CW00819023

FROM SOUP TO SUPERSTAR

FROM SOUP TO SUPERSTAR

The Story of Sea Turtle Conservation along the Indian Coast

KARTIK SHANKER

First published in India in 2015 by Litmus
An imprint of HarperCollins *Publishers*

P-ISBN: 978-93-5177-232-3
E-ISBN: 978-93-5177-233-0

2 4 6 8 10 9 7 5 3 1

HarperCollins *Publishers*
A-75, Sector 57, Noida, Uttar Pradesh 201301, India
1 London Bridge Street, London, SE1 9GF, United Kingdom
Hazelton Lanes, 55 Avenue Road, Suite 2900, Toronto, Ontario M5R 3L2
and 1995 Markham Road, Scarborough, Ontario M1B 5M8, Canada
25 Ryde Road, Pymble, Sydney, NSW 2073, Australia
195 Broadway, New York, NY 10007, USA

Typeset in 11/14 Electra LT Std by
R. Ajith Kumar

Printed and bound at
Thomson Press (India) Ltd

To Satish and Rom,
who inspired me to work on sea turtles
and
my parents, who gave me the freedom to be inspired

CONTENTS

PROLOGUE

PEOPLE ACROSS THE WORLD are fascinated by sea turtles today. Biologists, naturalists, lay persons, school children. The Annual Symposium on Sea Turtle Biology and Conservation, organized by the International Sea Turtle Society in various parts of the world, is attended by nearly a thousand people from over seventy countries. Thousands more are involved with sea turtles around the world. Given that there are only seven species of sea turtles, this group attracts more attention than almost any other. Consider that these creatures appear only briefly on land to nest, and then disappear quietly into the oceans for years. The enigma, the mystery, the frustrating inability to find out more.

While some of us remain passionate about understanding sea turtle biology, others dedicate their lives to saving sea turtles. But the domains of biology and conservation intersect more significantly in the sea turtle world than perhaps for any other animal group. By this, I mean that many if not most turtle biologists are involved in, or interact with, some aspect of conservation, and vice versa. In India, sea turtle biology and conservation began almost simultaneously in Odisha and Chennai. While the former was a government-driven programme with the forest department at its helm, the latter was initiated by wildlife enthusiasts and has remained largely a voluntary, non-governmental effort.

Five species of sea turtles are found in Indian waters. These include the olive ridley, green, hawksbill, loggerhead and leatherback turtles.

Olive ridleys nest along both coasts of mainland India and on the offshore islands, including the Lakshadweep Islands in the Arabian Sea and the Andaman and Nicobar Islands in the Bay of Bengal. A few thousand ridley turtles nest annually in the states of Tamil Nadu and Andhra Pradesh; and over a hundred thousand turtles nest most years during mass-nesting events, or arribadas, in Odisha. The lagoons of the coral atolls of the Lakshadweep are excellent foraging habitats for green turtles. They also nest in large numbers on uninhabited islands such as Suheli. The Andamans provide excellent feeding and nesting habitat for both hawksbill and green turtles, while the Nicobars have some of the best leatherback nesting beaches in this part of the Indian Ocean.

On the Indian coast, as elsewhere, sea turtle populations are threatened by a combination of fishery-related mortality, habitat destruction (through sand mining, beach armouring and light pollution) and depredation of eggs. The increase in fishing intensity throughout India has resulted in large-scale incidental mortality of sea turtles, drowned in fishing nets. From December to March each year, thousands of turtles wash up dead on the Odisha coast. At the same time, a large number of ports are being built along the coast of Odisha and other states. These, along with other forms of coastal development, probably pose even greater threats to the sea turtles which nest here. And then, there is the ever pervasive threat of climate change, which can affect sea turtle populations not only through rise in sea levels and extreme weather events that affect nesting beaches, but also through warming that can alter hatchling sex ratios.

It is against this background of constant threat from fisheries, coastal development, climate change and egg depredation that biologists and conservationists have worked on sea turtles over the past few decades. The histories of both disciplines are rich not only with stories of frustration and failure, but also extraordinary effort and the occasional success. And the satisfaction of a biological story discovered. This book pays homage to that history, focusing on the conservation of olive ridley turtles in particular.

In this narrative, I hope to provide a historical and contemporary account of sea turtle conservation and biology in India. All the references and citations are listed at the end of the book (marked by regular numerals). The minority that are academics and scholars (read geeks like me) will find this useful. I advise the rest of the sane world to read without interruption, but some of the articles that I have cited and quoted are fascinating and worth following up on later. Most of the literature on sea turtles in India is available on the website 'Sea Turtles of India' (www.seaturtlesofindia.org). I have also provided some notes (marked by roman numerals) at the end of each chapter which provide additional detail or interesting diversions. Each chapter is organized in roughly the same way, starting with a personal anecdote, followed by a detailed narrative of the subject of the chapter, an account of some of the key characters, some general theory of biology or conservation and a conclusion. The author is not responsible for deviations from this format; turtles are responsible for all digressions in this world.

BREAKING THE SURF

Juvenile frenzy

A TINY HATCHLING SCRAPES open its egg shell with its little egg tooth and a sharp thumb claws on each front flipper, and scrambles out into a sandy world. Two inches long and weighing less than twenty grams, it is one-and-half feet below the sand and finds itself in the company of a hundred other pushing, heaving hatchlings. The hatchlings wait quietly as their siblings proceed to shed their calcareous coats. A while later, the temperature drops, and so it must be night in the outside world, dark enough for the freedom run to the sea. As they start moving about energetically, the sand slowly filters down, the nest collapses, and the hatchlings find themselves moving towards the surface together. They emerge en masse and wait but briefly to determine the brighter horizon, the sea with the moon and stars reflecting off its surface, and then rush and tumble down the slope, feel the beating of the waves ahead of them and all of a sudden, they are in the water where they belong.

In seconds, other instincts kick in, and the hatchlings dive under the incoming waves and ride out with the current. Their first response is to swim against the waves, and this ensures that they reach offshore waters. The yolk sac that they have absorbed in the days after hatching and prior to emergence from the nest gives them the energy to swim without pause for a few days, and this 'juvenile frenzy' increases the chance that they reach safe havens in the sea. By the time they catch offshore currents and find seaweed rafts, the hatchlings have become oriented to the earth's magnetic field and this enables them to maintain their direction. Those who survive the dogs, crabs, ants and numerous other predators on the beach, and then the fish and seabirds that have waited to feast on them in nearshore waters, will essentially be at the mercy of oceanic currents for the next ten to twenty years of their lives. Perhaps one in a thousand will survive to adulthood and join other adults at a feeding ground thousands of kilometres from the place where they hatched. When they have matured, the males and females will use the earth's magnetic field and other cues to migrate back to their natal beaches to breed and nest and start the cycle again, as they have done for millions of years.

In 1988, more than twenty-five years ago as I write this, I was a second-year Zoology student at the Madras Christian College, quite unaware of the reptile rituals that occurred just a few metres from my house by the beach. I heard about the turtle walks and, after learning that this was not some peculiar collegiate ritual, decided that this was a diversion worth pursuing. We started each night on Besant Nagar beach and walked along the seashore looking for turtles or their tracks, reaching Nilankarai, 7 km south, by midnight or later. That year, the forest department had a hatchery on the beach where we slept till the morning.

During the first year, I saw no turtles on the few occasions that I joined the patrol walks. Several months later, in October, we formed a students' group to continue sea turtle conservation along the Madras (now Chennai) coast. Before we had any money, we had a society and a letterhead, with advisors, a president, a secretary and so on. In December, we started the regular patrolling between Besant Nagar and Nilankarai. Barely past the beachside temple at Besant Nagar, we were still chatting and not quite looking out for tracks yet, when we saw her. My first turtle! An olive ridley, three feet long, amongst the smallest of sea turtles. We crawled up slowly behind her and waited, while she, blithely oblivious to us, scooped out sand with one rear flipper, then the other, raising her head occasionally to take a sighing breath. When her flask-shaped nest was complete, she laid 126 eggs, soft, round and plopping three or four at a time into the nest. Once the laying was finished, she shovelled sand into the nest with her rear flippers, and then rocked from side to side, thumping the sand down on top of the nest. Finally, with her front flippers, she threw sand around in various directions to fool predators such as me and, with an ungainly waddle back to the water, returned to her world.

Centuries of fascination

Adumbakodi sidhaiya vānghi kodunkazhi
Kuypai venmanal pakkam sērthi
Niraitchool yāmai maraithēnru puthaitha
Kottovattu vuruvin pulavunārum muttai
Pārpida nāgum alavai pakuvāi
Kanavan ōmbum

Kumizhi Gnāzhalār Nappasalaiyar (Tamil Sangam literature, circa 4th century AD)

> The laying turtle collects and brings a bundle of Ipomea creepers
> Keeps them beside the heap of white sand to conceal
> Eggs, white as elephant tusks and round and foul smelling
> With open mouth, the male awaits the hatching of the young ones

Our fascination with marine turtles is not new; human cultures have interacted with them for tens of thousands of years. One of the earliest known written records of marine turtles from India is a remarkable short poem from Tamil Sangam literature (circa 4th century AD)[1]. This is a poetic and near accurate description of a nesting turtle, 'eggs, white as elephant tusks . . .', barring the crocodilian behaviour of the male. In sea turtles, there is no parental care, and neither parent awaits the hatching of the young ones, with open mouth or otherwise. Along the southern Indian coast, olive ridleys do, however, often nest in *Ipomea* (the goat's foot creeper with pink flowers), though it is far more likely that they are dragging the creepers by chance rather than intent.

From much earlier in Hindu mythology, turtles have been revered as an incarnation of Vishnu, one of the gods of the Hindu 'trinity'. There is even a temple dedicated to the 'Kurma avatar' at Srikurmam near Srikakulam on the Andhra Pradesh coast. Though mythology is unclear on this point, one might argue that the turtle that supported the mountain during the churning of the ocean must have been a marine turtle; furthermore, it must have been an olive ridley, given that only ridleys nest at Srikakulam, and in such large numbers at the nearby mass-nesting rookeries in Odisha.

The earliest documented reports of turtles at Gahirmatha, the famous mass-nesting beach on the coast of Odisha, are in a book, *A New Account of the East Indies*, based on the travels of Captain Alexander Hamilton[2]. He wrote in 1708 of the sandy bay 'between Cunnaca and Balasore Rivers' where a 'prodigious number of sea tortoises resort to lay their eggs'. More than a century later, in 1846, Andrew Stirling wrote of the value of the 'excellent turtle' off False Point in Odisha[3].

Several accounts from the nineteenth century deal with sea turtles in Sri Lanka. For example, there is a remarkable description of how the animals were tagged with brass rings during the Dutch occupation in the late eighteenth century by a district officer to check if descaled hawksbill sea turtles visited the same cove for nesting again. A hawksbill turtle tagged in 1794 was recaptured in 1826 and brought to J.W. Bennett, a member of the Ceylon administration, also known for his book on and illustrations of the fishes of Ceylon[4]. The '400 pound' turtle had apparently revisited the same cove to nest for thirty-two years. This may have been the first ever tagging programme in the world. Ideas about site fidelity in sea turtles existed even at that time. James Emerson Tennant, who was an Irish politician and spent some years as the colonial secretary of Ceylon, said[5]:

> In illustration of the resistless influence of instinct at the period of breeding, it may be mentioned that the same tortoise is believed to return again and again to the same spot notwithstanding that at each visit she had to undergo a repetition of this torture.

In the 1800s, sea turtles begin to make an appearance in British accounts of expeditions to the Andaman and Nicobar Islands and in sociological accounts of the aboriginal communities. One of the first of these is a book by Frederic J. Mouat, a British surgeon, where he recounts the tale of their expedition to the islands in 1857[6]. Apart from Mouat's accounts of turtle capture in the late 1700s, there are descriptions of sea turtles in an appendix by Edward Blyth, curator of the Museum of the Royal Asiatic Society of Bengal, on the zoology of the Andaman Islands. Blyth documented 'hawk's-bill' turtles and says that tortoiseshell was known to have been collected on the islands. He also recorded the green turtle and mentioned that *Sphargis coriacea*, (the earlier scientific name of the leatherback, now called *Dermochelys coriacea*) had been sighted. It is interesting that leatherback turtles were widely known to be open ocean wanderers by this time, but perhaps not greatly surprising given that so much of this information came from incidental sightings while travelling by sea. Blyth also mentions the loggerhead turtle, but it is likely that he meant olive ridleys – the two species were often confused in literature[i]. Thus, all four species known on the islands today had been recorded by this time.

In 1864, Albert Karl Ludwig Gotthilf Günther, FRS, published his *Reptiles of British India*[7]. Günther was a German-born British zoologist, who specialized in the study of fish and herpetofauna (amphibians and reptiles); he was a prolific herpetologist, describing more than 340 new reptile species. Günther included several species of sea turtles from India in his book. He referred to the ridley as the olive backed loggerhead which he believed was restricted to the East Indies. Echoing recent debates, Günther also commented on the differences between Atlantic and Pacific green turtles, which some authors at the time considered to be different species[ii], and said that it was known to nest on the sandy beaches of some 'sequestered island'. The eggs were collected by fishermen, whose 'expert eye baffle[d] the pains with which the turtle conceal[ed] her eggs'. Günther provided further evidence that the natal homing instinct of sea turtles was known at the time and wrote of the impact of directed take on abundance:

> As, however, turtles always resort to the locality where they were born, or where they have been used to propagate their kind, and as their capture is very profitable, they have become scarcer and scarcer at places where they are known to have been abundant formerly.

Günther also wrote of the practice of removing the shell of the turtle while heating it over a fire, but expressed doubts that epidermal shields would regenerate. Finally, he recounts a description by Major Tickell about an encounter with a leatherback, the largest of turtles, at the mouth of the Ye river on the Tenasserim coast, Burma (now Myanmar). The fishermen who tried to capture the turtle were nearly dragged into the sea, and it took ten to twelve men to drag the animal into the village. But, as in most parts of the world, the Burmese found the meat unpalatable.

The twentieth century history of sea turtles in India can be divided into roughly three periods. Before Independence, there are a few reports by British and Indian authors, mostly recording nesting events and sightings. Some anthropological accounts refer to sea turtle consumption by aboriginal tribes in the Andaman and Nicobar Islands, but there is little else. As later accounts show, there was egg and turtle consumption in many parts of the mainland, but little is written about it. Post Independence, the period till the mid-1970s contains literature on sea turtles largely as a fishery. Beyond that, sea turtles become a major subject of research and conservation.

In 1902, Alfred Alcock published his book, A *Naturalist in Indian Seas: Or, Four years with the Royal Indian Marine Survey Ship 'Investigator'*[8]. He wrote of seeing 'shoals of porpoises and turtles' near South Sentinel Island in the Andaman group. They also delighted at the sight of 'the entry of a brood of turtles into the world'. As people across generations appear to be tempted to do, they attempted to face them in a landward direction, and noted their 'unerring instinct' for finding the sea.

Alcock mentioned sea turtles in the Diamond Islands, Andamans and Lakshadweep, but not in Odisha. He reports leaving Minicoy 'taking with [them] good store of the turtles for which the island is so justly famous . . .' Though the ship visited the Odisha coast between December 1888 and March 1889 (currently the breeding season for olive ridleys on that coast), including Gopalpur, Ganjam and Chilika on the southern coast (where one of the current mass-nesting beaches, Rushikulya, is located), they did not appear to have encountered sea turtles there.

Recently, Jack Frazier, veteran sea turtle biologist, conservationist and historian unearthed another obscure reference to sea turtles on the mainland coast, this one from the *Guide to the Marine Aquarium*, in Chennai[9]. Started in 1909, this was then the only permanent aquarium on the 'Asiatic Mainland' and appears to have been quite popular, attracting over 1,50,000 people in 1919-20. In a description of the 'Turtle Tank' by J.R. Henderson, superintendent of the Government Museum, four species are described, including 'the Green or Edible Turtle (*Chelone mydas*), the Loggerhead (*Thalassochelys caretta*), the Hawksbill (*Chelone imbricata*), and the Leathery Turtle (*Sphargis coriacea*)', in order of commonness along the Chennai coast, which could have meant the erstwhile Madras Presidency rather than Madras (what is now Chennai) town, or maybe the Coromandel coast of Tamil Nadu and Andhra Pradesh. As Frazier points out, the commonness of green turtles and loggerheads is of interest, as is the absence of any mention of olive ridleys. Though ridleys and loggerheads were often confused, the mention of loggerhead size as comparable to that of green turtles suggests that the identification was accurate. Hawksbills were described as being common in the Gulf of Mannar.

In the early part of the twentieth century, there are several records of sea turtles on the Indian coast, many of them notes in the *Journal of the Bombay Natural History Society (JBNHS)*, which had been launched in 1886. In 1921, one Mrs Mawson documented her encounter of a green turtle[10]. She and her husband accompanied some local villagers, who collected the eggs of the turtle while it was laying. She writes, '[my husband] got comfortably seated on her back... she continued her walk as if nothing had happened'. The note is signed 'Malad-Marva, Salsette'; the former is a beach within what is currently Mumbai, while Salsette refers to the island on which Mumbai and Thane lie. Few or no green turtles nest in Maharashtra any more, but they are believed to have nested in the past along these coasts. About a decade later, Greaves wrote of the nesting of an olive ridley turtle again at Malad-Mervé beach[11]. He said:

> I was idly watching the sea when a curious upheaval, which subsequently resolved itself into something which looked like the Loch Ness monster, showed up near the edge of the water, and proceeded to steadily move towards the shore.

In 1942, one finds what is likely the first article on sea turtles in *Current Science*, India's leading science journal. P.I. Chacko[12] documented the nesting of olive ridleys on Krusadai Island in the Gulf of Mannar (though he referred to it as the olive loggerhead turtle). He noted that green turtles were common in the waters around the island, but did not nest on it. According to him, the female ridley turtle 'bites off pieces of plants found in the neighbourhood, such as Ravana's moustache, *Spinifex squarrosus*, the herb, *Launaea pinnatifida*, and the Ground-Glory, *Ipomea biloba* which grow wild on the island, and covers the top of the burrow'. This curiously mirrored what the fourth-century poetess had to say about ridley nesting, but since it has never been observed by scores of naturalists and biologists, it is likely to be local legend. He wrote that crude oil was extracted from the turtles for use in country craft, but did not mention large-scale capture of adult turtles, which he was likely just unaware of.

In 1958, Sanjeeva Raj, later Chair of the Zoology department at the Madras Christian College where I had studied, wrote in *JBNHS*, comparing the Tamil poetess' account to other natural history notes from the subcontinent at the time[13]. As he says, the observations from one-and-a-half millennia earlier were as good as any available at the time.

Sea turtles were also mentioned in passing in accounts of fauna, such as in Biswas and Sanyal's report on a Zoological Survey of India expedition to Great Nicobar Island[14]. They recorded a hawksbill turtle head at Campbell Bay (earlier mentioned by Edward Blyth[15]), and green turtles at Galathea, where we worked several decades later. While detailed accounts on sea turtles are fairly sparse in India, P.E.P. Deraniyagala, Sri Lanka's palaeontologist and zoologist par excellence, carried out extensive taxonomic studies and investigations of reproductive and developmental biology, which are recorded in minute detail in his many books[16,17].

Deraniyagala wrote a lucid account of the 'The Nesting habit of Leathery Turtle *Dermochelys coriacea*' in 1936, replete with beautiful illustrations made by his own hand[18]. Apart from his remarkably accurate description of the process, the manuscript is revealing in other ways. He wrote of the nesting turtle having a fishy odour, indicating an absence of the niceties that govern the research community's descriptions of nature today. When she had finished nesting, he says: 'At this stage, I struck her a sharp blow on the head with a stick and sat upon her, but undeterred she continued to churn up the sand . . .'

Historical records of leatherback turtles on the Indian mainland coast are of particular interest, as there have been so few in the last century, and no leatherback nesting occurs anywhere along the coast today. One of the first records is in the guide to the marine aquarium in Chennai, which noted the capture of a leatherback in Guntur in Andhra Pradesh. In the early 1900s, T.H. Cameron, the DSP (District Superintendent of Police) of Quilon (now Kollam), was attempting to collect specimens of the giant turtle which he had heard visited the coast to lay eggs[19]. Stationed at Quilon on and off since 1906, he had little luck till 1923, when he first encountered a nest. Finally, a turtle was caught at sea and brought to Cameron. Measuring seven feet, Cameron was able to identify it as a leatherback turtle, and sent it to another office to show to some other Europeans. Before it could get there, it was sold for Rs 60 and in all probability, consumed shortly thereafter. The fishermen told Cameron that leatherback turtles were common in these waters, and about forty would be caught each year during the breeding season. However, the numbers had already declined, and they were mostly seen near Tangacheri reef.

Jones, of the Central Marine Fisheries Research Institute, then recorded the nesting of a leatherback turtle in Calicut, Kerala, in

July 1956[20], the last confirmed record of nesting on the mainland coast. Since then, there have been just a handful of stray records of leatherback turtles on the mainland[21]. In 1976, a leatherback was recorded in Visakhapatnam and died of injuries or unknown causes, and it is not clear if it came ashore to lay eggs[22]. Another dead turtle was washed up in Kovalam, near Chennai, in 1982 and was reported by 'Ms Reliable Ferret' in the newsletter (now journal) *Hamadryad* of the Madras Crocodile Bank[23]. Most of the other records are notes in the *Marine Fisheries Information Service*, a generally valuable source for stranding records of sea turtles and marine mammals throughout the country. Like the turtle that Cameron reported, some recently caught leatherbacks have been sold for consumption in Kerala and southern Tamil Nadu (and some rescued by tourists and conservationists), which is noteworthy because leatherback meat is eaten at very few locations around the world.[iii]

A *brief history of research*

As much as the 1960s was a landmark decade for counter-culture and music – a.k.a. sex, drugs and rock and roll – in the USA, it is perhaps best remembered in India for the birth of the Green Revolution and the beginnings of self-sufficiency in food production. Back in the early 1970s, the mood in the wildlife community was that of exploration and exploitation, with the first murmurings of conservation, which was very much on the fringe of public consciousness. Enter Robert 'Bob' Bustard, an FAO consultant exploring saltwater crocodile populations for commercial exploitation. The late J.C. Daniel, who served for many years as the director of the Bombay Natural History Society (and editor of the *Journal of the Bombay Natural History Society*), and S.A. Hussain had, in 1973, visited Odisha and heard about a large sea turtle rookery there. Daniel's and Hussain's knowledge of this rookery is repeated by senior herpetologists, but does not appear to have been recorded anywhere. Bustard, while surveying Bhitarkanika, visited Gahirmatha and encountered a sea turtle skull, which he identified as an olive ridley. He discovered and wrote about the mass-nesting rookery, calling it the 'the world's largest'[24]. Over the next few years, he initiated a research programme with officers from the forest department, most notably the recently deceased C.S. Kar who worked for his Ph.D on olive ridleys in Gahirmatha. Kar tagged more than 10,000 nesting turtles during 1975–1980, and carried out extensive research on olive ridleys[25]. Bustard also persuaded the government to stop the collection of eggs, a ban which would never be revoked.

Coincidentally, around the same time, Romulus Whitaker was working in Chennai and, along with a few colleagues, started monitoring the beaches along the coast to protect the nests from takeover by local communities[26]. Whitaker had started the Madras Snake Park in 1969, and a few years later, moved the park to a location not far from the sea. In 1973, they began a conservation programme by relocating eggs to a hatchery, which was just the backyard of a friend's house. A couple of years later, Whitaker started the Madras Crocodile Bank Trust, which supported surveys of sea turtles across India, including the Lakshadweep and the Andaman and Nicobar Islands, documenting several sea turtle nesting sites for the first time. Strangely enough, these sea turtle projects began almost simultaneously in two different locations; the two groups were largely unaware of each other for the first few years, and charted very different courses. And though paths have crossed, and people have crossed over, the effect of these starting conditions can still be felt many decades later.

In the late 1970s, the Central Marine Fisheries Research Institute (CMFRI), led by E.G. Silas, also initiated studies in Odisha and Chennai. A research centre was established at Kovalam on the outskirts of Chennai, where eggs were incubated and hatchlings reared in a hatchery[27]. Around the same time, research was initiated at Utkal University in Odisha on temperature sex determination in olive ridley turtles which remains the only study of pivotal temperature for this population[28]. Priyambada Mohanty-Hejmadi and her colleagues also worked on various aspects of reproductive biology and physiology of olive ridleys[29,30].

As a Master's student at the department of Zoology at the University of Madras, I engaged with Prof. T. Subramoniam, a fine reproductive biologist. Subramoniam had demonstrated a remarkable instance of sex reversal in the once ubiquitous mole crab[31]. He had demonstrated that the males of this species achieved reproductive maturity at a much smaller size than females, but continued to grow over time, and eventually became females themselves. We enthusiastically initiated research on olive ridley turtle eggs from the hatchery on the Chennai coast, intending to compare blood and egg proteins, but it remained unfinished as I left shortly after for a Ph.D on small mammals.

Little laboratory-based and experimental research has been carried out on sea turtles in the decades that have followed. At least some of this can be attributed to increasing conservation attention and consequent difficulty in obtaining permits for what would be considered 'manipulative' research. In 1977, all sea turtles were placed on Schedule I of the Indian Wildlife Protection Act, 1972. This meant that handling the eggs or animals was forbidden, except for exemptions that would be made for research. However, this had to be cleared by both the state forest department and the Ministry of Environment and Forests, making it generally more challenging to obtain permits.

While research on sea turtles was sporadic and often isolated during the 1980s, sea turtle biology in India received a fillip in the 1990s and 2000s with projects of the Wildlife Institute of India (WII), monitoring programmes by the Madras Crocodile Bank in the Andaman and Nicobar Islands and other research programmes that have sprung up since, including at the Centre for Ecological Sciences, Indian Institute of Science in Bengaluru, Karnataka. In the 1990s, the WII's programme in Odisha, steered by Bivash Pandav and B.C. Choudhury, led to the 'discovery' of the mass-nesting site at Rushikulya[32]. Pandav, as part of his doctoral research, tagged over 1,500 mating pairs (for the first time in India) and 10,000 nesting turtles[33]. They also documented a rapid increase in fishery-related mortality of ridleys in Odisha[34], leading to a number of NGO campaigns and an increase in media interest in olive ridleys[35].

In the late 1990s, I became involved in sea turtle research in Odisha through a project on molecular genetics that was supported by the WII, as a collaboration with the Centre for Cellular and Molecular

Biology, based in Hyderabad. We found that olive ridleys on the east coast of India appeared to be genetically distinct from other global populations, and even differed significantly from the adjacent population in Sri Lanka[36]. We proposed that Indian ridleys and the Kemps ridleys could be remnants of a global population which was otherwise extirpated following climatic changes prior to and after the closure of the Isthmus of Panama four to five million years ago. Thus the Indian Ocean region, in particular the distinct Indian population, may have served as a source for ridley re-colonization of other ocean basins following the extirpation of populations there.

Following the IUCN Marine Turtle Specialist Group's 'Northern Indian Ocean Sea Turtle Workshop' in Bhubaneswar in January 1997, the Ministry of Environment and Forests launched Project Sea Turtle in 1999. Shortly after, they received funding for a national project from the United Nations Development Programme (UNDP). The Wildlife Institute of India coordinated the project, with B.C. Choudhury at the helm, and in early 2000, I joined this project as a coordinator. The project carried out surveys of the entire coast for sea turtle nesting and mortality through partners and conducted workshops and training[iv]. The results of the project, as well as other contemporary material from India and other countries in the subcontinent, were compiled into an edited volume in 2006[37]. In 2003, we initiated a follow-up project, funded by the Convention on the Conservation of Migratory Species of Wild Animals (CMS), in select states through the Madras Crocodile Bank Trust[38].

Since then, major research projects have been carried out by the WII in Odisha[v] and long-term monitoring programmes have been initiated by Dakshin Foundation and the Indian Institute of Science for olive ridleys at Rushikulya, Odisha, and leatherback turtles on Little Andaman Island[vi]. In the Lakshadweep, the Nature Conservation Foundation (NCF) has been carrying out research on the foraging of green turtles and their impacts on sea grass and fishing. Rohan Arthur and his team at the NCF have also studied the conflict between green turtles and local fishers over their perceived impact on fish catch[39,40]. As part of a larger project on human-wildlife conflict[vii], Aarthi Sridhar and

I examined how conservation measures had resulted in conflict over shared spaces between fishers and sea turtles in Odisha[41]. In the last decade, a few research projects have also been carried out on incidental catch and turtle excluder devices (TEDs), mainly in Odisha[viii].

Over the years, a number of students have conducted research on sea turtles in India for their Master's and Ph.D dissertations, mostly in Odisha but in other states as well[ix]. Much of the work on sea turtles, including surveys, monitoring and targeted research, has been done in the context of conservation. And no man did more to initiate and inspire this body of knowledge in India than a dropout from the Indian Institute of Technology, Chennai.

The essential turtle walker

Any account of sea turtles in India must begin with Satish Bhaskar, a pioneer in every sense of the word. I first met Satish in 1988, the year we started the Students' Sea Turtle Conservation Network (SSTCN) in Chennai. He used to come with us on a few turtle walks, and would often walk a 10 km stretch south of the hatchery by himself. This was a season when we used to find a nest or two a night if we were lucky. On one particularly good day, 31 January if I remember correctly, we had found four nests and there was great excitement. At the hatchery, we were busy digging nest pits to relocate the eggs. We saw Satish walking towards us with a bulging backpack and his typical smile, saying, 'Hey, I got six nests, man!' Not expecting this bonanza, he had only carried a couple of the cloth bags that we often used to collect the eggs. The third nest was in one pouch of the backpack, the fourth separated by newspaper, and the fifth in his shirt, which he had removed to pack the eggs. I have no recollection of where the last clutch was, but he had brought them all back safely.

Satish was already a legend then. He had started work with Romulus Whitaker, the founder of the Madras Snake Park and the Madras Crocodile Bank. We also knew that Satish had worked in the Andaman and Nicobar Islands, the Lakshadweep, in Odisha and in fact, most parts of the mainland coast. He had visited Papua on a leatherback survey, and had along with C.S. Kar from the Odisha Forest Department, attended the first world conference on sea turtles, where he was taken by Jack Frazier to a bar, the details of which have never been entirely revealed.

Satish was a student of the Indian Institute of Technology in Chennai in the early 1970s. As an army child, he had lived in many different parts of the country and, when he finished school, was given the option of joining the army or taking up engineering. Satish chose the latter but never really engaged with it. He spent most of his day swimming on Elliot's beach and elsewhere along the Chennai coast, experimenting with body surfing and snorkelling. One day, he met Siddhartha Buch, a keen naturalist, who suggested that Satish visit some sanctuaries to pursue his interest in natural history, and gave him a letter of introduction to a forest department officer in Karnataka. Satish never made the trip as he had little money, and gave the letter away to friends. He met Buch again, and this time, was sent to meet Rom Whitaker at the Madras Snake Park.

It was around this time that Rom and his friends started the turtle walks in Chennai. Pretty soon, Satish was a part of their group and one of the first residents of the newly formed Madras Crocodile Bank. Rom recalls that Satish would stay in shape by carrying huge sacks of sand from one end of the campus to the other. Rom was a dedicated herpetologist, and since his primary interests were snakes and crocodiles, believed that someone needed to focus full-time on sea turtles. He persuaded Satish that he could become India's turtle man, and they started planning a series of surveys.

As a preface to his first report in the biannual journal *Hamadryad*, the editor wrote[42]:

> Between 1st and 22nd March 77 Mr Satish Bhaskar, field officer of the Madras Snake Park, joined a survey team from the Central Marine Fisheries Research Institute on a visit to several islands in the Gulf of Mannar. He wished to learn about the status of dugongs and sea turtles in this area.

What an understatement that would turn out to be. During this trip, Satish saw several dead turtles (greens, hawksbills and ridleys) and a couple of live dugongs[43]. In 1978, they started planning sea turtle surveys around the country. The first of these was a trip to the Lakshadweep, where Satish stayed for several months visiting a number of islands[44]. Satish hitched rides with a fishing crew, finding his way to remote uninhabited islands such as Suheli Valiyakara and Suheli Cheriyakara.

Of the Lakshadweep, he wrote[45]:

> From a nature lover's standpoint, India's Lakshadweep islands, which is 120 to 200 miles off Kerala's coast constitute a happy hunting ground, as do coral islands all over the world. The marine biologist, fisheries scientist, scuba diver and amateur snorkeller will find enough sea life to keep himself happily occupied indefinitely.

Certainly it was enough to keep him occupied. He wrote:

> A small, streamlined shark traced a tight circle around me and swam off as I snorkelled near the reef a quarter mile from Kavaratti's lagoon shore. Its speed, grace and almost imperceptible swimming motions are impressive. Off Suheli Cheriyakara, the crew of a fishing launch I was travelling in harpooned a manta ray with a 13 ft wingspan. Two turtles copulating at the surface narrowly escaped the same fate.

Satish was fascinated with Suheli, the two islands (Valiyakara, the big one, and Cheriyakara, the small one) with their relatively pristine lagoons and rich marine life, and exceptional densities of green turtle nests. The problem was that the main nesting season was during the monsoon and fishermen did not go there when the sea got rough. This did not deter Satish. He decided that the way around this problem was to get there before the monsoon and stay till after, a period of about five months. Rom Whitaker and the rest of their team helped make elaborate plans with regard to food, fresh water, medication and other eventualities but obviously not everything could be accounted for. He had no contact whatsoever with the mainland, but was given some distress flares by the navy. The navy also promised him some food supplies, an experiment in field rations, which did not arrive as expected before his departure. A week after he arrived on the island though, a ship anchored offshore, and two officers rowed ashore in their Gemini dinghy, somehow navigating the channel past the reef. They left him a good stock of rations, but he had to find a way to drag it all back to his camp, where he had rented a fishing hut from one of the fishermen at Rs 30 per month. Later, he dragged his supplies all the way back to the other end of the island when he found he was cohabiting with an entire family of scorpions including a brood of thirty to forty babies. At the northern end, he would see sharks swim close to the shore, within a few feet of him, especially when a dead whale shark washed ashore and provided a good and easy meal for them.

It is not entirely true that he had no contact with the mainland though. This report appeared in *Hamadryad* in 1982[46]:

> Satish is spending 5 months on Suheli on his own and reached the island in early May. His wife has had one letter from him; found in bottle by a Sri Lanka fisherman who forwarded it to her. Tracing its course we find that the 'bottle-letter' dated 3rd July has travelled a distance of about 500 miles in an estimated interval of 24 days. The letter reached Sri Lanka around 27th July.

Satish says that it was no walk in the park getting the bottle to float away from the island. The first few bottles that he threw into the sea kept washing ashore. Finally he started wading out, looking for rip currents that would drag the bottle out to sea. He would get into deeper water, holding onto coral or rock so that he wouldn't get swept out. He would put a bit of money in the bottle to ensure that the finder would mail the letter, but after he started digging into his Rs 10 notes, it became too expensive. He also started attaching the bottle to a float, so that it would be more buoyant. Eventually, a bottle, attached to a piece of Styrofoam went cartwheeling out to sea. Maybe it was that bottle, maybe another, but one eventually reached a fisherman called Anthony

Damacious in Sri Lanka, who sweetly mailed the letter to Satish's wife with a note, a picture of his family and an invitation to Sri Lanka.

Satish even considered swimming the 1.6 km between the two islands, with a long sandspit in between. He thought it would reduce the distance and that he could rest on the sandspit. Wisely, with sharks and currents all around, he did not attempt this. But in later years, he would swim across from his isolated South Reef Island to Interview Island when he ran out of supplies. Satish was ready to leave the islands by September, but the boat that was due to pick him could not leave due to bad weather. Fortunately, he had enough rations as he ended up waiting nearly another month. By the end, it was just a few days short of five months.

Even today, not many researchers organize trips to these islands, giving some indication of the amount of enterprise it took thirty years ago. Subsequent to Satish's work, there were no sea turtle surveys on these islands till 2001, when Basudev Tripathy worked there as part of a WII project, and found the same exceptional green turtle nesting densities in Suheli[47]. These islands are seasonally used by fishermen and remain prime habitats for sea turtles and other marine life. In January 2015, I finally visited Suheli along with my researchers as part of our reef and bait fish research projects. Seven hours from Kavaratti and inhabited by just a few fishermen and a police outpost, the islands are spectacular and the diving exceptional. But an old green turtle track on the beach brought back memories of Satish's exploits here. There is talk today of establishing tourism resorts at these islands, which may not have positive outcomes either for the people of the Lakshadweep or the biological diversity of these islands.

Satish then undertook a six-week survey in June-July 1978 of sea turtles in the Gulf of Kutch[48]. He visited several islands, including Bhaidar, which he was to visit again later. He wrote of seeing turtles from the thirty-foot-high dunes near Dwarka. Dunes are today making a comeback in public consciousness as important coastal habitats, perhaps critical in protecting coastal communities from storm surges and even tsunamis. However, during the interim period, dunes have pretty much been flattened along the entire mainland coast of India.

Satish returned to his favourite, Bhaidar Island, in 1981[49]. On a bleak and rainy day in August, he set sail for the island with rations for five days, and two jerry cans of fresh water. Since the shallows extended quite far, his crew abandoned him about 1.5 km from the island and he waded ashore slowly and painstakingly. On the island, he found a bottle. He wrote:

> A bottle containing a note embossed with the insignia for the Military sea lift command (Atlantic) of the U.S. Navy had been washed ashore a short while earlier. Half-expecting the bottle to be a device to test sea currents, with perhaps a reward awaiting its finder, I carefully photographed it and extricated the note which said:
>
> *Help me*
> *Get me off this ship I am about to go crazy. I need help, please. Please return this letter and where found. Tossed July 12th 1981.*
>
> The message bore a name and a Massachusetts address. While the wording and the smell in the bottle strongly suggested that its late owner had been celebrating, here was a chance to learn something about the direction and speed of the currents the turtles nesting at Bhaidar might utilize or encounter – I had found the note exactly a month after its being cast out and needed only to know the exact location where it had been dropped.

His long association with the Andaman and Nicobar Islands began shortly thereafter. Satish visited the islands many times between 1979 and 1995, covering almost all the islands in the process. After an initial eight-month stint in 1978-1979, when he covered a large number of islands including much of South Andamans, Little Andamans and central and Great Nicobar[50], he returned in 1981 and then again in 1983-84 to cover the remaining islands. Starting with an extensive survey of Great Nicobar Island with Manjula Tiwari in 1992[51], he would spend the better part of the next few years in the islands, including his legendary sojourns on South Reef Island. He worked extensively in the islands between 1993 and 1996, surveying a large number of islands in the Andamans group and spending weeks or months on South Reef to tag and study hawksbill turtles[52].

In a tribute to Satish and his work, Rom Whitaker wrote[53]:

> In 1979, Satish visited the Andaman and Nicobar Islands for the first time and, like so many of us, got hooked. Over the next few years, again thanks to WWF and other funds, he visited most of the islands' major sea turtle nesting beaches. A near-mythical Satish-exploit of this period is his many months sojourn, over several years, on tiny South Reef Island on the west coast of North Andaman. He was studying the hawksbill and green turtle populations there. It was tough, with no freshwater, and of course, no food. Satish would swim the half kilometre of vicious currents to Interview Island and back to collect freshwater in a jerry can. Once he ran into one of the notorious feral elephants of Interview Island, who promptly charged him. As he ran down the forest path, Satish threw down his shirt which fortunately distracted the angry pachyderm. The next day he swam back to Interview to retrieve his jerry can and found his shirt in three pieces. He posted the pieces to his wife Brenda (back in Madras) with a reassuring note!

In 2013, Satish showed me the remains of this shirt, still in several pieces and preserved in pristine elephant stomped and ripped condition.

In between, Satish also covered Gujarat (December 1980 and August to December 1981), Kerala (April–May 1981), Goa (July 1981), Odisha, Andhra Pradesh (January–February, 1982) and West Bengal and Sundarbans (February 1983)[54]. He helped the Students' Sea Turtle Conservation Network in Chennai during the winters of 1989 and 1990, and conducted surveys for the estuarine turtles, *Batagur baska*, during April–May 1990 with Ed Moll, earning him the nickname 'Batagur Bhaskar'.

In 1984 and 1985, Satish worked on the leatherback rookeries on the northern coast or Bird's Head Peninsula of Irian Jaya or West

Papua in Indonesia[55]. The beaches where he worked, Jamursba Medi and Wermon, were not logistically easy places to access. To keep in touch, Satish would swim out for 100 metres or so each time there was a longboat heading for Sorong, the nearest city on the coast; there was only one boat which passed by at intervals of about 20-30 days. During his first visit in 1984, Satish was marking these turtles with paint. But he eventually stopped as the paint did not stay on very long. He monitored a 17 km stretch singlehandedly, sometimes walking up and down, while on some occasions, one of his assistants would take him back to the camp in an inflatable. Satish counted more than 13,000 nests during that season, which is the most that has been recorded on that beach[56]. He wrote[57]:

> Mesak trapped a wild pig in the jungle just behind the bench, and its stomach contained leatherback hatchlings – it was eating turtle eggs every night, so pigs have to be destroyed – anyway these pigs had been introduced from another part of Indonesia many years ago by the Dutch, so no harm destroying them. Every day between 50 and 130 leatherbacks are coming to nest, and this is not the peak season which passed in June-July, so this beach is the best in the whole of Asia for leatherbacks, and the second or third best in the world. Best of all was we followed two of the females out to sea after they finished nesting using the boat and I got some fantastic pictures from 20 feet away – of one leatherback coming up to breathe in the sea – these may be the first close-up sea photos of leatherbacks. Since that turtle was scared of the boat, it held its breath for very long (to avoid us) and when it came up it was quite desperate for air – so it didn't surface quite as it usually would, but shot up at full speed – the photo should show almost half the leatherback exploding out of the water – hope it comes out well.

Satish returned the following year and started tagging leatherbacks; he tagged his first turtle in April and saw her again after fifty-four days and worried that he had missed her nesting in between, even though he walked the entire 18 km stretch every other night[58]. But he was also tagging turtles intensively on a smaller stretch of beach with little help. He wrote[59]:

> . . . so far I have only one forest department man to help out with things, but 2 PHPA gentlemen from Bogor are expected to stay 2-3 months with me soon. So far we have tagged only about 300 leatherbacks – we should be tagging something like 2,000 before the season is through if the minimum estimate I made last year (i.e. 3,000/season) is to be realized. At this rate we will fall short by perhaps 1,000 turtles but of course, the season is just getting underway now.

By the end of the seasons, he had tagged almost 700 leatherbacks and was encountering few untagged ones. I have been unable to ascertain how many leatherbacks Satish eventually tagged in Papua, but nobody else worked on these beaches till the 2000s when a local conservationist, Creusa 'Tetha' Hitipeuw, started monitoring them with the support of WWF and local Papuans. Sadly, Tetha passed away a couple of years ago after a sudden battle with cancer. The project was eventually supported by the National Marine Fisheries Service, USA, which also satellite-tracked the leatherback turtles from this beach. Again, Satish was there first and long before anyone else found the enterprise or motivation to work at what would turn out to be, according to his instincts, a very important beach for sea turtles.

When he started tagging turtles, it gave birth to a local legend that I heard when I visited the beaches twenty years later. It was said that a strange visitor from a foreign land visited the island and put metal on the turtles, and then using very powerful magnets, attracted these turtles away to his homeland. Hence the decline of turtles at these nesting beaches.

William Betz and Mary Welch, visited Jamursba Medi in 1991, and were told that recent nesting levels were only 25 per cent of that reported by Satish Bhaskar. They wrote[60]:

> The local people are concerned about the drop as well, but they still take as many eggs as they can. They fail to see the connection

between their actions and the decline in the nesting population. Furthermore, they have made Bhaskar their scapegoat, claiming that his tagging program scared the turtles away or that the tags were designed to make the turtles follow him back to India! They claim that none of the tagged turtles have returned since, but in fact we observed two tagged turtles during our short stay on the beach.

Satish published one of his first overviews of sea turtle conservation in India in the *Indian Forester*[61], based on his original surveys. He had attended the World Conference on the Biology and Conservation of Sea Turtles in 1979 and his paper at the conference was published in the compendium, *Biology and Conservation of Sea Turtles*. This volume, (edited by Karen Bjorndal, a leading sea turtle biologist) remains a landmark document in the history of sea turtle biology[62]. Bhaskar's paper with C.S. Kar on sea turtles in India, is a classic compilation on the state of knowledge on sea turtles in the Indian subcontinent[63]. Bhaskar published an updated version of the review at the Central Marine Fisheries Research Institute workshop in Chennai in 1984[64].

The extent of Satish's work has to be placed in perspective. In 2000-01, the Wildlife Institute of India coordinated a sea turtle project with over $50,000 allocated for surveys alone and a total of about ten organizations working along the coast. With all of this infrastructure and money, the project mainly provided updates on information that Satish had gathered twenty years earlier. Certainly, more areas were covered and fresh data collected, but two things stand out. Nothing startlingly new was learned, and most areas had little additional research between Satish and this project. That the scale of work should even be comparable is itself astounding.

I have since met many of Satish's college mates, who all enquire after him. There is a common sense of nostalgia and appreciation for someone who refused to follow the rules. Satish received a special mention as part of the Rolex awards for enterprise (he did get a watch), but little else in the way of formal acknowledgement of his work. In April 2010, during the 30th Annual Symposium on Sea Turtle

Biology and Conservation in Goa, the International Sea Turtle Society presented him a Sea Turtle Champions Award for his outstanding contributions to sea turtle biology and conservation through his surveys. More importantly perhaps, he is revered as a turtle walker extraordinaire amongst the people who had the privilege to know him or hear of his exploits. Satish now lives in Goa with his wife Brenda. His children (Nyla, Kyle and Sandhya) who once helped him collect eggs from sea turtle nests are grown up and pursuing their own adventures.

A primer on sea turtle biology[65]

The ocean's travellers

Sea turtles are fascinating creatures whose adaptation to the marine environment is near absolute. Their languid grace in the water is almost balletic. They have beautifully streamlined bodies, long fore flippers and a physiology that frequently defies biological wisdom. They can survive for long hours while diving, accumulating toxic waste in their tissues, enabling species like leatherbacks to dive several thousand feet – it is one of the deepest diving air breathing vertebrates. They undertake enormous oceanic migrations, navigating using geomagnetic stimuli. For example, green turtles are able to locate Ascension Island, a mere speck in the middle of the Atlantic, while loggerheads undertake migrations of over 10,000 km from Baja California to Japan.

Many researchers have been honoured (or burdened!) with the title of sea turtle expert; Jack Frazier, who has worked on sea turtles for over forty years, is quick to point out that there is no such thing. Sea turtles are indeed remarkable animals but their complex life cycles make them very difficult to study and understand. There are seven species of sea turtles in the world belonging to two families: Dermochelyidae is represented by a single species, the leatherback turtle (*Dermochelys coriacea*), with a cartilaginous carapace as the name suggests. This is the largest sea turtle, growing up to six feet in length and weighing over 500 kg. The cheloniids or hard-shelled sea turtles include the green turtle (*Chelonia mydas*), loggerhead (*Caretta caretta*), Australian flatback (*Natator depressus*), hawksbill (*Eretmochelys imbricata*), and the two ridleys, the Kemp's (*Lepidochelys kempii*) and olive ridley (*Lepidochelys olivacea*).

Sea turtles are widely distributed in tropical and sub-tropical seas throughout the world. Find a sandy beach in the tropics on any continent and it is likely that there is some species of sea turtle which nests there. Five of the seven species are globally distributed; the flatback is restricted to Australia and the Kemps ridley to the Gulf of Mexico.

Reptiles have generally been described as sluggish, cold-blooded ectotherms. This stereotype has been demolished by leatherbacks. These animals have been known to maintain higher body temperatures than their surroundings, which allows them to foray into the cold temperate waters of the northern Atlantic and Pacific oceans to forage for jellyfish. Recent research using stomach pill thermometers shows that leatherbacks are able to maintain their core body temperature at 10° to 12° C above the surrounding water[66]. They are believed to regulate their body temperature by generating metabolic heat, but also their large size, insulation, and a blood circulation mechanism known as a countercurrent heat exchanger, found at the junction of the flippers and the body. This mode of thermoregulation that can neither be classified as homeothermy (as in birds or mammals) or poikilothermy (as in fish, amphibians and most reptiles) has been termed gigantothermy[67]. These animals provide important clues to the evolution of traits thus far believed to be the exclusive domain of birds and mammals. They may also provide insights into dinosaur life, such as where they may have lived, and their physiology and metabolism.

Though their foraging areas span the globe, leatherbacks nest mostly in tropical areas like most other species of sea turtles. Atlantic populations are believed to be largely stable or increasing with large nesting populations in the Caribbean, the northern coast of South America (Suriname and French Guiana), and West Africa. In the Pacific, there are large nesting populations in Central America, and on the northern coasts of West Papua (Indonesia) and Papua New

Guinea. Their decline in the eastern Pacific from nearly hundred thousand to less than ten thousand was much written about in the early 2000s to indicate the possible extirpation of Pacific leatherbacks[68]. Another remarkable decline occurred at Terengganu, Malaysia, where the population went from a few thousand nests to none at all in just a few years, reportedly due to unsustainable egg harvest and incidental adult mortality[69].

In the Indian Ocean, there are significant nesting populations in the southwest Indian Ocean, particularly in South Africa, where George Hughes, one of the pioneers of sea turtle biology, initiated a monitoring programme more than forty years ago[70,71]. Some nesting occurs in Sri Lanka, but other than that, there are few nesting beaches in the northern Indian Ocean. In the eastern Indian Ocean (south-east Asia) again, there are many significant beaches. The beaches at Great Nicobar Island have been surveyed several times, and Galathea, on the south-eastern coast, was monitored regularly till the 2004 tsunami[72].

Other species of sea turtles feed mostly in warm tropical waters. Green turtles feed in shallow, nearshore areas abundant in sea grass and algae. They are largely herbivorous as adults and may take many decades to mature. Green turtles were perhaps once the most abundant sea turtle in the world. Some authors estimate that there may have been more than a million green turtles in the Caribbean alone a few centuries ago[73]. These numbers are believed to have been depleted over the years due to exploitation. Green turtle populations are almost too numerous and widespread to list. There are large nesting populations in the Caribbean, Central America, Brazil and south-eastern USA, West Africa, Ascension Island in the southern Atlantic, western Indian Ocean (particularly Seychelles and Madagascar) and all over Asia, Australia and the Pacific Islands. The populations in Tortuguero in Costa Rica, Ascension Island and eastern Australia have a special place in history because sea turtle research started at these sites. Archie Carr, considered one of the founding fathers of sea turtle biology and conservation, started his enquiries into the lives and future of sea turtles at Tortuguero in the 1950s, while Colin Limpus initiated his research on green and loggerhead turtles in Queensland in the 1970s.

In India, green turtles nest in reasonably large numbers on the Gujarat coast, in the Lakshadweep islands (mainly Suheli) and on several islands in the middle Andamans. Foraging populations are found in the Gulf of Kutch, Gulf of Mannar, in several lagoons of the Lakshadweep islands, and at several sites in the Andaman islands. Injured or dead juvenile green turtles are occasionally washed up on the mainland coast. Their numbers in Agatti island in the Lakshadweep were particularly remarkable in the early 2000s, with the lagoon chock-full of green turtles of various sizes. Equally remarkable is the recent decline of the population in Agatti with a simultaneous rise in their numbers in the lagoons of Kavaratti and Kadmat[74].

Hawksbill turtles are unique in their spongivorous diet and occupy reef habitats. They generally nest on narrow island beaches with vegetation. One can find them throughout the world's tropics, with large nesting populations in the Caribbean (Mexico, Central American countries and Caribbean Islands), Brazil and West Africa (Bioko), western Indian Ocean islands (particularly Seychelles), west Asia and East Africa (Oman, Iran and Eritrea), Maldives, western Australia, south-east Asia (Indonesia and Malaysia), and Pacific Islands (Palau, Solomons and Samoa). In India, hawksbills are known to nest mostly in the middle Andaman Islands. Satish Bhaskar's monitoring of hawksbill nesting on South Reef Island between 1992 and 1995 remains the only study of a hawksbill nesting population in India[75]. Occasional nesting occurs in other parts of the Andaman and Nicobar Islands as well as the Lakshadweep. There are a few records of juvenile and adult turtles being washed ashore on the mainland coast from both the east and west coasts. There is also a record of nesting in southern Tamil Nadu, but this is a rare (now probably absent) occurrence along the mainland coast.

Loggerhead turtles feed mostly on crabs and molluscs. Worldwide, the largest loggerhead nesting populations are in south-eastern USA, Brazil, northern Mediterranean (mainly Greece, Turkey and Cyprus), Cape Verde on the west coast of Africa (though other parts may be poorly known), South Africa, Oman (Masirah Island) and Yemen, western Australia and Japan. There are few authentic records from

India, mostly from the Gulf of Mannar. In adjacent Sri Lanka, however, they are known to nest and forage in the offshore waters, though not in large numbers.

Flatback turtles are found only in northern Australia, with nesting beaches in western Australia, Northern Territory and Queensland. They have received little attention in terms of biology or conservation.

There are two species of ridleys – the kemps ridley and the olive ridley. Both are unique amongst turtles for the phenomenon of mass-nesting or arribada, when thousands of turtles come ashore simultaneously to nest. The kemps ridley is, in fact, known from a single population at Rancho Nuevo, Mexico, which was depleted from about 40,000 animals to a few thousand in the 1980s, mostly due to indiscriminate poaching of adults and eggs[76]. Intensive conservation programmes for this species have led to its recovery[77]. The olive ridley is more widespread than its sister, nesting both solitarily and en masse right through the tropical world. Major mass-nesting beaches for olive ridleys are found in Pacific Costa Rica, Mexico and Odisha on the east coast of India. While some mass-nesting sites, such as Suriname, are believed to have collapsed, others seem to be forming in Panama and Nicaragua, as well as in the Andaman Islands. Olive ridleys feed in deep, soft-bottomed areas which are rich in crabs and other crustaceans.

While solitary nesting occurs along much of the mainland coast of India and islands, there are major mass-nesting beaches in Odisha, notably Gahirmatha and Rushikulya, where more than hundred thousand turtles are known to nest in a single arribada. Mini arribadas (a few thousand turtles) have been recorded from the middle and north Andamans, and probably occur at other sites in Odisha.

A *turtle's life*

A turtle begins life beneath the sand on a tropical beach. The female turtle migrates from her feeding ground, perhaps thousands of kilometres away, to the offshore areas of the nesting beach. The males also migrate and mating occurs in offshore waters; sea turtles follow a promiscuous reproductive strategy, with multiple females mating with multiple males. Once her eggs are fertilized and ready, the female comes ashore at night, and crawls above the high-tide line. The distance from the high-tide line varies substantially and is a trade-off between the likelihood of flooding and the distance that hatchlings have to travel before reaching the sea. Once the turtle has decided on a site, she clears a body pit, which varies in size between species, and digs a flask-shaped nest with her hind flippers. Sea turtles always use alternate flippers to scoop sand, a behaviour that is so stereotyped that if a flipper is injured or missing, the turtle will still make a scooping movement with her phantom flipper during the process. This once prompted Archie Carr to say: '. . . a person, asked what a sea turtle is, might truthfully reply: "A sea turtle is a kind of turtle that never puts the same back foot into its egg-hole twice in succession."'[78]

Once the nest is ready (and it may vary in depth from 1.5ft for ridleys to over 3ft for leatherbacks), the turtle lays 100–150 soft-shelled eggs in the nest; leatherbacks frequently lay less than 100 eggs. The turtle then covers the nest and spends some time throwing sand around the nest to camouflage it. Sea turtles differ in small ways from each other in the closing sequence; for example, ridleys thump their nest down with their body, rocking from side to side. And leatherbacks often spend an hour or longer camouflaging their nests, leaving large plowed areas on the beach.

The females then return to sea and may nest several times in one season before returning to their feeding grounds. The nesting interval is about nine to sixteen days, but can be longer, especially for turtles nesting in two successive arribadas. Ridleys nest two to three times in a season, but other species lay more nests during a season, with green and loggerhead turtles known to nest up to ten times, or on some occasions, even more.

After about seven weeks, during which the eggs are incubated by the sun and metabolic heat, the eggs hatch, and the hatchlings emerge en masse at night and scramble for the sea. The incubation time is seven to ten weeks and sex is determined by temperature; higher temperatures tend to produce females. Sea turtles do not as a rule produce nests that have equal sex ratios. More commonly, the nests laid during different times of the season or on different kinds of beaches are predominantly male or female, thus resulting in the production of both sexes in a population over time.

When the hatchlings crawl out of the nest they have to find the sea, which they locate by orienting to the brighter horizon. Under natural conditions, the reflection of the moon and stars on the sea and the silhouette of trees and sand dunes in the background would make this a brighter horizon for the turtles. This again is very significant as landward lights can misorient hatchlings completely, leading to large-scale mortality. Many turtle nesting beaches today are significantly affected by light pollution, and the problem is likely to increase on most other beaches.

As the hatchings swim rapidly through the surf, they begin to swim against wave direction. This enables them to swim steadily in an off-shore direction. By this time, they begin to orient to the earth's magnetic field and this enables them to maintain their direction out at sea. It is believed that turtles can detect the earth's magnetic inclination angle and magnetic field intensity, giving them a bearing which can lead them very precisely to geographical locations. There may also be other factors such as chemical signals which enable them to precisely identify particular beaches and leatherbacks are believed to use oceanic ridges to navigate across open oceans. The hatchlings are believed to spend much of their early lives in seaweed rafts and driftlines in pelagic waters, until, years later, they move to other shallow water habitats. Mortality is particularly high during this early stage.

All sea turtles have a pelagic hatchling phase, spending many years floating with seaweed rafts and fish aggregating devices on trans-oceanic gyres and currents. Many species of juvenile sea turtles move to nearshore shallow water habitats for the remainder of their developmental years. There is contention about how long sea turtles take to reach maturity. While some estimate that loggerheads, ridleys and leatherbacks take ten to fifteen years to reach adulthood, there are indications that loggerheads may take three decades or so, although this may vary with region and associated conditions (temperature, prey availability and so on)[79]. Green turtles, which are primarily herbivorous, may take even longer, possibly up to fifty years to mature fully.

Once they have reached adulthood, both males and females migrate back to their natal beaches. Most species do not nest annually, except ridleys, who do so occasionally. Green turtles, loggerheads and leatherbacks nest once in two or three to five years. The remigration interval can be influenced by factors such as the food availability in the feeding area. This makes it imperative to collect biological data on nesting frequency to estimate numbers of turtles from nest counts. Furthermore, the fact that they do not nest every year means that there is tremendous natural annual variation in nesting which makes it difficult to estimate trends.

Adult turtles of different species prefer very different kinds of beaches. Olive ridleys and leatherbacks like broad sandy beaches with deep water approaches, uncluttered with coral or rock. They often nest on sand bars near river mouths. In fact the best known leatherback nesting beaches in Great Nicobar are all near river mouths (Galathea, Dagmar, Alexandria), while all the mass-nesting beaches of olive ridley turtles in Odisha are also near river mouths (Gahirmatha at the mouth of the Brahmani-Baitarani, Devi and Rushikulya near rivers of the same name).

As Archie Carr noted, sea turtle reproduction is marked by a few differences, but many remarkable similarities. Both aspects are fascinating for biologists who study the ecology and evolution of these animals.

Summing up

Starting with the fascination of a few biologists on different continents in the 1950s and 1960s—Archie Carr in the Americas, Robert Bustard and Colin Limpus in Australia, George Hughes in South Africa—sea turtles have become the most global of subjects, with research and conservation programmes in over seventy countries. Today, biologists study various aspects of sea turtle biology including nesting biology, physiology, migration, foraging, genetics, evolutionary biology, ecology, and so on. Amongst other things, they are long-lived, slow-maturing species that all live in a wide range of habitats separated by thousands of kilometres. So much has been done, and yet so little is known. It is this that makes them not only fascinating to study, but also difficult subjects for conservation.

Sea turtle research came to India surprisingly early, in the 1970s, thanks to a Scotsman, Robert Bustard, who had spent the previous decade in Australia, and an American, Romulus Whitaker, who had spent his early childhood in India, both more interested in crocodiles at the time. The former was responsible for initiating sea turtle programmes in Odisha, always a hotspot for sea turtle activity, while the latter initiated a conservation programme on the Chennai coast, which is active even today, and also an exploration of the remote Andaman and Nicobar Islands, which has led to research and monitoring programmes at several sites on the islands.

Subsequently though, a number of Indian pioneers such as Satish Bhaskar, C.S. Kar, E.G. Silas and others would drive surveys and research throughout the country, laying the ground for a new generation of biologists and conservationists in the 1990s and later. Bhaskar's groundbreaking surveys, especially in the island groups, uncovered important nesting sites for many species of sea turtles, especially leatherback and hawksbill turtles in the Andaman Islands and green turtles in the Lakshadweep Islands. Kar's research with his extensive datasets on tagged turtles and mass-nesting events set the stage for the many research programmes that would follow. One turtle that he tagged in 1978 was recovered in 1999, twenty-one years later, and remains the longest known interval of nesting for an olive ridley, at least in India.

As elsewhere in the world, research and monitoring programmes in India were closely linked to conservation, as sea turtles, whether deservedly or not, inspire that sentiment. In the chapter that follows, I trace a broad history of the interactions of humans and sea turtles in the subcontinent, which, during the post-Independence period, morphed from use of meat and shell to talk of sustainable utilization to eventually, a strongly protectionist view of conservation.

Notes

i Blyth used the scientific name *Couana olivacea* for loggerheads—the species name *olivacea* refers, however, to olive ridleys. During earlier times, the loggerhead was confused in many accounts with the olive ridley[80], and misidentification of these two species appears to have been common. Though the adults look different, the species are related, and juvenile loggerheads can be confused with adult ridleys. Also, the olive ridley was then known as the olive backed loggerhead turtle, and some errors might just have resulted from the incomplete use of the name. Later, Günther referred to the ridley (*Caouna olivacea*) as the Indian loggerhead and described as congeneric with the loggerhead turtle, which Günther believed was confined to the Atlantic, while the ridley/Indian loggerhead was confined to the East Indies.

ii In fact, a few taxonomists considered green turtles from the Malabar coast as a separate species, and those from the Red Sea were considered by some to be different as well. This is perhaps not surprising, given the considerable variation in shell colouration patterns, but recent genetic analysis has shown that green turtles worldwide can be considered as the same species, though there is a deep divergence between Atlantic-Mediterranean and Indian-Pacific populations[81]. Günther referred the Indian form as *Chelonia virgata* (as opposed to *C. midas*, which today is *C. mydas*). Günther also described the hawksbill (*Eretmochelys imbricata*) as *Caretta squamata* or *Caretta imbricata*, a genus now ascribed to loggerhead turtles (*Caretta caretta*), though he also provides the genus name currently used, calling it *Eretmochelys squamata*. The turtle was first described by Linneaus, until Louis Agassiz separated the Indian Caret, as Günther refers to it, from the Atlantic form. Günther correctly expressed doubt about whether this distinction would hold.

iii In 2002, a leatherback was caught and killed for sale of meat in southern Kerala[82]. In this instance, a local diver retrieved the carapace from the harbour where it had been dumped, for display in the Central Marine Fisheries Research Institute office. In a couple of instances, however, the turtles were alive and released into the sea. In Vizhinjam, a male leatherback was accidentally caught in a shore seine in 2001. A foreign tourist then negotiated its release for the sum of Rs 2,000, preventing it from being carved up for sale in the market. The same turtle was captured a few days later, but the fishermen apparently released it voluntarily[83]. A year later, another leatherback was captured in a shore seine in southern Kerala, and released by the community[84]. Similarly, a leatherback was rescued in southern Tamil Nadu in 2009 and released with some fanfare by the forest department and conservationists[85]. Another leatherback was captured in a fishing net and released by forest officials and police personnel in Vizhinjam in Kerala in 2009[86].

iv The overall goal of this project was 'to develop a sustainable model for the conservation of olive ridley sea turtles and restoration/conservation of their habitats along the Indian coastline, through a holistic and people centered approach'[87]. The project was coordinated by the Wildlife Institute of India (WII). One of the objectives of this programme was to extensively survey the entire coast for sea turtle nesting and mortality with intensive sampling of key areas, and to build a coastal network of organizations and individuals involved or interested in marine conservation. These surveys were carried out by a range of non-government organizations including the state forest departments, national research organizations and NGOs. On the west coast, Wesley Sunderraj and Justus Joshua carried out the survey of the Gujarat coast through the Gujarat Institute of Desert Ecology (GuIDE), Varad Giri of the Bombay Natural History Society (BNHS) surveyed Maharashtra and Goa coasts, while B.K. Sharath of the University of Mysore covered the Karnataka coast. S. Bhupathy of the Salim Ali Centre for Ornithology and Natural History (SACON) surveyed the Tamil Nadu coast, while the Andaman and Nicobar Environmental Team (ANET) covered the islands, starting a tagging programme in Great Nicobar Island. Basudev Tripathy of WII carried out the surveys in Andhra Pradesh and Lakshadweep. The NGOs Thanal and Nature, Environment and Wildlife Society (NEWS) covered the Kerala and West Bengal coasts respectively. Odisha was reported on by the state forest department. The project included a number of workshops and training programmes, as well as study tours to Australia and Malaysia for state forest and fisheries department officers. The first satellite telemetry study was also initiated through this project, where four transmitters were fitted on ridleys in Odisha in April 2001. A review of legal instruments was commissioned from the environmental lawyer, Sanjay Upadhyay, as also a review of community-based conservation in Kerala, Goa and Odisha by Roshni Kutty through Kalpavriksh. A small Geographic Information Systems (GIS) study was carried out on the nesting beaches of Odisha by the Odisha Remote Sensing Application Centre. The project also produced a series of manuals on sea turtle conservation.

v The Wildlife Institute of India returned to research on sea turtles in Odisha, led again by B.C. Choudhury, with support from the Directorate General of Hydrocarbons (DGH), a part of the Ministry of Petroleum and Natural Gas[88]. This was in the context of exploration for natural gas in the offshore areas of Odisha by public sector companies and private corporations. The project, conducted over several years, carried out satellite telemetry of over sixty olive ridley turtles, and monitored nesting populations and offshore distributions at multiple sites along the coast of Odisha from 2007–2011.

vi With support from the Marine Turtle Conservation Act Fund of the US Fish and Wildlife Service, Dakshin Foundation and the Indian Institute of Science, Bengaluru, launched long-term monitoring programmes for olive ridleys at Rushikulya, Odisha, and leatherback turtles at Little Andaman Islands. Both projects monitor a number of parameters including offshore and nesting population abundance, nest predation, hatching success as well as the effects of climate change on hatchling sex ratios. In addition, the leatherback project has satellite-tracked over ten turtles during their post nesting migrations in the Indian Ocean.

vii In 2007, the Indian Institute of Science initiated a large-scale project on human-wildlife conflict, in collaboration with the Norwegian Institute of Nature Research, with support from the Royal Norwegian Embassy and the Norwegian Research Council[89]. The project examined a number of cases of conflict with different taxa (carnivores, herbivores, elephants- and so on) in a variety of landscapes (forests, grasslands, human-dominated areas, seascapes)[90].

viii Much of the research has been carried out by the Central Institute of Fisheries Technology (CIFT), which modified the Georgia Jumper

TED for use in India, and designed the CIFT-TED[91]. This device won an award from the Southeast Asian Fisheries Development Centre (SEAFDEC). CIFT conducted trials with this device in Odisha and in Andhra Pradesh[92] to estimate catch loss due to the use of the TED. Subsequently, trials were conducted by WII in the early 2000s[93], and by WWF in 2009 and 2010, to promote the use of TEDs in Odisha[94].

ix M. Rajagopalan of the Central Marine Fisheries Research Institute was one of the first to work on a Ph.D on sea turtles in India[95]. He worked on ecophysiological studies on olive ridley turtles, while his student, S. Venkatesan, worked on similar aspects on green turtles[96]. Another student, P. Kannan, carried out his doctoral research on incidental catch of sea turtles along the Indian coast[97]. P.S. Rajasekhar also submitted a doctoral thesis on sea turtles in Andhra Pradesh[98]. C.S. Kar was the first of many to conduct doctoral research on olive ridley turtles in Odisha[99]. Following Bivash Pandav's pioneering work in the 1990s, Karthik Ram studied offshore distributions of sea turtles in Gahirmatha[100] at a time when conservationists were actively campaigning the state forest department to patrol offshore waters and curb mortality. Basudev Tripathy carried out an M.Phil. dissertation on olive ridleys in Rushikulya and returned for his Ph.D in 2004[101]. He studied nesting populations as well as offshore distributions of mating turtles in Rushikulya. Tripathy's work was supported in part by conservation grants, which he used to strengthen a local conservation group that had been formed by Pandav's field assistants at the site. In the late 2000s, Suresh Kumar worked as a research assistant on the DGH project with WII and completed his Ph.D on offshore distribution and migration.

Over the years, a number of students have also conducted research on sea turtles for their Master's dissertations. Arjun Sivasundar, from the Department of Ecology (formerly the Salim Ali School of Ecology) Pondicherry University, was one of the first to work on leatherback turtles in the Andamans[102,103]. K. Banugoppan[104] and J. Subramanean[105], also from this department, studied nesting along the Pondicherry and northern Tamil Nadu coasts. In recent years, Divya Karnad, from the National Centre for Biological Sciences Master's programme in

Wildlife Biology and Conservation, studied the impacts of lighting and temperature on olive ridley turtle hatchlings in Rushikulya[106,107]. A year later, M. Muralidharan from the Wildlife Institute of India studied the impacts of predation and nest site selection in ridleys at the same site[108]. In 2010, Ema Fatima, from Guru Gobind Singh Indraprastha University, conducted a population genetic comparison of the olive ridleys on the east and west coasts of India[109], through Centre for Ecological Studies (CES), Indian Institute of Science. The same year, Sasmita Mallick, a student of Forest Research Institute, compared the attitudes and perceptions of communities to sea turtle conservation and management along the Rushikulya coast[110]. Aparna Lal, a Master's student at the Wildlife Institute of India, carried out her research on the foraging ecology of green turtles in the Lakshadweep and their impacts on sea grass beds[111,112]. A few others have since carried out Master's projects on marine turtles in India.

SEA TURTLES: FROM FISHE TO FLAGSHIP

Kachua kababs

A GROUP OF US visited Indira Point on Great Nicobar Island in early 2002, three years before the tsunami. Meera Anna Oommen and I were based at the camp at Galathea, about 10 km away, studying tree-shrews and leatherback turtles respectively. Suvir and Sudhir were visitors from Pune, filming and exploring the islands. On our way, we had passed the Nicobari village of Chinghen (Chingam). Across the Galathea river, we first came across a cluster of huts in a clump of trees. We passed some olive ridley carapaces lying in the grass. A Nicobari woman squatted, removing the meat from a ridley turtle which had just been killed. She had already removed the ovaries and oviducts, from which both shelled and unshelled eggs lay neatly piled in another carapace. Further on, we came to the village itself, another cluster of typical Nicobari huts by the sea. As advised, we had presented the village head, Captain Sitaram, with a bottle of whisky and had spoken to him of common friends such as Ravi Sankaran, Manish Chandi and Rauf Ali. Another ridley – this one alive – was tied to a coconut tree.

The Nicobaris have been eating turtles for as long as they can remember. They catch them when they come ashore to nest. We are told that they like green turtle meat so much that they cannot wait for it to be cooked, and often eat it raw while they are chopping up the turtle. Unfortunately, regular take at the nearby beach has led to the decline of green turtles and they get few nowadays. But olive ridleys will do.

We went onwards, past where the road ended, and onto the lighthouse at Indira Point. The road is now gone, and the lighthouse partly submerged. En route, we looked for the coconut crab, a magnificent animal that shreds coconuts open with its enormous mandibles. But the last one on that stretch of beach seemed to have been eaten. Or so we assumed; it is unlikely it would have died of old age in that part of the world.

Indira Point has great political and geographical significance. Fifty kilometres from Campbell Bay along the North–South Road, it is the southernmost tip of India. No land lies directly south all the way to Antartica. Sumatra lies only 140 km away to the south-east, but there is just a vast expanse of ocean to the south and west. I had read Satish's account ('A Hawksbill nested on 4-5 April 79 at Pygmalion Point') and was on the beach trying to get a sense of the place as he must have seen it more than twenty years earlier. And as if time had stood still, I saw a hawksbill turtle nesting under a scaveola bush. I considered that a special connection to Satish's time there.

On the return journey, the onlive ridley at the end of the rope was missing. Instead, there were succulent pieces of meat being roasted slowly over a wood fire. A group of Nicobari youngsters sat around the fire waiting for the meat to be cooked. Subsistence take of turtles for meat occurs across the world. Many communities will talk about the taste of turtle meat, and the special place that a turtle hunt or turtle feast has in their culture. However, in the conservation community, a belief has come about that the consumption of turtles is inherently egregious, or that it is inevitably harmful to populations. Even in India, a dramatic transition occurred from viewing sea turtles as a fishery to according it a special place as a charismatic mega vertebrate, almost, if not actually, on par with tigers and elephants. Not all the consequences of this change were good. How did this change come about and what did that mean for sea turtle conservation in India?

Centuries of consumption

In many parts of the Indian subcontinent, adult sea turtles were not intentionally harmed because of religious beliefs that turtles are an incarnation of Vishnu, though they were killed for oil in Gujarat and Lakshadweep, and for meat and blood (as well as exported whole) in Tamil Nadu, West Bengal and the Andaman and Nicobar Islands, and millions of eggs were taken from beaches in Odisha. Many communities believed that the meat (and blood) had medicinal value. Muslims in the region generally do not eat turtle meat products because Islamic custom does not approve of it, although eggs are often consumed and other products may be used for a variety of purposes. In much of south India, turtles are considered as bad omens and cannot be brought into the house. There is a saying that has been recorded in coastal Tamil Nadu as well as in Lakshadweep that 'the aamai (=turtle; Tamil) is closely followed by the ameen (=tax collector)'. Christian and ethnic tribal communities do eat turtle meat and eggs. In Kerala, Christian communities even consume the occasional stranded leatherback, a species whose meat is minimally palatable and rarely eaten in any part of the world. The tribal communities in the Andaman and Nicobars are exempt from the provisions of the the Indian Wildlife (Protection) Act, 1972 and consume green turtle and olive ridley meat with great relish. In the past, when turtle eggs were exploited, many communities would leave about two to five in the nest, in theory to ensure the perpetuation of the species, though it is unlikely that so few eggs would hatch or emerge or contribute to the survival of the species. In any case, this tradition appears to have disappeared.

Sea turtles have been exploited over many centuries, probably millennia, for their meat, shell and eggs. Jack Frazier, a prolific writer and pioneer of sea turtle biology and conservation and a self-proclaimed trouble-maker at large, has produced a wonderfully exhaustive account of sea turtles in historic and pre-historic cultures, particularly from the western Indian Ocean and from the Caribbean and Yucatan peninsula[1]. He finds that sea turtles are known from various archaeological sites in the Middle East from 5000 BCE onwards. Marine turtle remains, mainly fragments of limb bones and carapace, have been recovered from numerous sites in the Arab world, including UAE, Saudi Arabia, Oman, Kuwait, Bahrain and on the Israeli coast. Zooarchaeological remains are also known from the south-eastern US, Caribbean Islands, and Yucatan peninsula. Similarly, many archaeological remains have also been found in South America and other parts of the world. Various cultural artifacts have been found at these sites, in addition to marine turtle remains, indicating that marine turtles have been part of human cultures for thousands of years. Sea turtles are represented in cultural artifacts from 2000 BCE onwards in seals from the Middle East, and in ceramics, and petroglyphs from Central and South America. Tortoiseshell products are known from various parts of the world from ancient times, including those from India and Sri Lanka.

Some of the oldest texts referring to marine turtles and their trade are also from several thousand years ago. There are accounts of turtles inscribed on cuneiform tablets from Sumerian cities from about 2000 BCE. Texts from Mesopotamia imply that marine turtles were used for medicine, food and in rituals. There are pictograms of turtles in Chinese writing. In ancient Greece, some authors wrote about the 'turtle eaters' (chelonophagi) who were apparently a primitive group of people who lived on islands in the Red Sea and caught and ate the large turtles that were found in their waters[2].

In about the first century BCE, a traders' handbook for the Indian Ocean written anonymously, called *Periplus Maris Erythraei*, described the trade from the Indian Ocean to the Mediterranean and Egypt[3]. In the *Periplus*, tortoiseshell trade figures prominently, including from India, Sri Lanka, Malay and Sumatra[4].

Strabo, a Greek historian and geographer from the first century CE, writes that large quantities of tortoiseshell and ivory were exported from Sri Lanka and India[5]. By this time, tortoiseshell was well known as a luxury item and used for a variety of purposes. The Roman historian, Pliny (24–79 CE), reported that turtle fishing was active in Taprobane, as Sri Lanka was then known. In the sixth book of a 1601 translation of Pliny's *Natural History*, the following lines appear[6]:

> They take also a great pleasure and delight in fishing, and especially in taking of tortoises: and so great they are found there, that one of their shels will serve to cover an house: and so the inhabitants doe employ them in steed of roufes.

Hawksbill scutes were sent abroad as gifts by Sinhala kings, perhaps as early as the first century BCE, when ambassadors were sent from Sri Lanka to Rome. Even at that time, the artisans of Galle and Jaffna were known for their use of hawksbill scutes.

Subsequent accounts in the ninth and tenth centuries CE confirm the extensive trade in tortoiseshell across the Indian Ocean. Though marine turtles are likely to have been part of literary accounts from ancient cultures such as Greece, China, Arabia and India, not much is known, perhaps because the literature has not been adequately explored in this context.

Thomas Bowrey writes in A *Geographical Account of Countries Round the Bay of Bengal, 1669–1679*[7] about Coromandel:

> Metchlipatnam affordeth many very good and fine Commodities . . . as also Chaires and tables of that admirable wood Ebony, Chests of drawers, Screetores finely wrought inlaid with turtle Shell or ivory, for which a Very great trafficke is driven into most parts of India, Persia, Arabia, China and the South Seas, as well as into England and Holland.

He writes later of palanquins being inlaid with ivory and turtle shell. Clearly, there was knowledge of some aspects of the natural history of turtles at this time, as at a later point in the book, in writing about the 'alligators' (actually crocodiles), he says that 'they lay their eggs in the Sand (Even as Turtle doe), and hatch with the heat of the Sun.'

The most significant mention of sea turtles from this period though is A *New Account of the East Indies*, subtitled 'Being the observation and remarks of Capt. Alexander Hamilton from the years 1688–1723', and further described as 'trading and travelling, by sea and land, to most of the countries and islands of commerce and navigation, between the Cape of Good Hope and the Island of Japan'[8]. Hamilton's account covered the 'treats of the coasts' of Guzerat, Bombay, Goa, Canara and Malabar on the west coast, as well as much of the east coast from Adam's Bridge to Fort St. George, and the Coromandel coast, from Fort St. George to Ganjam. He then writes about his travels in the ancient kingdom of Orixa, ending at Ballasore at the northern end. At the end of his first volume, he says:

> About 12 leagues to the Northward of *Cunnaca*, is the River's mouth of *Ballasore*, where there is a very dangerous Bar, sufficiently well known by the many Wrackes and Losses made by it. Between *Cunnaca* and *Ballasore* Rivers there is one continued sandy Bay, where prodigious Numbers of Sea Tortoises resort to lay their Eggs . . .

Cunnaca refers to the kingdom of Kanika in north-eastern Odisha, and the Rivers Cunnaca and Ballasore to the Rivers Maipura (to the south) and Budhabalanga (to the north). The reference was first noted by Priyambada Mohanty-Hejmadi[9], who pointed out that absence of any reference to trade or consumption (while there are repeated references to the consumption of fish and game throughout the account) implied that the use of turtles or eggs had not yet started in this region.

From the 1800s, there are several accounts of hunting of sea turtles by the tribals of the Andaman and Nicobar Islands. Colonel Colebroke's account (retold by Frederic Mouat[10]) from the late 1700s stated that more than a hundred turtles were captured by a British crew, which appears to have been a common source of turtle exploitation at the time. Ships would often capture turtles and keep them in the hold as emergency food supply. They were also maintained in pens at the seashore in Port Blair and slaughtered for food when required. Anthropological accounts of aboriginal communities by E.H. Man[11] and M.V. Portman[12] in the Andaman islands and Baden Kloss[13] in the Nicobar islands dealt mainly with the hunting and consumption of turtles and the rituals associated with them. Apart from the detailed descriptions, what is remarkable about all these writings is the superior attitude towards 'savage' and 'primitive' tribes, a view held not just by the observant administrator or curious explorer, but by anthropologists. They are spoken of in remarkably demeaning terms as if they were animals, and yet, in some of the accounts, there seems to be a grudging respect for their simple life.

Anslem de Silva, one of Sri Lanka's most published herpetologists, has written a fine historical account of marine turtles in the edited volume on *Marine Turtles of the Indian Subcontinent*[14]. By the nineteenth century, there was considerable documentation of sea turtle use in Sri Lanka. At the time, hawksbill turtles were caught and flipped over on their carapaces and left on the beach. Later, the turtles would be suspended by a pole slipped through their limbs over a flame to heat and remove the scutes, usually the costals and vertebrals. The turtles were then released as they believed that the scutes would regenerate. Also, the artisans believed that the scutes taken from dead animals did not possess quite the same quality[15].

Turtles were caught while nesting as well as from the water by harpooning and with nets. These turtles were slowly butchered in the markets while the animals were still alive. Tennant decried this practice, calling it a 'repulsive exhibition' and a 'frightful mutilation', while describing it in great detail[16]: '. . . the plastron and its integuments having been previously removed, and the animal thrown on its back so as to display all the motions of the heart, viscera and lungs.' A graphic description of the disemboweling followed, and customers ordered the parts that they were interested in:

> Each of the fins is thus successively removed, with the portions of the fat and the flesh, the turtle showing, by its contortions, that each act of severance is productive of agony. In this state, it lies for hours, writhing in the sun, the heart and head being usually the last pieces selected.

Till the end, Tennant wrote, the opening and closing of the eyes and mouth showed that 'life is still inherent'. Not just did the British observers find this practice barbarous, the humane society of England protested against this practice, and requested the Ceylon government to pass an Act against cruelty to animals, perhaps providing one of the first documented instances of animal rights versus use.

From consumption to conservation

In the twentieth century, the two main centres of turtle trade were the Gulf of Mannar and Odisha. There was also possibly a supply of green turtles from the Andamans to Kolkata at the turn of the century. According to Frazier[17], there was a regular trade in tortoiseshell from Zanzibar to India. But there is not much documentation of the volume of take and trade. Most early accounts deal with sea turtles in the context of their consumptive value. The earliest of these are two papers in *Journal of the Bombay Natural History Society (JBNHS)* in 1950 entitled 'Edible chelonians and their products' by Acharji of Zoological Survey of India (ZSI)[18], and 'Turtle Fishing in the Sea around Krusadai Island' by Kuriyan[19] of the Marine Biological Station, in Pamban, Gulf of Mannar. Archarji's article started with: 'On account of the food shortage that is at present prevailing in India …'

By this time, the fishery departments of the coastal states were mandated to make special efforts to increase fish production. Acharji suggested that not enough attention had been given to chelonians as a source of food and protein. He wrote: 'One great advantage in the fishery of these animals is that they can be kept for a considerable time out of water, and do not suffer any appreciable deterioration in their edible qualities.'

Turtle farming in other parts of the world was known and recurred in many articles over the next two decades as a possibility for India. In addition to a few freshwater and land turtles, Acharji gave a brief description of green, hawksbill and leatherback turtles, and potential uses for food, tortoiseshell and oil, respectively. In his description of the species, he calls the leatherback 'a great wanderer, travelling from ocean to ocean', indicating again that the transoceanic migrations of these animals was already part of scientific lore. Surprisingly, he makes no mention of olive ridleys (most likely because there was widespread confusion about its taxonomic status). At this time, turtles (probably green) were sold for approximately Rs 30–40.

In the Gulf of Mannar, green turtles were taken in large numbers both on Sri Lankan and Tamil coasts. Interestingly, in writing about the capture and consumption of green turtles in Jaffna, he wrote that Hindus were the chief consumers of meat. Kuriyan, however, in his article about turtle fishing practices around Krusadai Island in the Gulf of Mannar stated that it 'has long been conducted by non-Hindu fishermen' and 'Hindu fishermen do not generally fish for turtles; their indifference being attributed to their religious belief that the second of the *dasavatars* had the form of a turtle'. But Acharji, in a footnote, documented the consumption of *Kachuga* (a freshwater turtle) meat, saying that 'in the ancient law books of the Brahmins, the meat of tortoises is not forbidden'.

His source for this is unclear, as any meat consumption is today strictly forbidden amongst Brahmins in south India. In Bengal, however, where most of the turtles from Odisha were shipped, the Brahmins do eat fish, which they consider the 'fruit of the sea'. Perhaps turtles were also considered fish, as they are in some other parts of the world[i].

Kuriyan detailed the kinds of net that are used to capture turtles in the region, including nets made of the bark of *Acacia*. He indicated that green turtles were mostly captured for meat, while other turtles were used for oil and their shell. Green turtles were graded according to their size, and the price varied accordingly. There was a significant rise in the price after the war. In 1957, Krishnamurthy included a paragraph on the turtle fishery in his article on the fishery resources of Rameswaram in the *Indian Journal of Fisheries* (survey in 1952)[20]:

> There are four merchants at Pamban and two at Rameswaram whose chief occupation is the collection and the export of the turtles that are landed in the Island. Annually 2000 turtles are being exported by the four merchants from Pamban and 400 turtles by the two merchants from Rameswaram. All the turtles are being exported to Ceylon, by country crafts, 11 in number, at an interval of 15 days.

In the next reasonably comprehensive account of turtles as fishery, Jones and Fernando of the Central Marine Fisheries Research Institute (CMFRI) wrote in 1973 that about four to six thousand turtles were taken annually in the late 1960s in southern Tamil Nadu, with about three-quarters being green turtles[21]. They also wrote that olive ridleys were considered to have a 'disagreeable flavour' while the demand for hawksbills had declined due to the availability of plastic substitutes for tortoiseshell and because the meat was sometimes poisonous. Here, as elsewhere, leatherbacks were not consumed but their fat used for oil for caulking the bottoms and sides of wooden boats. They estimated that about 3,000 to 4,000 turtles were caught from Pamban to Kanyakumari and about 1,000 turtles from Palk Bay between April and September, the peak being from May to August. They suggested that there may have been a decline after the cyclone in 1964 which may have affected sea grass beds. The turtles were caught using special nets (amaivalai) made of cotton or nylon. There was a collective slaughter every Sunday at Tuticorin. The meat was exported to the UK and Germany (1,000 to 1,500 kg per year), and also from Pamban to Jaffna, but this stopped after the Sri Lankan government imposed restrictions.

Jones and Fernando wrote: 'While estimating the turtle fishery potential of the area, the results of a series of experimental fishery carried out between Kuttankuli and Cape Comorin, during 1960–62 by some merchants when 3500–4000 turtles were caught of special interest. It is reported that each year between December to February 1000 to 1500 were caught and no sign of any decline was noticed during that period.'

In evaluating the fishery, they indicated that there was scope for expansion, but restrictive measures were required for effective management. They noted that the capture of turtles while laying eggs could lead to a decline. There were also self-regulation mechanisms; due to a large mesh size, small ones were seldom caught, and when they were, the fishermen released them into the sea (sometimes with compensation of one rupee from the merchants). They concluded that a detailed study was required 'for optimum utilisation of the valuable seafood as information on the same is very meagre at present'. In any case, protecting small juveniles most likely did not help the populations sustain; large juveniles or sub-adults would have to be protected.

There is not much data on the earlier trade and export in sea turtles from this region. In 1996, Agastheesapillai compiled the data on export from 1945 to 1963-64 from port office records[22]. He estimated about 1,000–4,000 turtles in most years, with green turtles comprising about 90 per cent of the catch. The average weight of the turtles was about 51 kg based on 2,186 specimens weighed at Mandapam station between 1971–1978. At the time, meat, shell and oil were exported to the US, the UK, Germany, Bahrain, Singapore, Netherlands and Belgium.

There are also several accounts of poisoning from eating turtle meat during the late 1970s. Tennant and Deraniyagala had reported cases from Sri Lanka from the nineteenth and early twentieth century where people had died after eating turtle meat. Tennant wrote that flesh of the turtle was poisonous during certain seasons and recorded an instance from 1840, where twenty-eight persons were taken ill after eating turtle meat and eighteen died[23]. Deraniyagala wrote that the flesh of leatherbacks and hawksbills may be toxic because of the food they consumed, particularly some toxic marine algae or ascidians or jellyfish[24]. He said that experienced fishermen would test the meat by throwing raw liver to crows.

In India, E.G. Silas and Bastion Fernando compiled the cases of turtle meat poisoning from southern Tamil Nadu from 1961 to 1983[25]. Most of these were in Tuticorin and were attributed to hawksbill meat, though green turtle meat was also implicated in a couple of cases. In each of these cases, a few people died, and in a couple of instances, more than ten perished. Many cases were probably not documented. Bhaskar noted in *Hamadryad* in 1978 that 'on the South Coromandel coast, deaths from this phenomenon occur every two to three years. Live turtles were brought from nearby villages and slaughtered in the market place; fresh blood is an elixir, and the meat sold at Rs 4–5 per kg'.[26] He found green turtles, hawksbills and ridleys being sold, and noted that a 'crude test' was carried out to ensure the edibility of the meat. In many of these cases, breastfeeding infants were also reported to have died.

The fishery at Tuticorin continued despite the introduction of the The Indian Wildlife (Protection) Act, 1972 and the protection of sea turtles. Later surveys by Bhupathy (a leading herpetologist who died tragically in a field accident in 2014) and others in the early 2000s however indicated that the proportions of sea turtle species had changed dramatically[27]. Once dominated by green turtles, catches were now dominated by olive ridleys with greens forming less than 20 per cent of the catch. Turtle slaughter was mostly clandestine and the fishermen were wary of sharing information with strangers.

At one point in the 1980s, local youth had formed an association called the Young Christians Blood Drinkers Association as a protest against the protection measures. In 1982, after a survey of the area along with B.C. Choudhury and Anne Ahimaz, Zai Whitaker wrote in *The Indian Magazine*[28]: 'Fresh turtle blood was being drunk as an elixir at one rupee a glass and old men and women gulp it down, blood at the corners of their mouths like so many draculas.'

When I visited the Tuticorin harbour in 2001, there were boards that announced the protected status of sea turtles, and the fishermen were reluctant to talk about turtle consumption. Despite this, two olive ridleys were tethered to the dock in plain view. The lads who had caught the turtles were soon tracked down, and they appeared after confirming that I did not work for the forest department. Soon it transpired that they had a large green turtle stashed away as well and they were much more reluctant to show me this catch. Though both species are on the same schedule of the Wildlife Act, the relative rarity of green turtles by this point appeared to make it a more valuable item but also apparently 'more illegal' to catch.

The other major turtle fishery occurred in Odisha. Prior to the 1970s, there was no organized fishery for adult turtles in Odisha, but whenever live adult sea turtles were found in fishing nets, they were collected and transported to the nearest railway station from where they were booked to Kolkata. The late 1970s saw heavy directed take of olive ridley turtles in Odisha, fuelled by demand in West Bengal and recently introduced mechanization in boats. Thousands of turtles were shipped from the coast of Odisha to Kolkata, piled high on their backs in the open coaches of goods trains, mostly alive, requiring no refrigeration. Subsequently, the trade declined due to enforcement of laws by the Indian Navy, Indian Coast Guard and the Forest Department of Odisha.

Over the 1950s and 1960s, several articles appeared in the journals, *Indian Seafoods* and *Seafood Export Journal*, on the prospect of turtle fisheries and farming. In 1975, Santharam wrote an article titled 'The marine edible turtles' in the *Seafood Export Journal*[29]. And in 1976, T.S.N. Murthy, a herpetologist with ZSI and A.G.K. Menon documented the export of turtle meat and tortoiseshell to Japan, Europe (East and West Germany, Switzerland, Italy, the UK, and so on) and the US[ii]. However, by the end of that decade, though articles still referred to turtles as resources, the notion of sustainability and conservation had entered the lexicon. By 1981, Murthy concludes his article by saying[30]:

> Thanks to the IUCN, sea turtles are listed as species threatened with extinction and the government of India has made them protected species. Unless the breeding beaches are properly policed, the depredation of turtles and their eggs by the beasts and humans alike continue unabated. India, with a coastline of over 4,000 miles and about 1,280 islands, is actually aware of the need to draw maximum food yields and in this context turtle culture can offer a partial solution to the protein food shortage. We cannot afford to ignore turtles any more.

There is inherent self-contradiction in the demands for culture and consumption, and policing and protection. Clearly, no one had ignored turtles. This was the language of takeover by scientists and government.

Several publications from CMFRI in the 1980s reflected this transition[iii]. In 1984, they conducted a workshop in Chennai[31], where Satish Bhaskar shared his, by now, extensive knowledge of the sea turtles of the Indian coast, and Chandrasekhar Kar his research in Odisha. Most of the other presenters focused on various aspects of threats, recovery and management. Rajagopalan alone spoke about the economic and cultural value of sea turtles, but more in the past than in the future tense[32]. The papers are published in a special publication of CMFRI, which is even more valuable because the post-presentation discussions have been transcribed into it.

In 1968, the CMFRI had held a 'Symposium on Living Resources of the Seas around India' (the proceedings were published in 1973)[33], which included just one paper on turtles i.e. on the turtle fishery in the Gulf of Mannar (by Jones and Fernando[34]). By 1985, when the Marine Biological Association conducted the 'Symposium on Endangered Marine Animals and Marine Parks', there was an entire section on sea turtles[35]. Several international biologists – including Brian Groombridge and Ed Moll – presented papers at this symposium. All the papers were on aspects of biology and conservation. Brian Groombridge provided an early overview of 'India's sea turtles in world perspective' and pointed out the importance of the arribada beaches in Odisha and the nesting grounds of leatherbacks and other sea turtles in the Andaman and Nicobar Islands[36]. Groombridge touched upon the 'contentious issue' of sustainable use and argued strongly that 'there should be no question of legal utilisation of this resource in the immediate future' till there was more 'refined protection' of the beaches. He did allow that the potential existed for India to support one of the largest ridley populations and simultaneously exploit the resource for eggs and meat, 'for communities not averse to these foodstuffs'. Only Whitaker took a positive view of use, and devoted a small section to farming (giving the Cayman Islands example) and recommended the use of eggs from arribadas where many get destroyed by erosion. Interestingly, his original title 'Rational use of estuarine and marine reptiles' was published as 'National use . . .' instead, an irony at many levels[37].

The tide was turning.

Following the national preoccupation with the food crisis in the 1960s, some discussion of environmental conservation and wildlife had begun to enter the rhetoric of natural resource use. Though the Indian Forest Act had been passed much earlier, in 1927, a seminal event was the enactment of the Indian Wildlife (Protection) Act, 1972.

This paved the way for the listing of 'endangered' species in various schedules, Schedule I according the highest protection. In 1977, all the species of sea turtles found in India were listed under Schedule I of this Act. The Act also enabled the creation of sanctuaries and national parks, though these were mostly created in terrestrial areas. There is considerable debate today about the absence of a scientific basis for the creation of these parks. Most were created on the basis of historical contingency and political factors, rather than biological or ecological factors. Mostly, charismatic mammals, such as elephants, tigers and so on, were used as flagships or umbrellas for the creation of these protected areas. More importantly, these did not account for communities who may already have been living in these areas for generations, assuming that relocation was an inevitable consequence. After thirty years of debate, controversy and bitter struggle, the Scheduled Tribes and Other Traditional Forest Dwellers (Recognition of Forest Rights) Act, 2006, was passed, recognizing the rights of forest-dwelling communities. The positive and/or negative consequences of this Act are yet to play out.

Sea turtle conservation in Odisha began in the early 1970s when the large-scale legal/incidental take of turtles from Gahirmatha was widely reported[38]. In the early 1980s, numerous petitions and letter writing campaigns were supported and endorsed through the *Marine Turtle Newsletter*, an international newsletter[39], and several hundred letters were in fact written to Prime Minister Indira Gandhi[40]. J. Vijaya, a young, adventurous field biologist, conducted surveys in the early 1980s and reported on the large numbers of turtles being sold in fish markets near Kolkata[41], and this, along with her photographs of hundreds of turtle carcasses (published in *India Today*[42]), brought even more attention to the extraordinary numbers of turtles being killed in Odisha. Prime Minister Indira Gandhi's support, and her initiative to involve the coast guard in protecting the marine area at Gahirmatha, helped drastically reduce the direct take to a point where it was thought to be negligible.

While the Odisha programme was coordinated by the forest department, the turtle hatcheries in Chennai were operated first by the Madras Snake Park, followed by the CMFRI and Tamil Nadu Forest Department[43]. By the late 1980s, the state hatcheries in Tamil Nadu were closed due to lack of funding, and the programme returned to the realm of NGOs. Since 1988, the Chennai hatchery has been operated by the Students' Sea Turtle Conservation Network (SSTCN). The state forest departments established hatcheries in several other states in the 1980s, including Karnataka, Gujarat, Odisha and West Bengal. Student and NGO programmes were also initiated at a number of these sites in the 1990s and are now active in almost every coastal state. Most importantly, these programmes served as powerful tools of education, spreading awareness about sea turtles and coastal conservation. These programmes are diverse in geographical location, cultural representation, conservation issues addressed, and also in their impacts on turtles and the respective societies involved.

While sea turtle conservation began to receive more attention at a countrywide level through the efforts of NGOs and national-level projects, Odisha remained firmly in the centrestage. Following reports in the 1990s that thousands of dead turtles were washing ashore[44,45], various projects such as Operation Kachhapa were initiated[46], mainly aimed at reducing trawler-related mortality through support to the forest department. In the last decade, a number of large NGOs such as Greenpeace and WWF have been involved in sea turtle conservation in Odisha, but many small local NGOs have also been active at the different mass-nesting beaches. In 2004, traditional fishworkers, local conservation groups and national conservation agencies came together as the Odisha Marine Resources Conservation Consortium to pursue common objectives for the conservation of marine resources, including marine turtles, while promoting the livelihoods of the fishing communities[47].

In their own ways, both Romulus Whitaker and Robert Bustard left a legacy of research and conservation. Bustard's 'discovery' of Gahirmatha had far-reaching impacts in initiating a research programme and inciting several decades of conservation and rhetoric. Odisha has been very much the focus of sea turtle research and conservation over the last few decades. Whitaker's turtle walks developed into a regular turtle monitoring programme on the Chennai coast, and eventually inspired NGO-based conservation along the entire mainland coast. The Madras Crocodile Bank Trust, also established by Whitaker, initiated surveys in the 1970s not only along the mainland coast, but also on the offshore islands.

Bustard's encouragement of research and his collaboration with the government led to state-centred conservation action (fuelled, of course, by activists), and multiple research projects over the years, mostly by large national institutions. It took several decades for local community groups to become involved with sea turtle conservation in Odisha. On the other hand, the turtle walks in Chennai, though managed by the state for a few years in the 1980s, always involved local non-government organizations, and eventually passed to a students' group, paving the way for many more such initiatives along the mainland coast.

Of course, these trajectories are not entirely explicable by the initial conditions or the philosophies of those who started the programmes. The states are very different from each other economically, socially and culturally. The Chennai programme was based out of a city and had more access to young, educated environmental enthusiasts. Odisha, with its globally significant mass-nesting populations, was always likely to attract outsiders, both national and international conservationists, as well as political attention, as the conservation stakes were higher. Nevertheless, both programmes are central to the history of sea turtle conservation in India. But let us turn to someone who was responsible for the change in perspective within the fisheries community to a more holistic and conservation-centred view of sea turtles.

The visionary

Started in 1837, the Madras Christian College has many distinguished alumni. Less glamorous than some of the well-known colleges in the country's capital, this 170-year-old college nevertheless has produced its share of luminaries over the years. More to the point, students of this institution have played a fairly key role in the trajectory of sea turtle conservation in the country. P.J. Sanjeeva Raj, who wrote in

1958 about the Tamil poem on sea turtles, was faculty and head of the Department of Zoology for many years. And during my time there as a student of zoology, a group of us from the same department, along with other friends and enthusiasts, founded the Students' Sea Turtle Conservation Network (SSTCN). Down the years, many students of this college (and others in Chennai, of course) have organized or participated in the group's activities.

But one of the first to influence sea turtle conservation was E.G. Silas, former director of the CMFRI. Silas is a well-known figure in sea turtle circles, as he promoted both research and conservation of sea turtles through CMFRI in the 1970s and 1980s. Having read many of his papers, and heard his oft-quoted prediction in 1983 that the nesting beaches of Odisha would turn into 'the largest grave yard of olive ridleys',[48] I decided to seek him out.

I met him at his residence in Kochi, where I had traced him through colleagues in the fisheries world. To my delight, I discovered that Silas was a student of Madras Christian College from 1946–1950, exactly forty years before me. After graduating, he joined Madras University to work on the biogeography of fish in central, south and south-east Asia. In 1955, his paper on 'Speciation among the freshwater fishes of Ceylon' was published in the *Proceedings of the Symposium on Organic Evolution*, conducted by the National Institute of Sciences, New Delhi (now the Indian National Science Academy, or INSA). This symposium was considered a landmark event, signalling the growth of evolutionary science outside the 'West', and had papers by J.B.S. Haldane, Peter Medawar, Sunder Lal Hora and others.

Silas himself had worked under the guidance of Sunder Lal Hora, a biogeographer of considerable influence. Hora's Satpura hypothesis, which suggests that the Satpura range served as a bridge for faunal dispersal between north-eastern India and the Western Ghats, has great appeal even today. Considering that many of my own students work on various aspects of biogeography (including on Hora's hypothesis), I must admit that I felt a certain thrill at having found a link to the man.

After completing his Ph.D, Silas moved to Bombay (now Mumbai) to work as the registrar and curator of the Bombay Natural History

Society (BNHS) at Salim Ali's invitation in 1954. Soon after, he went on a Fulbright fellowship to the Scripps Institute of Oceanography as a research fellow. Silas returned and continued as the curator of BNHS till 1959, when he took a position with the CMFRI in Mandapam, Tamil Nadu. Mandapam was the headquarters of CMFRI till it moved to Kochi in 1971. He got the position of scientist in 1963, and became director in 1975.

Silas told me that a turning point came when he attended the United Nations Conference on the Human Environment in Stockholm (also called the Stockholm conference) in 1972. This was the first UN conference on environmental issues and a landmark event for international policy on the environment. There was, however, little awareness of environmental issues in India at the time, though Prime Minister Indira Gandhi was one of only two heads of state present, along with the host, Olaf Palme. When he returned to India and took up the directorship of CMFRI, he discovered that there was little concern about conservation and management in the institution. These issues were not even, he said, 'in the dictionary of fisheries at that time'.

Silas decided that there needed to be a focus on two or three major issues, amongst which were marine mammals and sea turtles. Silas organized a system to collect data on the accidental catch of marine mammals and sea turtles, as well as records of dead animals which washed ashore (strandings) through the fifty-three regional centres of the CMFRI. He also started involving the younger staff of CMFRI, including Rajagopalan, who would go on to work principally on sea turtles through his career. In addition, Silas started the research station at Kovalam, near Chennai, where eggs were translocated to the hatchery for research.

Of course, there was not a lot of support for this at the beginning. CMFRI, started in 1947, became a member of the Indian Council of Agricultural Research (ICAR) in 1967. Initially, there was a lot of criticism from ICAR for taking up these programmes, as they were focused entirely on increasing fish catch. But Silas believed that all these were interconnected and interlinked. He was able to convince ICAR that, in order to fully understand fisheries, one had to look at the whole ecosystem, including other species such as whales, dolphins and sea turtles. Silas had attended the World Conference on Sea Turtle Conservation in 1979. There he encountered some of the leading sea turtle conservationists and biologists, from whom he got ideas for programmes in India. Frazier recalls a conversation with Silas and a Sri Lankan colleague at this meeting, where he tried to promote the idea of a binational marine reserve in the Gulf of Mannar, an idea they must have believed was far-fetched, given the politics of the region.

Silas's interest in research was fundamental. He was greatly interested in the life cycles of sea turtles; questions such as the fate of hatchlings fascinated him. Was it plankton, was it ocean currents? Speculating on the distribution of leatherbacks, he said: 'At one time, the leathery turtle was breeding along this coast. And we had at least some information on nesting along Calicut and in these areas, and they coincided with the explosive blooming of medusae along the coast, and medusa is a feed for the leathery turtle.'

In his fascination with the breeding habits of various species of sea turtles on Indian coasts, Silas was instrumental in influencing Satish Bhaskar to spend an entire season on Suheli island in the Lakshadweep, now part of the Satish legend.

Silas was greatly interested in the instances of poisoning from turtle consumption that occurred along the southern Tamil Nadu and Kerala coasts. He visited Quilon (now Kollam) to look at the post-mortem records, and wrote detailed reviews of the numerous cases of turtle poisoning from southern India and Kerala. He said: 'And then this also made me think about the turtle poisoning that had taken place and people had died along the Kerala coast, particularly in Quilon and other places. I was able to do some work on that at the time and found out how this poison was systemically entering into the milk of the mother when she was feeding the baby, affecting both.'

In the 1980s, Silas also pushed for more research on the olive ridley arribada populations in Odisha. They conducted status surveys in Odisha and West Bengal, wrote about the 'large and mini arribadas' of olive ridleys, and documented incidental mortality. Silas and his colleagues had already started documenting the stranding of sea turtles on the Odisha coast. He recalls seeing 7,000 dead turtles on one trip.

In 1983 itself, Silas and his co-authors wrote: 'We feel that unless urgent action is taken in regulating or adopting new modifications in fishing gear, the nesting beaches along the Odisha coast may turn to be indeed the graveyard – the largest graveyard of olive ridleys – anywhere in the world,' a warning that has been remarkably prophetic.

Jack Frazier recently remarked to me: 'As an objective, responsible scientist, Silas clearly recognized that the massive incidental catch and mortality of olive ridleys in the waters near to the Gahirmatha arribada beaches had to be mitigated, and one of most direct ways of achieving this was through gear modifications – particularly turtle excluder devices (TEDs). Hence, Silas took a position that no one else in fisheries in India would countenance; but he had the integrity and vision to state what was needed for the sake of fisheries as well as marine ecosystems, being decades ahead of his times.'

Silas warned of the bureaucratic difficulties in implementing conservation through the state departments. Despite providing training to many of the state fisheries offices in special training programmes, he found that the staff would remain committed for as long as they were in that particular position, but would get transferred after two or three years. This is a problem that persists today in both the state fisheries as well as forest departments.

In 1984, Silas organized the first workshop in India on sea turtle conservation and management. This workshop, held in Chennai, had over forty participants and included such luminaries as Jack Frazier, Romulus Whitaker, C.S. Kar, Satish Bhaskar, Shekar Dattatri and M. Rajagopalan, all of whom presented papers on various aspects of sea turtle biology and conservation. Remarkably, many of the issues raised then are still problematic now – incidental catch, egg predation, loss of habitat through erosion and so on.

Silas felt that the 'complicated life history' of sea turtles, particularly the distances between feeding and breeding grounds, rendered both studies as well as management difficult. In 1985, as the president of the Marine Biological Association of India, Silas also conducted a

'Symposium on Endangered Marine Animals and Marine Parks' in Cochin, to mark the twenty-fifth anniversary of the association. The meeting was attended by 170 participants from sixteen countries. This was an event well ahead of its time, given how little attention marine conservation has received in India in the following decades. The inaugural speech given by Salim Ali, doyen of Indian ornithology, for whom Silas had the greatest admiration. Jack Frazier could not attend the meeting, but sent a typically thought-provoking and philosophical paper on the meaning of the term 'endangered' in conservation. Silas himself was nothing if not pragmatic. In his presidential address, he questioned the basis on which management decisions were made. On the 'destruction of virtually millions of sea turtle eggs' by erosion, he asked: '. . . could we utilize the "doomed eggs"? Should emotion override the development of sane management reasons?'

Echoing this sentiment, Salim Ali pointed out that while it was difficult to persuade people that they were over-exploiting a resource 'when they see thousands of individuals of species like gulls and terns and ridley turtles . . .', going to the other extreme of indiscriminate exploitation was equally futile. He said: 'As I have often stressed to wildlife enthusiasts of the sentimental and emotional variety, meaningful conservation does not mean the complete denial of these very valuable natural resources which have helped to sustain man for thousands of years.'

Silas served as the director of CMFRI from 1975 to 1985. During this decade, he managed to inspire and instigate a great deal of research on topics that had, till then, been ignored. He proposed many ideas, such as the inventorying of genetic resources, that were novel at the time. More importantly, he brought attention to conservation and sustainability and integrated ecosystem approaches. Silas inspired many of his colleagues such as Bastion Fernando and Krishna Pillai to actively conduct research and collect information on sea turtles in their regional stations in Tuticorin and Vizhinjam, respectively. He also motivated M. Rajagopalan to pursue doctoral research on the eco-physiology of sea turtles and in setting up the centre at Kovalam, Chennai.

Still relatively active in local CMFRI events, Silas commands a large amount of respect and affection in the fisheries community in India. Nothing can sum up his contribution and love of research more than his parting present to me, two papers published on biogeography in the early 1950s and a review on the future of research on fishes (replete with accounts of the latest techniques) published in 2010.

A primer on sea turtle conservation

Sea turtles are probably the most widespread icons for conservation. They might not be as recognizable as the panda, they may not have received as much publicity as whales and dolphins, they may not have received quite as much attention as tigers and elephants in India, but they are not far behind all of these, if indeed they are behind at all. This is best illustrated by the participation in the Annual Symposium on Sea Turtle Biology and Conservation. Over the last decade, the symposium has attracted 500 to 1,000 participants at each of its annual meetings. More significantly, these participants come from sixty to eighty countries around the world. All of these countries have sea turtle populations. This remarkable global distribution of sea turtles has also created a worldwide community of biologists and conservationists working on a range of local and larger issues.

From the earliest days of the modern conservation period (*circa* 1950–1960), those who worked on these species believed that many sea turtle populations were being depleted. The reasons for these were varied. In the initial days, the main concern was the take of sea turtles and depredation of eggs by natural and feral predators, as well as by humans. Many early conservation programmes sought to reduce the killing of sea turtles on the beach as well as in the water. Much of this was carried out through the respective states by strict, even draconian, enforcement. The move towards working with communities has come about in recent years.

In the last few decades, incidental mortality from drowning in fishing nets and other gear has become an increasing threat. Globally, this coincided with the rapid development in the 1960s of mechanized fishing, bottom trawling in particular. In nearshore areas in many parts of the world, bottom trawlers caught large numbers of sea turtles, which drowned and were thrown overboard as they could no longer be utilized due to protective laws. For example, tens of thousands of loggerheads have died as incidental catch on the eastern coast of the US[49], and more than hundred thousand ridleys have been killed on the Odisha coast and elsewhere[50]. Sea turtles are also killed in nearshore gill nets, especially since mechanized winches and boats and better material led to the increase in the length of these nets[51], and in pelagic longlines (loggerheads and leatherbacks in particular)[52].

In response to the threat from trawling, TEDs were developed in the US[53], and a long period of research and negotiation with trawl fishermen followed[54]. TEDs comprise a metal grill at the back of the net, which allows fish and/or prawns to pass through, but not larger animals such as turtles or marine mammals, and a trapdoor through which these animals can escape. While effective, most fishermen believed that TEDs would result in a substantial loss of catch, and were reluctant to use them in the US and wherever else they were introduced over the following years[55].

In addition, coastal development has remained a pervasive threat to sea turtle nesting habitats[56]. In some cases, beaches have been completely destroyed due to sand mining or construction. In many parts of the world, coastal development has led to erosion, in response to which beach armouring structures (sea walls) are established, which result in the loss of sandy beaches or prevent access to them for sea turtles. Development has come from a variety of sectors, including residential housing, industry, roads and highways, tourist resorts, and so on. In some parts of India and south-east Asia, extensive exotic tree plantations (ostensibly planted with good intention) have resulted in the erosion and loss of nesting beaches[57].

Coastal development has another nasty effect on sea turtles: misorientation of hatchlings. Since sea turtle hatchlings locate the sea using the brighter horizon when they emerge from nests, any illumination on the landward side can misorient them so that they are unable to find the sea and die due to predation or dessication as they mistakenly wander inland. This alone has probably resulted in the mortality of many millions of sea turtle hatchlings around the world. Sea turtle biologists in Florida have studied the impact of lighting on sea turtles for many years and even produced a manual on turtle-friendly lighting[58], but unfortunately, its use is not widespread. In the water, pollution has been a widespread threat, but the impacts of this are little known. There are a few studies to suggest that diseases such as fibropapillomatosis may be aggravated by immune suppression resulting from pollution[59]. In recent years, climate change has emerged as a significant concern to sea turtles. Rising sea levels can result in the loss of nesting beaches, and changing temperatures can affect nesting phenology as well as sex ratios of sea turtle hatchlings.

Sea turtles are now protected by several national level and international laws[iv]. Across the world, conservation programmes have worked tirelessly to address threats. While some have worked at the local level to address threats from fishing or tourism or depredation, others have attempted to bring about change in national and international law and policy. A standard conservation practice across the world is the establishment of beach protection programmes. These involve the protection of eggs on the beach either in-situ, through protective fencing of one sort or another, or the establishment of hatcheries. Hatcheries in particular have become a very widespread conservation tool for sea turtles, both for protecting eggs that would be otherwise doomed (to predation or erosion) as well as for education and awareness programmes[60]. Another reason why hatcheries are popular is that many beaches have some level of artificial illumination, and newly emerged turtles produced in hatcheries can be safely released near the water, ensuring lower mortality.

Some icons of the global conservation movement include Projeto Tamar in Brazil, the Turtle Conservation Project in Sri Lanka, Caribbean Conservation Corporation in Costa Rica (now called the Sea Turtle Conservancy) and so on. Many of these projects embody the movement towards community-based conservation, which eschews top-down draconian enforcement and attempts to work towards conservation objectives while also promoting community and livelihood benefits.

~

The World Conference on Sea Turtle Conservation was held in November 1979 in Washington DC. More than 300 participants from forty countries attended this meeting, presenting papers on the status of sea turtles in different parts of the world and on various aspects of sea turtle biology. The papers were published as the *Biology and Conservation of Sea Turtles* in 1982, a collection that was the Bible of sea turtle biologists around the world for over two decades.

In her original introduction, the editor, Karen Bjorndal, quotes David Ehrenfeld, one of Archie Carr's students, as saying '. . . a combination of our incomplete knowledge about sea turtles and the numerous constraints imposed by their biology dictates a very conservative conservation strategy'[61]. This was the period when conservation biology established itself as a discipline, and Ehrenfeld was among its pioneers.

However, the event that spawned the most significant annual meeting of turtle people was a small gathering of biologists at Jacksonville, Florida, in 1981. From this, the Annual Symposium of Sea Turtle Biology and Conservation evolved into a meeting of several hundreds by the mid-1990s. In 1998, the symposium moved away from the east coast of the US, and leaving the country for the first time, travelled to Mazatlan, Mexico. The event, incidentally my first symposium, was such a success that the symposium started to move outside the US on a regular basis. Between 2003 and 2009, the symposium travelled to Malaysia, Costa Rica, Greece, Mexico again (Baja California and later Oaxaca) and Australia. In 2010, the symposium was held in Goa, India, with more than 500 participants; it was at that point, one of the largest ecology or conservation conferences held in the country. The theme, 'the world of turtles', sought to focus on the people and habitats that turtles interacted with, rather than the animals themselves, 'to explore the world that turtles live in and interact with; the world of coral reefs, seagrass meadows, open seas and sandy beaches; the world of people, living and working on the coast or at sea and of fishing cultures and livelihoods'.

The meeting differs from most taxon-centric meetings (like cetacean or carnivore meetings) in that it includes and provides an opportunity for participation to a widely diverse group of stakeholders, including academics, activists, volunteers, practitioners, community representatives, government representatives, and so on. Jack Frazier, for example, though affiliated with the Smithsonian Institution, has been largely nomadic, conducting projects all over the world and providing invaluable advice in Central and South America, South Asia, and elsewhere. Maria Angela 'Neca' Marcovaldi, the iconic leader of Projeto Tamar in Brazil, established one of the best-known conservation programmes in the world, a flagship for community conservation. Alejandro Fallabrino has served to rally the Latin American community around sea turtle conservation. Lily Venizelos (married to the grandson of Eleftherios Venizelos, a major political figure and prime minister of Greece in the 1920s and 1930s) started out as a concerned citizen and has been a leading campaigner for sea turtle conservation in Greece for over twenty-five years. She played a critical role in the protection of nesting beaches at Zakynthos and was awarded the ISTS Lifetime Achievement Award in 2015. Roderic Mast, as vice president (sojourns) at Conservation International, travelled the world with Lukas Walton (of Walmart), Eddie Vedder (of Pearl Jam) and Harrison Ford, diving in the Bahamas and Raja Ampat and persuading donors to fund conservation. He has also served as auctioneer at the symposium, helping sell a variety of turtle paraphernalia to trinket addicts to raise money for students. And Jean Beasley, one of the most loved figures in the sea turtle community, runs a sea turtle hospital in North Carolina and has inspired rescue centres around the world. These are but a few of the remarkable, committed and dedicated people who make up this community.

~

The Marine Turtle Specialist Group (MTSG) was established in 1966 by Archie Carr at the behest of Sir Peter Scott, one of the founders of WWF and vice president of IUCN. The group was established under the Species Survival Commission of the IUCN to provide information about the extinction status of different species of sea turtles. The categories of species status at the time included 'giving cause for very grave anxiety', 'considerable anxiety' or 'some anxiety' which sound almost amusing in comparison to today's more grave sounding categories of 'critically endangered', 'endangered' and 'vulnerable'. Carr was joined in this endeavour by other leading turtle biologists of the time such as Leo Brongersma and John Hendrickson, and youngsters Peter Pritchard and Robert Bustard. Carr headed the group for eighteen years, with assistance from Tom Harrisson, Nicholas Mrosovsky, George Balazs, and Karen Bjorndal, who served as deputy chair. Peter Pritchard also served as coordinator of marine turtle conservation programmes from 1969 to 1973. Eventually, Carr passed the mantle on to his student, Karen Bjorndal. The MTSG attained a membership of over 200 sea turtle experts from about fifty countries by the 2000s, with many accomplishments over the decades of its existence[v]. Over the years, several of its members have criticized the MTSG for its lack of transparency and for serving the interests of a few, much like its parent organization, the IUCN.

∽

The Marine Turtle Newsletter (*MTN*) was started in 1976 by Nicholas Mrosovsky. Considering the threats to sea turtles around the world, and the growing global community, Mrosovsky started the newsletter as an informal tool for rapid communication between biologists and conservationists[62]. Importantly, the newsletter also supported conservation campaigns in Mexico, India and Ascension Island. By 1980, the newsletter already reached seventy countries. A variety of topics had been debated and discussed including research techniques such as tagging and telemetry, and controversial conservation issues such as the consumptive use of turtles. Several leading sea turtle biologist couples, Karen and Scott Eckert, Annette Broderick and Brendan Godley, Lisa Campbell and Matthew Godrey, have edited the newsletter in the last two decades, while Michael Coyne ensured that all issues went online. Over the years, the format of the newsletter has become a little more formal, and it has recently dispensed with the hard copy, which reached over 2,000 subscribers in 100 countries, and is now an online only publication. The online version is of course accessed by thousands of researchers and conservationists across the world.

In India, we launched the newsletter, *Kachhapa*, in 1999 as a part of Operation Kachhapa, a project for sea turtle conservation in Odisha, supported by the Wildlife Protection Society of India (WPSI), New Delhi[63]. Modelled on the *MTN*, *Kachhapa* was also initiated to provide a forum for exchange of information on sea turtle conservation in the Indian subcontinent. I started and edited the newsletter between 1999 and 2003 with Belinda Wright of WPSI. In 2004, to broaden the scope of the publication, I initiated the *Indian Ocean Turtle Newsletter*[64]. Today the newsletter is distributed free of cost to over 1,000 subscribers, including government and non-government organizations and individuals in the entire Indian Ocean region. I co-edited the newsletter with Chloe Schauble from 2009–2011, and we recently handed over the editorship of the newsletter to Andrea Phillott and Lalith Ekanayake.

~

In the last decade, the advent of the internet has created many virtual communities. The email listserv, CTURTLE, maintained as part of the Archie Carr Centre for Sea Turtles at the University of Florida, Gainesville, has served as an important communication tool for sea turtlers around the world. However, the virtual centre of the sea turtle universe is seaturtle.org, created by Michael Coyne over a decade ago[65]. Today, this website serves as a repository for vast amounts of news, information, articles and images about sea turtles, and provides a large number of tools for conservation and research. It has one of the largest directories of people and groups working on sea turtles, and includes databases on tagging, tracking, sightings and strandings.

The excellent fishe

In the early 1960s, Archie Carr authored a book titled *So Excellent a Fishe*[66]. The term comes from a law passed by the first ever session of the House of Assembly of Bermuda in 1620. The assembly, concerned with the indiscriminate take of sea turtles, passed an 'Act agynst the killing of ouer young tortoyses' prohibiting the killing of sea turtles less than eighteen inches in length[67]. Here's a quote from the Bermuda assembly, 1620:

> And at all times as they can meete with them, snatch & catch up indifferentlye all kinds of Tortoyses both younge and old, little and greate and soe kill, carrye awaye and devoure them to the much decay of the breed of so excellent a fishe the daylye skarringe of them from of our shores and the danger of an utter distroyinge and losse of them.

The law prohibited the killing of young turtles within five leagues of the islands, and the penalty for the offence was fifteen pounds of tobacco, to be shared between the public and the informer.

Carr began his work in the Caribbean and the tale of the discovery of sea turtle beaches and the threats affecting them are recounted in a number of beautifully written books. The first of these, *The Windward Road*, describes his adventures in the Caribbean during his first years there[68]. Much that he writes about the region and its people and its turtles is illuminating. However, what is intriguing is his own ambivalence about the consumption of turtles and other animals. In the foreword to the 1979 edition, Carr reflects on this at some length. He considers that in the early1950s, when the book was written, he greatly enjoyed the wild meat fare that was served him but had 'transitory pangs of conscience'. These developed into a full-blown personal dilemma over the next two decades. On the one hand, Carr could not resist, even while discussing the rights and wrongs of the issue, providing a full-blown description of wild meat consumption in the Caribbean, including manatee, armadillo, peccary, turtle, iguana, spider monkey, tepescuintle (a rodent) and hiccattee (freshwater turtle). The following lines epitomize this ambivalence: 'In fact, I will say without hesitation that clear green turtle soup is the finest gastronomic contribution of the English people.'

For those who have sampled various world cuisines and, consequently, have a less appreciative view of English culinary skills, this may not be saying much. But Archie clearly liked his turtle soup. He goes on to say: 'Giving it up was my greatest sacrifice to the religion of turtle preservation. I have almost completely given it up . . .'

But he did go into supermarkets and confiscate illegal cans of turtle soup, and having berated the manager, consumed them in the privacy of his room. That, he says, 'epitomises the predicament of the conservationist devoted to eating the object of his concern.'

Previously, in the foreword, Carr reproduced a 'panicky' letter he wrote to Sir Peter Scott, the Chair of the Species Survival Commission, about the predicament of the Kemps ridley. Speaking of 'dramatic drops' and 'terminal declines', Carr warned that the species might go extinct unless the Mexican government acted immediately. Perhaps he was right about this, but the letter strikes the same tone that activist petitions take. In any case, the contrast between the ambivalence over use and the zeal with which he advocated state protection is striking.

Globally, conservationists have fallen into two broad camps. One group believes strongly that any kind of use of wildlife can only have negative consequences, or that the animals that they work with are close enough to collapse, that they should receive the highest degree of protection. Others believe that the only way to build a large enough constituency for conservation is to involve communities, and in some cases, that may involve consumptive use of the resource. While there are relatively few sustainable use initiatives for terrestrial animals (with some exceptions, such as crocodiles), conservation in the fisheries context is largely driven by the idea of sustainable use or resource management.

Nicholas Mrosovsky, another legendary turtle biologist, visited Carr in Tortuguero to explore the possibility of post-doctoral research on sea turtles. He narrates how, as rite of passage, Carr would make his students and colleagues eat a raw sea turtle egg on their first visit to the nesting beach. Mrosovsky would do the same to his students at a time when such practices had become taboo in the sea turtle community, of course, only reinforcing his reputation as the black turtle of the conservation community. While many in the next generation would go the way of the new Carr philosophy and fight for the preservation of sea turtles, Mrosovsky became a lifelong advocate for the conservation of turtles through use[69,70]. He often cut a solitary figure in this endeavour, with only a few of his students, some crocodile biologists (and yours truly) supporting this perspective, or demanding a less emotional assessment.

Mrosovsky and Carr disagreed over the issue at meetings of the Marine Turtle Specialist Group, which they even co-chaired for a brief period, leading to an increasingly acrimonious debate in the turtle world. Carr had three female students, known as Archie's Angels, in Anne Meylan, Karen Bjorndal and Jeanne Mortimer. All of them, in one way or another, became strong advocates of protection and opponents of the use of turtles. They were joined by a near-entire generation of sea turtle conservationists through the 1980s and 1990s, when the very topic of use became anathema[71]. Few of them shared the confusion that Carr himself professed on the issue, and where he was troubled by the nuances (at least at one time), they seemed troubled only by turtles.

Much of this was based on the debate about the appropriate baseline against which to evaluate current populations. Sea turtles in many parts of the Caribbean were exploited by Amerindians even before the arrival of Columbus, but are believed to have been heavily depleted in the centuries after[72]. The exploitation and extirpation of the Cayman Island population is well documented. Jeremy Jackson, a marine biologist, used exploitation records to estimate the pre-exploitation population of green turtles in Caribbean at over thirty million animals[73]. Later, he and Bjorndal estimated the possible carrying capacity at over 100 million green turtles, based on the availability of food resources[74]. In a similar fashion, they estimated the possible hawksbill turtle population at half a million. Regardless of the precision of these estimates, clearly sea turtles had once been abundant in the Caribbean and were not any more. The debate over the issue of use continues, but has been somewhat overtaken by the increasing recognition of the rights of local communities and community-based conservation. While not quite gaining acceptance, the bitterness over the issue appears to have died down, at least for now.

Carr passed away in 1987, just a few months before I discovered sea turtles. His contributions to sea turtle conservation and biology are legendary. Carr was awarded the Eminent Ecologist Award by the Ecological Society of America in 1987. The Archie Carr National Wildlife Refuge in Florida, and the Dr Archie Carr Wildlife Refuge in Costa Rica, were both established in his honour. Beyond his academic life as a fine herpetologist and naturalist, where he guided many students and published many papers, he wrote charming (and award-winning) books about sea turtles that inspired an entire generation.

In establishing the MTSG and bringing together many of the world's leading sea turtle experts, Carr was able to highlight conservation issues in political contexts. His contribution to Caribbean sea turtles lives on through the Caribbean Conservation Corporation, which he founded and directed for nearly thirty years. The organization was born from the Brotherhood of the Green Turtle, established in 1959, by a New York publisher's representative who read *The Windward Road*. The brotherhood had the aim of 'restoring green turtles to their native waters, and insuring to Winston Churchill his nightly cup of turtle soup'. In the first they have succeeded remarkably, even if they have not restored Jacksonian numbers, but the second objective has succumbed to death by philosophy.

Notes

i For example, in Mexico, there was large-scale turtle consumption during the month of Lent, when red meat consumption is forbidden. However, since turtles are considered fish, they are consumed extensively. J. Nichols, a conservationist and biologist working in Baja California, petitioned the Vatican for many years to have turtles classified correctly as reptiles within the Catholic faith, which might have reduced local consumption. The Vatican, which had made dramatic advances during that decade by apologizing to Galileo and pronouncing that the teachings of Darwin might not be entirely in conflict with the Book of Genesis, declined to take a position on the taxonomy of turtles.

ii The export was valued at about Rs 50,000–1,00,000 per year for meat, and upto Rs 1,00,000 per year (roughly about $15,000 to $25,000 at the time) for tortoiseshell, though it is not clear how closely these approximate the real figures. It is estimated that India exported roughly 1,00,000 kg of sea turtle products valued at about $100,000 from 1963 to 1974. In the 1970s, exports of tortoiseshell were growing dramatically, especially from countries like Indonesia, Thailand and Philippines (in Asia), and Indian exports are believed to have reached 80,000 kg by the late 1970s[75].

iii In the 1980s, the CMFRI produced two bulletins on the marine resources of the Andaman and Nicobar Islands[76] and Lakshadweep Islands[77], both of which featured chapters on marine turtles. Though the articles were on 'sea turtle resources' the focus of Bhaskar and Whitaker's piece was on the surveys in the islands for these species[78]. Three of their four recommendations were towards protection and law enforcement, while the fourth suggested that mariculture should be investigated. Lal Mohan, a CMFRI officer, recommended turtle farming in his chapters, but even he talked mainly about threats to sea turtles and how they might be mitigated.[79] There was a small paragraph on the existing fishery and a longer one on conservation in the Lakshadweep Islands, where he called for Suheli Valiyakara to be declared as a sanctuary for the green turtle, and said that 'the turtles in the lagoon should also be protected from being killed' and 'the trade of products of turtles like turtle scutes, turtles meat and turtle oil should be prohibited under appropriate provision of the Indian Wildlife Act, 1972'. In an issue of Marine Fisheries Information Service (MFIS) in 1983, Silas and co-authors wrote an overview of the conservation and management of sea turtles, and spoke of the need to develop a good monitoring system[80]. At a national workshop on data acquisition and dissemination, they introduced the idea of monitoring cetaceans and sea turtles. The workshop recommended that data on mortality and resources of endangered marine mammals and turtles be collected because their populations were in decline, and stressed that 'it is essential to conserve these species'.

iv In July 1975, all seven species of sea turtle were listed on Appendix I of the Convention on International Trade in Endangered Species of Wild Fauna and Flora (CITES). Most species of sea turtles are listed as threatened (critically endangered, endangered or vulnerable) in the Red List of the IUCN, with only the flatback turtle listed as data deficient. Sea turtles are also listed on Appendices I and II of the Convention for the Conservation of Migratory Species (CMS). Under the auspices of CMS, several regional agreements were signed, including, in 1999, 'The Memorandum of Understanding concerning Conservation Measures for Marine Turtles of the Atlantic Coast of Africa', which covers much of Western Africa. In the early 2000s, two more agreements, 'Inter-American Convention for the Protection and Conservation of Sea Turtles' and the 'Memorandum of Understanding for the Conservation and Management of Sea Turtles in the Indian Ocean and Southeast Asia (IOSEA)' were signed by signatory states in their respective regions to promote the conservation of sea turtles.

v The MTSG conducted a series of workshops in the 1990s assessing global and regional conservation needs. These had significant impacts in many regions, not least in South Asia, where the Northern Indian Ocean workshop was held in January 1997. At around the time, the MTSG also produced a manual on 'Research and Management Techniques for the Conservation of Sea Turtles' which has been widely used in all parts of the world[81]. Recently, the group has been defined by a series of 'Burning Issues' meetings that have resulted in publications on defining and prioritizing sea turtle populations[82].

FROM THE PORTS OF ODISHA TO THE PANS OF KOLKATA

The eagle eye

INDIRA GANDHI'S INTEREST IN wildlife and conservation is well documented. From the letters that her father Jawaharlal Nehru wrote to her, the theme of nature and environment recurs in the political family's values and actions. In order to get a first-hand account of her interest in sea turtle conservation in the early 1980s, I met Samar Singh one morning at the India International Centre in New Delhi. Singh had worked in various government departments on environment and wildlife.

The Indian Wildlife Act had been passed in 1972, but environmental issues still did not get the attention they deserved. Both forests and wildlife were part of the Ministry of Agriculture, where Singh joined as the Director of Wildlife Preservation in 1980. While the Department of Environment was formed in 1980, forests and wildlife remained with the Ministry of Agriculture till 1982, when Mrs Gandhi shifted wildlife to the Ministry of Environment, directly under her charge, to give it more importance. After briefly reuniting forests and wildlife under the Ministry of Agriculture, Prime Minister Rajiv Gandhi created the Ministry of Environment, Forests and Wildlife in 1986 (currently known as the Ministry of Environment, Forest and Climate Change).

Singh recounted with great nostalgia working with Mrs Gandhi, whose interest in environmental issues and decisiveness he greatly valued. Singh recalls a briefing meeting sometime in the early 1980s, along with the minister for environment and some bureaucrats. In the middle of the meeting, Mrs Gandhi wrote something on a slip of paper, folded it up and flicked it at Singh. He says that he opened it with 'trembling hands, wondering "what has come towards me like a missile"'. The note said: 'We have to send a baby elephant for the Japanese General. Find one. We will name it *Moti*. It can be male or female.' Many years earlier, when Nehru had visited Japan, he had presented an elephant to the Ueno Zoo in Tokyo Zoo for the children. When that elephant died, they had requested another elephant.

Aside from Mrs Gandhi's personal interest in wildlife, she was also tuned in to the international community, both to legal instruments, as well as to world leaders who endorsed wildlife conservation, who were her friends and acquaintances. She hosted the Conference of Parties of the CITES (Convention on International Trade in Endangered Species of Wild Flora and Fauna) in 1981, which brought many issues to prominence in India including the turtle take in Odisha. She took time off from the Commonwealth Conference she was hosting to attend the tenth anniversary of Project Tiger, which was being attended by Prince Philip. Singh recalls that she was more than willing to meet a variety of foreign visitors connected with wildlife conservation, even if it was just for a few minutes.

Indira Gandhi is credited with having initiated or supported much of the wildlife preservation of the 1970s and early 1980s. This included the enactment and enforcement of the Wildlife Act, the establishment of national parks and sanctuaries and protection afforded to large mammals such as elephants and tigers. The Indian Board for Wildlife had been formed in the 1950s and was chaired by Dr Karan Singh during her first tenure as prime minister. When she returned to power, the board was reconstituted with her as the chairperson, and she presided over it till her death in 1984. She insisted on chairing the standing committee of the board during this period, as also the steering committee of Project Tiger. According to Singh, who served as the member secretary of the committee, despite the burdens of prime ministership, she did not miss a single meeting of these committees and gave all the issues the attention they required.

Mrs Gandhi had given general instructions to her staff to mark news items about wildlife, so it was no surprise that she should have encountered and acted on a piece about sea turtles. Sometime in 1982, with a great deal of media attention on turtles, she wrote a memo on a news clipping and sent it to Samar Singh. Singh says, 'There was a short news item about the problems of Gahirmatha, and she wrote in her own hand, "Who is keeping an eagle eye on this? I.G."'

From intentional to incidental catch

The Gahirmatha beach is on the eastern edge of the Bhitarkanika mangrove forests. Till 1952, Bhitarkanika was part of the Kanika *zamindari* (private landholding), when the system was abolished by the state government and the land transferred to the state forest department. Not much seems to be known about the rulers of Kanika before British occupation of Odisha. The 'kingdom' appears to have been established in the thirteenth century by the brother of a ruling chief of Mayurbhanj. The descendants of this chief ruled the region till the early 1800s. These 'rajas' were semi-independent rulers of a small principality, and possessed sovereign rights, but had allegiance to the kings of Odisha, the Mughals, and to the Marathas over several centuries. The region also had several subsidiary chiefs belonging to the Nag dynasty. Kanika was finally occupied by the British in 1803 who needed to connect their territories to the north and south. In the eighteenth century, they had already established commercial establishments at Jagatsinghpur and Balasore in northern coastal Odisha. After attempts to acquire the state by peaceful means, the British under Lord Wellesley finally annexed the state by force, including the zamindari of Kanika[1].

The earliest reports of Gahirmatha are in Captain Alexander Hamilton's book which has already been mentioned[2]. This site has obviously been a ridley rookery for centuries if not longer. The rookery is also briefly mentioned in *Orissa – Its Geography, Statistics,*

History, Religion and Antiquities by Andrew Stirling[3] (Persian Secretary to the Bengal government), published in 1846*. Stirling wrote:

> The value of the excellent *Turtles, Oysters, Crabs and Prawns*, found of *False Point*, and in other parts, was unknown to the natives prior to their subjection to the British rule, but they are now of course eagerly sought after, to supply the stations of Balasore, Cuttack and Juggernauth.

This confirms, however, Mohanty-Hejmadi's comment that the turtles were probably not being utilized in the early 1700s[4]. By the mid 1800s though, use of the turtles had begun. Apparently, one more thing we owe the British. By the 1900s, there is considerable evidence of widespread utilization of eggs.

Boatloads of eggs

It is not clear when egg collection started, but until 1952, the local zamindar (landowner or collector of land revenue) levied a duty (called *andakara*) for the collection of eggs from Gahirmatha. Eggs were sold in the riverside villages of Bhitarkanika and also transported in large numbers to Calcutta (now Kolkata). Most Oriyas did not consume turtle meat or eggs, except perhaps the poorer and lower caste communities. Locally, in Bhitarkanika, turtle eggs were dried in the sun and preserved in large quantities to be used as cattle feed[5].

The management of the area was transferred to the revenue department, which continued to issue licenses till this moved to the purview of the state forest department in 1957. The Odisha Forest Department issued licenses for egg collection at the rate of Rs 15 per boatload of eggs, with each boat containing about 35,000 to 100,000

---

* An anonymous internet comment on this book says, 'This book is quite rude by modern standards. It is contemptuous of Hinduism in general and Orissa in particular.'

eggs. The estimated legal take for just the 1973 season was 1.5 million eggs, but given the account of boatloads being shipped out, this likely represents only a small fraction of the total take[6].

In 1975, after Bustard had discovered the rookery, he recommended to the state government the banning of egg take and licences, advice which was duly followed[7]. Though his intentions were only to temporarily stop egg take until the populations could be assessed and the take regulated, this action spelt the end of egg utilization in Odisha. He wrote:

> Now, following my advice, the government of Orissa has completely banned the taking of eggs and revoked all egg licences. Furthermore, special staff have been posted to protect the nesting females. In April 1975 this nesting rookery was included inside a sanctuary known as the Bhitarkanika sanctuary. The main purpose of the sanctuary is protection of the saltwater crocodile (*Crocodylus porosus*), and the Pacific Ridley sea turtle.

Boatloads of adults

From Stirling's brief comment, one might infer that adult female turtles were being caught and consumed. However, there is little evidence that turtles were consumed as meat in Odisha. On the other hand, there are numerous first-person accounts of ridleys in the meat markets of Kolkata.

From at least the 1960s until the early 1980s, thousands of adult ridley turtles were shipped from the ports of Odisha to the pans of Kolkata by both road and rail. In the 1960s, the turtles were largely taken as by-catch and shipped along with fish. During this period, large numbers of ridley turtles were recorded in the Kolkata markets by observers like Anne Wright, one of the first and a long-serving member of the National Board for Wildlife, and by Indraneil Das in the 1970s. Turtle meat (both freshwater and marine species, apparently) was widely consumed in West Bengal. Das recalls visiting the markets as a schoolboy and finding a vast array of turtles on display next to the fish section. Many were being stripped of their shells and meat when they were still alive.

The introduction of mechanization in the local fishing industry in the 1970s led to a dramatic increase in turtle take. S. Biswas of the Zoological Survey of India (ZSI) wrote a comprehensive review of olive ridley turtles in the Bay of Bengal[8]. He wrote that olive ridleys were shipped mainly from Puri to Kolkata by rail: 'The turtles are despatched by train upside down, both pairs of fore and hind limbs tied together with an identification mark and a number inked on the plastron.'

Each turtle was sold for Rs 20 at Puri and Rs 60 in Kolkata in the mid-1970s, and about Rs 100–150 in the early 1980s. At roughly Rs 10 per kg, this was still cheaper than mutton and fish. But there was little local consumption. Biswas wrote: 'Though turtle meat is not popular, it is also used by the local low caste people. I have seen turtle meat being sold in the Chandbali market of Orissa and Namkhana (Sundarban), West Bengal.'

In southern Odisha, Andhra fishermen (nolias) were mainly involved in turtle fishing; these migrant fishermen from Andhra would live in temporary settlements on the Puri coast. Turtles caught at Konark were shipped on the top of buses to Puri or Bhubaneswar, while those caught in Gahirmatha were taken to Chandbali by boat, and from there to Bhadrak by bus or truck. In northern Odisha, Bengali fishermen from West Bengal or Bangladesh were responsible for turtle fishing.

In 1974-75, the month-wise despatch of sea turtles from Puri railway station to Kolkata shows that 6,339 turtles were transported between November and January. In Digha, however, 21,361 turtles were recorded in three months between mid-October 1978 and mid-January 1979. Turtles were collected at Digha from several seaside depots and sent by truck to the Kharagpur railway station for the Kolkata markets.

From about 1975 till about 1985, more than 1,000 fishermen from Digha, Medinipur and 24-Parganas in West Bengal, and Balasore in Odisha, were involved in turtle fishing for nearly six months of the year (October to March). Turtles were usually caught while mating when they are relatively easy to approach and capture. They were encircled with nets or hauled aboard by hand. Turtles were even caught using standard gill nets (normally used for seer, pomfret and other pelagics) from country boats (either Kakinada boats or catamarans) operating as teams. Each boat would contain eight to ten fishermen, and five to eight such boats would be towed by a mechanized boat, once the fishing operation was completed. Each boatload would contain about a hundred turtles. At the peak, eight to ten such teams operated in the Gahirmatha area. Given the surfeit of catch, injured or dying turtles were thrown overboard and washed up later on nearby beaches. Apparently, female turtles were preferred, and males were often released. The fishermen often formed fishing units of 50–150 members on a contract basis for the season. The trade, as for fish, was controlled by middlemen, called aratdars (godown-owners) at Digha and Kakdwip.

If these groups fished even once a week during the peak season, catching about hundred turtles per boatload, they would have captured about 80,000 to 1,00,000 turtles per season. Based on the fact that six-seven truckloads, each with about 150 turtles, arrived in Kolkata every day during the peak nesting season, Indraneil Das also estimated that 80,000 turtles may have been caught annually during the late 1970s and early 1980s.

In 1977, when sea turtles were listed in Schedule I of the Wildlife Act, the forest department became involved in curbing the take of turtles. With the formation of the Wildlife Wing of the Odisha Forest Department in the 1970s, Indian Railways eventually stopped transporting sea turtles, though it continued for a couple of years as a fishery product. The trade continued by road, particularly from Digha and other ports in West Bengal, where the turtles were taken from the offshore waters of Odisha.

At CMFRI, E.G. Silas had also become interested in the issue of sea turtles. As director, he was able to direct time and effort towards research and status surveys. In 1981-82 and 1982-83, CMFRI teams visited the Odisha and West Bengal coasts to assess populations and threats[9]. The team included Silas himself, M. Rajagopalan, who would go on to do his Ph.D on sea turtles, Bastion Fernando, who carried out much of his work on the southern Tamil Nadu coast, and others. Though turtles were transported from Puri to Kolkata by rail, it was not possible to estimate numbers from their records, as they were generally booked as fish, due to lower freight charges. CMFRI scientists noted that, by 1982, the Southeastern Railways had stopped booking sea turtles from Puri to Kolkata, and the fishery from central Odisha had virtually stopped. There is a story, however, of an unknown vintage of a railway employee, who while on a 'fish pilfering spree . . . was relieved of a couple of toes' when he accidentally stepped on a turtle. This led to the insistence of the Railways that the turtles should be caged, which the traders were understandably reluctant to do because of the associated costs[10].

At the same time, several conservationists started to note the take of sea turtles. Anne Wright, from her position on the wildlife board of the state, and Bonani Kakkar of the World Wide Fund for Nature (WWF), who had been instigated to action by the young Indraneil Das, protested to the Railways that they had to stop the transportation of the Schedule I species. By this time, there was considerable awareness in the fishing communities of the ban imposed by the forest department. However, the fisheries department continued to be preoccupied with promoting the fishing economy, leading to what Silas et al. called 'an anomalous situation . . . where the Department of Forest and not the Department of Fisheries is empowered to control a resource caught in the sea'[11]. This situation, which continues till today, is viewed by many conservationists as a major obstacle to the effective conservation of sea turtles.

Silas et al. also wrote that the chief poachers of turtles were fishing vessels from Thailand and Taiwan[12]. Not much else is known about the extent of turtle fishing in Indian waters by foreign vessels. In their conversations with skippers of fishing vessels in Paradip and Visakhapatnam, they discovered that the area north of Gopalpur was a mating ground for turtles, suggesting that the Rushikulya rookery had significant nesting even before it was 'discovered' a decade later.

The ban on the booking of turtles by rail which led to the decline in take from southern Odisha, combined with an increase in mechanization, resulted in substantially increased take from

Gahirmatha in the mid- to late-1970s. In the early 1980s, a series of events led to increased vigilance and enforcement, and by 1982-83, the operations became much more clandestine.

In the winter of 1983, Ed Moll, a leading international freshwater turtle biologist, J. Vijaya, a young herpetologist with the Madras Crocodile Bank, and Satish Bhaskar were conducting a survey across India. Vijaya had already covered Kerala, Andhra Pradesh and Uttar Pradesh, and along with Moll and Bhaskar, was visiting Madhya Pradesh and Uttar Pradesh again, along with Odisha and Bihar[13]. They arrived in Odisha in February just after the mass nesting had occurred, and their disappointment at missing the event was apparent. They did, however, participate in the release of ridleys that had been confiscated from poachers in an operation involving the navy and coast guard. While the presence of trawlers and incidental mortality was already on the horizon as a serious threat, the nesting beach was being protected by forest department personnel, both by bicycle and on foot. Moll and Bhaskar were joined by Biswas from Zoological Survey of India (ZSI), and continued onwards to survey the Mahanadi.

Vijaya continued separately to check on the situation at Digha, where she had reported the slaughter of ridleys the previous year[14]. This time, she found that the fishermen had become secretive; turtles were unloaded in thc late afternoons and evenings and moved to holding tanks that were half a kilometre away from the beach in the fields. The peak season had apparently passed, but several truckloads a day were still being transported.

Silas's team and others documented the clandestine turtle trade in West Bengal in the early 1980s[i]. They noted that several hundred turtles were rescued by the police and forest department of West Bengal in 1983, and more turtles were seized from trucks at various police checkposts in West Bengal in 1983 and 1984. There were also numerous press reports about the illegal trade and rescue of sea turtles at the time, particularly in the *Dainik Chetana*, a Bengali daily. In one instance, a truck with ninety turtles was caught at a forest department checkpost, and produced before the sub-divisional magistrate court at Contai, which appears to have been fairly active in dealing with cases of 'turtle poaching'. The driver and assistant were released on bail (Rs 500) and the turtles were released[15]. On another occasion, the newspaper reported that a businessman and his driver were caught 'when they tried to transport turtles on the night of Christmas day'. Having paid once, the lorry was stopped again and 'a bribe of Rs 750 had to be given to an influential man' and 'local inhabitants complained that illegal business is flourishing well in the darkness of night'[16]. An article in the Oriya daily, *The Samaja*, reported that 'some dishonest businessmen from West Bengal have been making huge profits by catching lakhs of turtles from the sea' as 'turtle meat and eggs are served as costly items in big modern hotels in Calcutta'[17].

In 1985, the team saw turtles being transported openly in a cycle-rickshaw on Sagar Island. On another occasion, a turtle was loaded onto the roof of a passenger bus, but rescued by the intervention of a forest official. One article from *Dainik Chetana*, dated 14 January 1985, reports on the seizure of a truck with more than fifty sea turtles headed from Digha to Kolkata by 'rickshaw labourers'[18]. The article says that 'the truck driver attempted to run away with the truck but the public and rickshaw labourers prevented it'. The driver admitted that two persons from a nearby village had purchased the turtles in Digha to transport to Kolkata. The article stated that 'sea turtles are reducing fast in number and on the way to extinction . . .' This report, as well as one in *Dainik Teerbhumi*, pointed out that it was legally prohibited to capture and sell sea turtles[19].

A day later, the same newspaper reported a follow-up to the case. The driver and trader appeared in front of a judge (Alok Kumar Mukhopadhyay), who ordered the release of the twenty-nine live turtles into the sea, which was duly carried out 'on the holy Makar Sankranthi day in the presence of many bathers', representatives of a wildlife organization and the police. The driver was released on bail, but apparently allowed to leave with twenty-one dead turtles. According to this article, over 250 sea turtles had been similarly rescued and released on earlier occasions over the previous two years[20].

The strong enforcement and consequent decline in turtle take was due to the efforts of a number of people, but two remarkable, strong-willed women played key roles. One is amongst the best known political figures from India. The other was known mostly to her friends and a small community of herpetologists. But both of them were central to saving ridley turtles back in the 1980s.

Black and white

J. Vijaya, a young, adventurous field biologist, first visited Digha in January 1982. At the time, the beaches were strewn with turtles, and they were openly sold in the markets. Her passionate reports, along with her black-and-white photographs of hundreds of turtle carcasses, brought even more attention to the extraordinary numbers of turtles being killed in Odisha. The report was picked up by *India Today*, which ran a two-page article titled 'Massacre at Digha', written by Dilip Bobb in March 1982[21]. Perhaps it was this article that caught the attention of those who mattered in high places in New Delhi.

In the early 1980s, numerous petitions and letter-writing campaigns were endorsed through the *Marine Turtle Newsletter* (*MTN*), supported by the editor, Nicholas Mrosovsky. Following Bobb's article in *India Today* and other reports, Mrosovsky first wrote to the readers of the *MTN* in December 1982 (in the same issue in which Vijaya's brief note appeared), urging them to write to the prime minister of India about the issue[22]. In a lightly sardonic tone, he reminded readers that the newsletter was free and nothing was asked of them except their interest in the subject. He went on to outline the issue of the olive ridley slaughter that Vijaya had reported, commenting briefly that utilization and conservation need not be incompatible. In March 1983, he wrote again, urging readers who had not yet written to the prime minister of India to do so[23]. His note said: 'With a Prime Minister favourable toward conservation, and still turtles there left to conserve, it is important that the authorities there be made aware of world-wide efforts to conserve sea turtles and of what has happened to the arribadas of olive and Kemp's ridleys in Mexico.'

Several hundred letters were written to Prime Minister Indira Gandhi. Later that year (July 1983), Mrosovsky wrote another editorial, this time to provide an update on the campaign, which had resulted in coverage in *The New York Times*, *The Times* (London), *The Times of India*, *The Globe and Mail* (Toronto), BBC Radio, and a TV network in Belgium[24]. In the Netherlands, the journal *Lacerta* had taken the campaign forward. Mrososvky received fifty-nine letters in response to the petition and heard that over 200 letters were received by the prime minister.

Typically reflective, Mrosovsky wondered if the campaign had helped, insightfully acknowledging 'the danger that outsiders may do more harm than good when they try to intervene in situations whose complexity they may not fully understand'.

How one wishes that international campaigners (no doubt well-intentioned) thought this way more often. It was clear by this time though that the government was concerned, whether due to the campaign or not. Mrosvosky was curious; he wrote: 'So when I spoke recently with Mr Samar Singh, Director of Wildlife Preservation, in the Department of Environment, India, I asked him directly, "Has the letter writing been helpful?" He replied thoughtfully, cautiously, "Yes", he said, "it has strengthened our hand". That seems a fair assessment.'

Having visited Gahirmatha and being aware of the issues, Singh composed a long note on the issue highlighting the inability of the local authorities to enforce the law. During this time, Singh had held meetings in Bhubaneswar to push for better enforcement, including seeking the help of the police. In response, Mrs Gandhi only said, 'Get the Coast Guard to help. Report compliance in 24 hours.'

This caused considerable consternation in the Ministry of Agriculture (where wildlife was vested) as the ministry had no control over the coast guard. Especially since the prime minister had asked for a report within twenty-four hours. Eventually, they approached the Ministry of Defence, whose initial response was that the coast guard was not for wildlife conservation. Singh went across to the Ministry of Defence himself and was at the receiving end of a diatribe that the coast guard had better things to do and 'what were these turtles anyway?' Singh explained to them the issue over turtle protection in Odisha, but more importantly, showed them the note signed by the prime minister herself. Twenty-four hours later, the coast guard was patrolling the Odisha coast[ii].

At the other end, the officer-in-charge received the government order, instructing the coast guard to patrol the coast and protect turtles. Commandant Anil Panwar, posted at the regional headquarters in Chennai in later years, recalls receiving the orders, which were well within the mandate of the coast guard. He remembers the coast guard being involved in other wildlife issues such as the protection of dugongs in Palk Bay, and in some of the first sightings of the whale shark, which were communicated to the Zoological Survey of India.

The protection programme, christened 'Operation Geeturt', involved the coast guard, navy, and state agencies, including the forest and fisheries departments and the police in West Bengal and Odisha. In 1984, Archie Carr, chairman of the Marine Turtle Specialist Group, received a response to his request to Mrs Gandhi to help protect turtles in India[25]:

> I have received your letter of the 11th March. There have been other similar letters about the protection of marine turtles. Even before these reports were received last year, I called for immediate action through the Orissa State Government and the Coast Guard of the Indian Navy to prevent the hunting of these turtles or for collection of eggs by beachcombers . . . You will be glad to know that our States have also started taking steps to collect the eggs and get them hatched in a central hatchery and release the young ones into the sea. We are aware of the importance of the endangered species to our eco-systems. Our concerned Ministries here and in the State Governments have been asked to take the required measures to see that the olive ridley turtle, which is an endangered species, is looked after.

Operation Geeturt led to the decline of turtle take to minimal proportions. The operation was discontinued in the early 1990s when a new coast guard headquarters was established at Paradip, but following a request from the chief wildlife warden of Odisha to the director general of the coast guard, the operations were resumed.

While the presence of the coast guard was a deterrent to turtle fishing, it proved less effective against the trawlers which fished in near-shore waters, where the larger coast guard vessels could not operate. In the last decade, the coast guard has been willing to assist, but their involvement has been sporadic for a variety of reasons, related perhaps to the changing politics of sea turtle conservation in Odisha. .

The pioneers

The Wright way

In the early 1970s, Robert Bustard broadcast the discovery of the Gahirmatha rookery to the scientific world. However, concern over the ridley turtles of Odisha began a decade earlier. And the theatre of conflict was not Odisha, but Kolkata. In the 1960s, Anne Wright was one of the few prominent people involved in wildlife conservation in India[26]. Anne had grown up in India in a variety of picturesque locations, riding horses and elephants as a child, and learning to appreciate nature around her. She lived in Kolkata after marrying Robert Wright, also a Britisher who was born in India. Like many in their generation, they were very much part of the hunting tradition, but turned to conservation when they realized the critical need for it.

Anne Wright was one of the founding trustees of WWF India in the 1960s, and a member of the Tiger Task Force commissioned by Prime Minister Indira Gandhi to select tiger reserves for the launch of Project Tiger in 1973. She served on the national board for wildlife for nineteen years and was on the wildlife board for several states including much of north-east India, West Bengal, Bihar, Odisha and the Andaman and Nicobar Islands. Wildlife board meetings were then held twice each year and she spent a considerable amount of time travelling to these meetings. Wright played a little known but crucial role in the drafting of the Indian Wildlife (Protection) Act, 1972. She obtained, through the good offices of the Kenyan polo team, which was visiting India at the time, a copy of the Kenyan Wildlife Act, retyped it and passed it on the bureaucrats in New Delhi who were framing the Indian Act. Anne was one of the first to raise the issue of tiger poaching and exports; her activism resulted in the initiation of tiger conservation in the country.

Anne was also amongst the first to take note of the huge numbers of sea turtles that showed up in the Kolkata market and to realize that this could be hurting the population. She visited Gahirmatha several times and highlighted the issue in meetings of the wildlife board, which provided the impetus for conservation action in the late 1970s and early 1980s.

Anne first heard about Bhitarkanika, not because of the turtles, but because of salties (salt water crocodiles). She said: 'I first heard of Bhitarkanika because there was a monster crocodile there, twenty-six feet or something! And everybody wanted to shoot it. And we were thinking, "Oh, let's hope that it doesn't get shot," but it did, because there was an Anglo-Indian captain of a ship who was determined to shoot it. It was a record for that kind of crocodile.'

Subsequently, she visited Gahirmatha on a few occasions with the forest department and coast guard. Anne remembered one of her first visits to the rookery: 'I was camping on the beaches by myself, in a tent. There was a marvellous young man who was doing some research there. Suddenly, on a moonlit night, he came tearing across the sand. He had nothing on, except for a little lungi, and was shouting, "C'mon, c'mon, c'mon, the turtles have arrived!" So we both ran, and there we were, amongst a sea of turtles, all coming out of the water.'

Anne was one of the first people to campaign against the planting of casuarina on the beach. They persuaded the forest department to remove the trees and widen the beach at Gahirmatha, but that policy did not stick, and the beaches there have more casuarina than they should.

Anne also recalls a trip to Gahirmatha, where on the long boat ride, they saw a boat coming the other way, and according to her 'out of it popped this chap with a long beard, long hair, and he looked like a – you know, the old man of the deep, or something, coming out of the boat'. The old man of the deep turned out to be a young Jack Frazier, visiting Gahirmatha for the first time. Needless to say, Anne was a great influence on her daughter Belinda, who would become very much part of the turtle conservation melting pot in Odisha a couple of decades later. In the 1980s, Anne Wright was awarded the Order of the Golden Ark and the Member of the Most Excellent Order of the British Empire, or MBE, for her conservation efforts in India.

Bob the builder

Robert 'Bob' Bustard was the first of a series of influential biologists to work on sea turtles in Odisha. Scottish by birth, he spent his early years in Hong Kong and western Australia, and graduated from the University of St. Andrews and the Australian National University in Canberra. Bustard actually started out making natural history films for television in the UK as he realized that was the way to 'get to' people en masse; he appeared on Sir Peter Scott's natural history programme, *Look*, and was shortlisted as best amateur natural history filmmaker. He then applied for a post in the BBC natural history unit and found that the person in charge had a Ph.D and decided to pursue an academic career. He arrived in Australia to pursue a Ph.D on the ecology and population biology of four species of geckos and a predatory skink that resulted in the award of the very prestigious Queen Elizabeth fellowship, usually just one a year in biology. The fellowship could be held anywhere in the Commonwealth of Australia, and Bustard chose to stay at Australian National University (ANU) at Canberra and became the first staff member of the newly formed Research School of Biology. He spent over a decade working on sea turtles, crocodiles and lizards, before 'giving it up' in the early 1970s to work with indigenous Australians, particularly the Torres Strait Islanders.

Colin Limpus, currently Australia's foremost turtle biologist, first met Bustard at Heron Island in 1968, while he was reinventing himself as a biologist after undergraduate studies in physics. Limpus offered to show Bustard flatback turtles and took him to Bundaberg where he had worked. When he told Bustard about the large amounts of data that he had on loggerhead sea turtles (relative to the small amount of flatback data), Bustard's response was, 'There isn't one thing that you can teach me about loggerheads.' A month later, they had their first tag return from a flatback, then considered to be a non-migratory species, and they wrote their first paper together. Two months later, they had another tag return. The saga was to continue. Bustard took Limpus under his wing and mentored him.

Bustard's tagging programme for green and loggerhead turtles at Heron Island continued between 1964 and 1972, and they tagged pretty much every nesting turtle through the season. Heron Island was an ideal base since it was a small coral cay with a beach perimeter of a little over a mile and the team could be certain of seeing every nesting turtle without exception every night for three months. As research funds grew, they purchased a pearling lugger and visited islands that hadn't been set foot on by Europeans since the Royal Navy had first mapped them a century or more before. They also borrowed his dentist's light aircraft and flew the whole of the Queensland coast (2,250 miles) plus the inner and outer Great Barrier Reef recording turtle nesting.

His work in Queensland would be carried on by Limpus, whose multi-decade monitoring of sea turtles in that region is considered one of the finest long-term research programmes on sea turtles anywhere. Bustard also helped formulate the Queensland sea turtle legislation, which was at the time one of the most significant laws for sea turtles in the world.

During this time, Bustard decided to apply his knowledge of population ecology to conservation. For example, with sea turtles, he knew that a component of population regulation was the destruction of earlier laid nests by subsequent nesters. Hence, it was possible to remove a percentage of earlier laid nests in certain locations without in any way affecting natural hatchling production. The removed eggs could be used for other purposes – for food or farming enterprises, for instance.

After the Queen Elizabeth fellowship ended, Bustard became a fellow in the research school at ANU, and set up the Applied Ecology unit there. Around the same time, the Commonwealth was keen on improving the employment prospects of native Australians. Bustard applied for projects through this unit and received large grants that facilitated and supported the farming ventures for crocodiles and turtles.

Bustard's fundraising skills greatly benefitted from a meeting with John Hendrickson, a legendary turtle biologist and senior member of the MTSG, who had worked extensively on green turtles in Malaysia. Hendrickson's advice to Bustard, and a couple of nights of work on a portable typewriter in a hotel room, culminated in two proposals which were funded and set the stage for larger grants in Australia and later, India.

Bustard initiated a series of turtle farms – perhaps over thirty – in the Torres Strait Islands. Christianity had come to the Torres Strait Islands in the 1870s and by the early 1900s, it was being celebrated as the 'coming of the light'. Bustard arrived in 1970, about a hundred years later, with his plans for farms and jobs and money. This generated so much enthusiasm that it came to be known as the 'second coming of the light'.

However, the farming enterprises in Queensland became highly politicized, resulting in conflict between the Commonwealth governments and Queensland politicians. Bustard returned to Scotland for a sabbatical, but was invited to India almost immediately afterwards by Prime Minister Indira Gandhi to advise on the conservation of the gharial. Australia's loss was India's gain.

Robert Bustard now lives in the Isle of Man in the Irish Sea and visits India occasionally, spending a few days in Odisha as a state guest on each trip. We did not meet, but we did exchange emails. Commenting on his work, Bustard said: 'The project was originally to catch the last remaining wild gharial and put them in New Delhi Zoo. I was not about to do that but I had to convince the government that we could set up protected areas and I would help them to ensure that the gharial could be fully protected in these.'

The project soon expanded to include all crocodilians. But then, he discovered sea turtles. In 1974, he was on a countrywide survey of crocodiles for the government of India and heard news of an olive ridley turtle rookery. He first visited Gahirmatha with Saroj Raj Choudhury and the following year with B.C. Kanungo, both forest officers with whom he worked closely. Having confirmed the existence of the rookery, he returned in 1975 to initiate research and spent eight years in India. Memorably, and with long-lasting consequences, he branded Gahirmatha as the 'World's Largest Sea Turtle Rookery?'[27]. In this article, he wrote: '. . . during a three week period in February, 1976 a total of 1,58,171 nesting female turtles were marked. Since it was not possible to apply monel tags this year, there may be some repeats among those marked. Nevertheless, the rookery is enormous and a most remarkable find for a species listed in the Red Data Book … of IUCN.'

While advising the government on sea turtle conservation, he also initiated research programmes in Odisha and Uttar Pradesh. Bustard's Ph.D students in Odisha at the time included L.A.K. Singh, Sudhakar Kar and B.C. Choudhury, known as 'gharial' Singh, 'crocodile' Kar, and 'mugger' Choudhury, after the three crocodilians they were going to study, namely the gharial, saltwater crocodile and freshwater crocodile or mugger. They would be joined later by Chandrasekhar 'turtle' Kar, who initiated his research on olive ridley turtles in Gahirmatha. In addition, two students worked on different aspects of the gharial in Uttar Pradesh: Sushant Choudhury and A.K. Srivastava. All of them would go on to become accomplished researchers and powerful spokespersons for conservation.

Bustard played a strong role in influencing policy both with the Central as well as the state governments. With regard to the sea turtle egg harvest, he said: 'Having requested the Odisha F.D. to stop selling *Lepidochelys* eggs on a royalty basis (which they quickly did) I turned my attention to the adult ridleys. On my return to India, I saw turtles being carried up the beach at Gahirmatha to be taken to the rail head at Puri, for on forwarding to Kolkatta (sic). I photographed all this with L.A.K. Singh. Working together we had great achievements as did B.C. and I in the seventies and early eighties.'

Bustard's book, *Australian Sea Turtles*, provides an excellent account of his research there[28]. He published extensively with his Indian students, mostly on crocodiles, and took care to publish in Indian journals such as *Indian Forester* and *JBNHS*, so that it would be accessible to audiences here. His legacy remains in the form of the programmes that he initiated in the 1970s, and the researchers he trained. Bustard remained an active supporter of sustainable use, but remained cautious about legalizing take which could not be regulated or controlled. Nevertheless, it is ironic that his original advice to the Odisha government to stop egg utilization led to a ban, which still stands today.

Bustard was also a great believer in 'taking conservation to the people'. At Heron Island, he and his students would take out tourist parties to see nesting turtles every night to 'build up a huge body of "pro-turtle" people'. He wrote many popular articles in wildlife magazines, gave radio and TV talks and wrote books including *Kay's Turtles*, which was on the Canberra best-sellers' list for weeks and won a top literary prize in Victoria in Australia.

Neil for turtles

In the late 1970s, another Kolkatan was visiting the markets, usually in the company of his parents in Gariahat and others around Kolkata, looking for turtles. Indraneil Das, a schoolboy at the time, was horrified at the turtles being killed, some stripped of their shells while still alive, including species listed as protected by Indian laws. Neil started by keeping a few of the 'rescued' turtles as pets at home and started visiting the Zoological Survey of India, 'pestering' senior scientists like S. Biswas and D.P. Sanyal for information. He also started making his own sketches of turtles, and as early as 1984 (when just out of college), published his first field guide to the turtles and tortoises of India, based on these amateurish sketches. Neil would often skip college and go to the markets looking for interesting specimens – *Hardella*, *Kachuga*, *Geoclemys* and others, especially at the fish wholesale market of Howrah, a major transhipment point for turtles captured from the rest of India to eastern India and across the border, to Bangladesh.

Neil also became very interested in sea turtles at the time, especially their conservation, after a visit to the fish market at the town of Canning, in the Sundarbans, where ridley shells were stacked over wooden huts to function as roofs! He had contacted Rom Whitaker at the Madras Crocodile Bank and visited them shortly thereafter. He met Vijaya and found that they shared an interest in turtles, and in addressing the astronomical take from the Odisha coast. Neil got Bonani Kakkar of WWF Kolkata interested in the issue, and they visited Gahirmatha together, and alerted the turtle conservation community of the impending missile-testing range that was being set up nearby. Kakkar, amongst others, was involved in pressuring the Railways to stop the transport of the sea turtles, helped with the campaigns against ports near the Odisha sea turtle beaches, and started an education programme for sea turtles, featuring native scroll-painter-singers.

Indraneil Das has become one of the world's leading herpetologists, an expert on the reptiles of south and south-east Asia. Based at Universiti Malaysia at Sarawak, he has published more than a hundred papers on the taxonomy of amphibians and reptiles and also served as the editor of *Hamadryad* for more than a decade.

Sepia tones

In the 1980s, one person who had a lasting impact on sea turtle conservation in Odisha was J. Vijaya. Known as Viji, she was a pioneering young herpetologist, perhaps India's first woman herpetologist, at the Madras Snake Park and then at the Madras Crocodile Bank in the 1970s. She joined the snake park in 1975 as a volunteer and spent her weekends doing all the things that volunteers do – cleaning cages, filing papers, managing visitors and so on. Along with Shekar Dattatri, she went on field trips to the countryside around Chennai looking for snakes and lizards.

In the early 1980s, Viji became interested in freshwater turtles, and soon after, accompanied Edward Moll, chairman of the IUCN's Freshwater Chelonian Specialist Group (as it was then titled), on his nationwide survey of freshwater turtles. They travelled from West Bengal to Assam, and eventually estimated that over 50,000 flapshells, 5,000 to 10,000 large softshell turtles, and 15,000 hardshell turtles were shipped to Kolkata each year for slaughter and consumption. Viji also visited some of the most dangerous regions in northern India, including Chambal, known for its dacoity and lawlessness.

Her visit to Digha in West Bengal and the consequence of her actions are well documented. Vijaya wrote[29]:

> At 7a.m. on a Thursday morning January, 1982 we arrived at a market in Calcutta, carrying out routine survey work for the Freshwater Chelonian Group of the IUCN. Several Pacific Ridleys were on their backs, eyes bulged from the pressure of being overturned for several days with flippers wired together. Three or four customers wanted sea turtle meat so a female was slid across the slippery, gouged concrete floor next to the scales. The young cutter drained his tea cup, and picked up the just sharpened knife. He bent over and deftly cut around the margin of the plastron, avoiding the flailing flippers and the sudden desperate attempts to reach and bite the knife hand. The dark blood overflowed onto the cement as the plastron was ripped off, all of the pulsating innards exposed. The flapping and biting action continued, but feebler now as the reptile was eviscerated and the important meat carved out for weighing. The female ridley didn't die for 10 minutes . . . No one was concerned about the suffering, nor was there any worry about the Indian Wildlife Act, under which sea turtles receive the 'highest' protection!

Viji's most significant contribution was her rediscovery of the forest cane turtle. The species had been recorded in 1911 by Henderson at the Madras Museum, and had not been seen in over sixty years[30]. Viji set off to look for the turtles with the Kadar tribals. She visited her first Kadar settlement in June 1982 in Vazhachal near Chalakudy, and wrote[31]:

> 'The Moopan', or headman, was appointed to accompany me as he was the oldest man available to accompany a girl into the forest. Moopan, whose actual name I was never allowed to address, was a dignified man, four-and-half feet tall, with a serene face. Rain or shine, we would go out with his big umbrella and his sickle, which he used to chop off plants to make way in the jungle. Though we

never found a single turtle, he took the trouble to teach me every tree of significance and every hill in the vicinity.

When she found the turtles, she wrote proudly in *Hamadryad*[32]:

> After a period of 67 years, the forest cane turtle *Hoesemys silvatica* has been rediscovered during a recent survey in the Kerala forests by Snake Park . . . According to the Katumaran tribals, forest cane turtles are difficult to find, because of their small size and their habit of hiding in groves of eeta or reed bamboo, under dead leaves, logs, etc.

Viji went on to study various aspects of the behaviour and biology of forest cane turtles and travancore tortoises[33,34]. Living out of a cave in the forest, she conducted mark recapture studies by notching the carapaces of the turtles. She attempted to study movements by attaching threads to their carapaces and following them. She also established captive breeding populations for the species at the crocodile bank[35]. In 2006, the species that she had rediscovered was named after her and is today known as *Vijayachelys silvatica*.

In 1984, Viji went to the US to Eastern Illinois University to pursue her graduation under Ed Moll. During this period, Viji had a nervous breakdown, suffering possibly from schizophrenia. She returned to India for her fieldwork in 1987, and was found dead in the Guindy forest several weeks after disappearing from home.

In 2006, the *Indian Ocean Turtle Newsletter* carried a tribute to Viji[36]. Her sister, Prabha, wrote a powerful and touching eulogy for this talented but troubled girl, herpetologist, friend[37]. Her opening lines were: 'When Viji's remains were found in the summer of 1987, life turned to a sepia-toned freeze frame for us, her family. For years, my sister and I shared an intense relationship, so intense that I felt all her pain within my flesh and blood, but was helpless to change anything for the better. She was three years older than me and I loved her beyond measure . . .'

Prabha writes of a strong-willed, mercurial young girl with an incredibly strong love of nature, and a strong sense of social justice.

On the one hand, their pets included monkeys (Massey, the bonnet macaque), white mice, a chameleon, sand boas, and once, even a little fox; and of course, turtles (Melvin the cane turtle and Emma the Travancore tortoise). Not to mention Millicent, 'the giant spider who birthed several hundred hairy children all over our bedroom'. On the other hand, she also was 'the only eight-year-old I know who invented three different class versions (super rich, middle class and dirt poor) of the same game.' For Prabha, she was simultaneously 'George in the *Famous Five* series, Jo in *Little Women* and Scarlett in *Gone with the Wind*.'

Indeed, Viji changed everything that she touched in her short life, her friends, her colleagues, the animals she worked on. From Rom Whitaker, who gave her a start, to Ed Moll, to her peers, Shekar Dattatri and Indraneil Das, everyone who knew Viji had great admiration for her. Not just for natural history skills, but for her perseverance and her passion for nature and conservation.

Turtle Kar

Once Robert Bustard had discovered Gahirmatha, he set about trying to establish a research and monitoring programme there. In 1980, he reported in the *Marine Turtle Newsletter* that he had seven Ph.D students working under his supervision on various aspects of crocodilian and turtle biology in India. 'Not training overseas,' he said pointedly.

One of these, Chandrasekhar Kar would go on to complete his Ph.D on marine turtles in Odisha. Chandrasekhar joined the forest department in Odisha as a research scholar in 1976 to work on crocodiles and sea turtles. It is remarkable that the forest department had such research positions at the time, as most states do not have such positions now. He was initially stationed at Nandankanan Zoological Park, but decided to shift to Gahirmatha as no one else was willing to work there.

In 1977-78, Bustard approached Madhab Chandra Dash at Sambalpur University, and asked to supervise Chandrasekhar's research formally. Chandrasekhar carried out fieldwork at Gahirmatha between 1977 and 1982, including a range of studies on the mass-nesting population. Bustard was accepted as a co-guide, despite substantial bureaucratic difficulty in getting a foreign supervisor approved.

Chandrasekhar worked under extremely taxing conditions for several years. The beach was a three-day trip from Bhubaneswar. Chandrasekhar would travel by bus to Gupti, stay overnight there, take a boat to Dangmal, stay overnight again at Sudhakar Kar's saltwater crocodile research camp there, and then go by rowing boat to Ekakulanasi, the beach at Gahirmatha.

When I spoke to him in 2011 about his work, he became very nostalgic about his field days, and bemoaned the fact that his work now mostly consisted of pushing paper. He said:

> The place where I was staying was Habalikatti. There was no habitation in a 10 km radius, no post office in a 35 km radius. Once in fifteen days I used to send one person, he used to walk 35 km, cross the river, because there was no post office even in Gupti at that time. And I tried my level best so that Satbhaya, Gupti and other places would have a post office. No books were available, we were not allowed to come to the university and refer to the books. No literature was available. There was no chance to meet the guide even once in a year. There were no cell phones, there was no computer, nothing of the sort. No solar lights, no VHF. And in such a situation we were staying 365 days in the field. During the entire period of five to six years, my parents couldn't visit my place, none of my relatives could visit. And we were not allowed to come frequently – not to the head office, not even to the regional office.

The nearest village was Satbhaya, 10 km away. When the villagers came to Habalikatti to catch crabs, Chandrasekhar would plead with them to stay overnight with him, but he would usually end up alone. He eventually did succeed in his attempts to get a post office at Satbhaya, for which the village was eternally grateful.

Chandrasekhar did receive many visitors, though his father failed to visit after several attempts, only reaching Gupti. He travelled to Gahirmatha with T.A. Davis, G.M. Oza and Bedi which led to the publication of a controversial article. They were accompanied on that trip by L.N. Acharjyo, a well-known academic in Odisha at the time. Anne Wright, Jack Frazier, Rom Whitaker and various others followed over the years.

Rom Whitaker wrote in *Hamadryad* following his visit: 'The Research Officer should be given further support in terms of equipment and personnel (i.e. he has no adequate camera, film, typewriter, two-way VHF radio or assistant officer).'

Chandrasekhar also discovered a second rookery at the mouth of the Devi river in 1981. Partly because of the isolation, Chandrasekhar had hired locals and conducted surveys with them. One of these, from the Devi region, told him about a large nesting ground for turtles there. After several unsuccessful attempts to reach the location, he finally made it in the winter of 1981, after many hours of travelling by bus and then foot. Arriving at the beach, they found evidence of recent mass nesting in the form of exposed eggs and egg shells.

As part of the forest department, Chandrasekhar was also closely involved in enforcing the Wildlife (Protection) Act to stop the large turtle fishery of the 1970s. Chandrasekhar took pride in the efforts of the forest department in stopping the transport of turtles by rail and then road. He also frequently represented the forest department in the 1980s on sorties with coast guard ships and aircraft.

Over the years, Chandrasekhar was involved with many research projects including the projects conducted by the Wildlife Institute of India and Indian Institute of Science. He was a mentor to Bivash Pandav and Basudev Tripathy, who found recognition as sea turtle biologists in Odisha in the 1990s. In 1982, Chandrasekhar left for Sambalpur University to work in a GoI-sponsored project under M.C. Dash. He later rejoined the forest department through the Odisha Public Service Commission as a research officer in 1989 and was promoted to senior research officer in 2003. Chandrasekhar retired in February 2014, and passed away suddenly shortly afterwards. Sudhakar Kar, his colleague and friend for nearly forty years, wrote a tribute highlighting his contributions to sea turtle conservation in Odisha[38].

Kar's book, *Gahirmatha: A Turtle Paradise*, co-authored with his supervisor M.C. Dash and dedicated to 'the memory of Smt. Indira Priyadarshini Gandhi and Dr Salim Ali', remains a comprehensive account of sea turtles in the region and his work there. With a foreword by Marie Dimond highlighting the importance of the information collected by Kar, the book provides particularly rich accounts of population dynamics and reproductive biology, based on surveys, monitoring and tagging studies. Kar marked over 10,000 female nesting turtles and at least one of these returned twenty-one years later to nest in Gahirmatha.

As a conservationist, Chandrasekhar was torn between the philosophy of his mentor and the ideology of the department he worked for. Bustard was a strong proponent of sustainable use. At the end of *Gahirmatha: A Turtle Paradise*, Dash and Kar devote a chapter to 'Conservation Strategy', where they discuss the utilization of 'doomed eggs' and the economics of take. They say that if appropriate management measures are taken, then 'a small number of the adult males as well as all the mutilated and deformed females can be cropped for human consumption'. At the end, they state clearly: 'However, this is suggested on an experimental basis under strict scientific management by the forest department and experts.'

The idea of use

In one of the first examples of the hyperbole surrounding sea turtle conservation in India, D.N. Ganguly of Calcutta University wrote an article titled 'An appeal to save Pacific ridley turtles from mass killing in West Bengal and Orissa'[39]. In his abstract, he outlined various threats and concluded that 'The species, hence, may soon be extinct'.

As in a number of accounts up to that period, the author incorrectly states, in a footnote, that this species is 'popularly known as the loggerhead or olive ridley'. Ganguly noted that 10,000 turtles were being dispatched from Puri to Kolkata each year. Apart from threats from depredation and hatchling predation, he also speculated that male–male competition due to shortage of females might result in some mortality. After recommending stringent protection measures, he does however recommend farming in the shallow waters for this 'meat-producing turtle'.

The articles that attracted the most attention, however, were an exchange that same year in *Environmental Conservation*, an international journal[40]. In February 1977, Gunavant Oza, from MS University in Baroda, Rajesh Bedi, a filmmaker, and T.A. Davis of the Indian Statistical Institute, Kolkata, visited Gahirmatha. They were joined by the wildlife warden, G. Chowdhury Mishra, two veterinary officers and a research scholar (probably Chandrasekhar Kar). G.M. Oza had been to Gahirmatha in previous years and had seen large numbers of turtles nesting. That January, he arrived in Bhubaneswar to participate in the Indian Science Congress symposium on the conservation of wildlife and forests, which itself was unusual for the time. He was, as they wrote, 'shocked to witness in the Railway Station living sea turtles carried away every night in several hundreds'.

Their opening lines: 'The once common and familiar Pacific Ridley seems to be losing its fight for survival – ironically enough in one of the world's largest rookeries' and '. . . this sea turtle is also nearing extinction in the Cuttack District of Orissa . . .'

They went on to say: 'These were brought on the roofs of passenger buses, in scorching heat with their bellies upwards, puffing and groaning in apparent agony. They were then dragged off the roads and sometimes mercilessly handled before being taken off on the train – though with doubtful legality.' They detailed through the rest of their article, titled 'Sea turtle faces extinction in India', the plight of sea turtles on the Odisha coast. 'Will it not be a sad day for all of us if they vanish because of commercial slaughter?' they lamented.

This article was to set the tone for two decades and more of hype about sea turtles in Odisha. A similar article in *Environmental Awareness* stated that 'though the '76 season (Jan–March) had brought over 1,58,161 nesting female *Lepidochelys olivacea* ashore, not a single turtle visited this beach in '77'[41]. This was duly reported in *Hamadryad* with editorial notes and reprinted in *Marine Turtle Newsletter* (with the heading 'India: mass slaughter of sea turtles')[42]. *Hamadryad* questioned the wisdom of publicizing habitats and locations of commercially viable species such as sea turtles, as had been done the previous year, when scientists became excited about Gahirmatha.

Jack Frazier, one of the stalwarts of global sea turtle conservation and an occasional visitor to Gahirmatha, was quick to respond. In a response to Davis et al. titled 'Crying wolf or saving sea turtles', also published in *Environmental Conservation*, Frazier wrote that many of the claims by these authors were unsupported and some had been refuted[43]. As he pointed out: 'Apparently, Davis et al. were only able to spend a few hours on the beach in 1977, and on the basis of their brief observations made inferences for the entire season – a very questionable procedure.'

Frazier agreed that the conservation of olive ridleys in Odisha was important, and that Davis et al. had raised critical issues, but their message had been undermined by refutable statements. He ended with the classic line: 'Pleas for rational management must themselves be rational.'

Shortly thereafter, Chandrasekhar Kar, the research scholar working in Gahirmatha, responded to these authors in an article in the *Marine Turtle Newsletter*[44]. He pointed out that he had been practically living at the rookery since 1976 and that their inference based on a single day's visit was incorrect, and that more than 1,00,000 turtles had nested during arribadas in both 1978 and 1979. In this article, Kar mentioned (perhaps for the first time) the emerging threat from trawlers. He said that they likely captured turtles only incidentally, but might be tempted to sell them because of the price they fetched in Kolkata. Kar then made the case for greater protection of sea turtles and their beaches on the Odisha coast.

Kar's article ended with an acknowledgement to Robert Bustard, his mentor. Ironically enough, the pages that followed in that issue of *MTN* accommodated a fine philosophical piece by Bustard himself on whether sea turtles should be exploited[45]. Ironic for two reasons: one, the mentor's article follows the student's and presents a radically different perspective; and two, Bustard was originally responsible for the ban on the collection of eggs in Odisha. Bustard's article highlighted a divide that was to polarize the international sea turtle community over the next three decades. Bustard wrote that 'unless wildlife is utilized, we will lose most of it'. Confessing that he was responsible for the cancelling of the licences, he recommended 'limited and scientifically managed exploitation'.

Echoing a battle that is still very much alive in India today, Bustard wrote: 'The international conservation fraternity's "protect everything" philosophy does real conservation – which surely includes sustainable yield/utilization – as opposed to mere preservation, a great disservice in that it makes the countries in the developing world feel that total protection alone represents advanced thinking.'

By the early 1980s, trouble was brewing over the continuing intentional take of adult turtles in Odisha. Vijaya's survey led to the article in *India Today* and the letter writing campaigns in *MTN*. Even here, though writing against the 'inhuman slaughter', Mrosovsky suggested that 'what is needed is not total prohibition but rational, or at least controlled, utilization'[46].

In 1982, Bobb's article in *India Today* quoted one conservationist as saying: 'With an efficient, logical programme for the Ridleys, the Bengalis can eat turtles forever. If the present slaughter continues, they have less than a decade to enjoy them.'

In 1991, Bhaskar wrote[47], 'One of the objectives of research must be to forestall these losses and to make turtle eggs, which are a valuable protein source, available for human consumption, *without affecting natural turtle populations* (his italics).'

Even Chandrasekhar Kar in his 1990 book, *Gahirmatha: A Turtle Paradise*, included a chapter on the economics of harvesting one thousand mutilated turtles and roughly fifteen per cent of the eggs that would have been destroyed in arribadas[48]. However, no talk of use has ever been raised since the mid- to late-1990s, coincident with the beginning of major conservation initiatives.

The issue of sustainable use has long been debated in the conservation community with differences based on geographical location, taxonomic group and so on. For example, some countries like South Africa and Australia have been at the forefront of use as a conservation tool. In 1980, George Hughes wrote an article titled 'Conservation, utilization, antelopes and turtles'[49]. As he pointed out, the take of turtles had been intensive in every sense of the word in areas like the Caribbean and in Seychelles, and yet the turtles had survived.

In 2001, Nicholas Mrosovsky, another longtime supporter of sustainable use, wrote in *Kachhapa* that the situation in Odisha was 'Triple Waste': the waste of eggs, the waste of adults, and the waste of an opportunity to help the people[50]. Many turtle biologists, however, both then and now, believe that turtle use is the thin end of the wedge, and that it would, in any form, lead to the inevitable decline and extirpation of sea turtle populations.

The general reluctance to consider egg harvests is probably due partly to the protectionist approach to turtle conservation, in particular the way the ridley has been used as a flagship creature for conservation. For the species to be used as an effective (= emotively provoking) flagship, it has been portrayed as highly endangered, which means of course that no use is possible, and strong protectionist approaches have therefore been advocated. This has precluded even considerations of other approaches to conservation, so much so that it is considered bad form to even raise the issue in public.

At the same time, there is no doubt that there would be practical problems in the implementation of a scientific, controlled exploitation of turtle eggs. Given limited personnel and enforcement, there would be innumerable challenges in ensuring that all the take is legal. Here, as elsewhere, management and enforcement authorities have made a choice: a blanket prohibition, with all of its political pitfalls, is still easier to enforce than managing different levels of legal exploitation.

Rights and wrongs

Both in India and internationally, sea turtles have been accorded the hallowed status of 'charismatic mega-vertebrate' in the same corner as tigers and elephants. Hence, the very thought of any form of consumptive use of sea turtles appears to have become anathema.

At the Annual Symposium on Sea Turtle Biology and Conservation in Kuala Lumpur, Malaysia, in March 2003, a group of us (including Matthew Godfrey, Lisa Campbell, Charles Tambiah and I) organized a special session on use, the stated purpose being to get beyond the poaching paradigm, to accept that use existed, and to explore its impacts and sustainability[51,52]. The keynote speaker and many others worked principally on crocodiles, a group that has had much more success in promoting conservation through use. Similar sessions on use, indigenous communities and fisheries would be held at symposia in Mexico, Australia and India from 2008 to 2010, but the international community remains lukewarm at best, mostly cold, and occasionally hostile to the issue. Most international treaties on sea turtles only deal with use in the context of reducing it, and providing alternative livelihoods. This largely West-driven philosophy of sea turtle conservation still pervades much of wildlife conservation in India. Let alone sea turtles, India does not even have a programme of commercial harvest of crocodiles, one of the flagships of conservation through use across the world. Though there are some voices of dissent, they are some distance away from having any impact.

Unfortunately, this is not an issue of marine turtles alone. The malaise runs much deeper, to the differences in the philosophies of animal rights on the one hand and environment and wildlife conservation on the other, with the concern for wildlife being only a superficial similarity between the two[53]. Conservation philosophies do not suggest that there is anything morally or ethically wrong with using animals, particularly as food, although this does not mean that cruelty is condoned. In fact, sustainable utilization can be a powerful tool in motivating communities to conserve a resource. It has been argued that many conservationists – termed preservationists in some circles – are opposed to any kind of utilization because their actions and ideas stem from an animal rights philosophy.

Philosophers have distinguished animal liberation and environmental ethics as being based on different intuitions, principles and behaviour[54]. Broadly speaking, our relationship with nature can be divided into two categories: the biocentric approach (on which animal liberation is based) posits that nature should be preserved because of its inherent right to exist, while the anthropocentric approach views nature in the context of its relationship and usefulness to humans. In general, pure biocentric approaches have been not been considered as useful for environmental conservation[55]. However, some have suggested that, as a compromise between the extreme positions of bio- and anthropocentrism, environmental ethics can invoke 'weak anthropocentrism', which invokes and uses the cultural value of nature in human society, in addition to its utilitarian value[56,57]. Animal rights is typically not concerned with human welfare, and its use in political spaces such as conservation has the potential to alienate at least some proportion of the people involved. Hence, animal liberation and environmental conservation may require very different approaches and actions. The onus is on environmental conservationists to carefully choose and implement approaches that have the best consequences, not just for individual species, but for ecosystems and diversity in the long run. Conservationists must seek not single-point-single-species successes, but success across geographical regions, over time and for many taxonomic groups.

Notes

i Silas's team visited Digha and other areas in West Bengal in 1982-83. During 1981-1982, they estimated that about fifteen fishing units were operating from Medinipur district, each with six country craft and a motor launch, catching about 6,000 turtles per season per unit. In December 1982, they found fifty-nine turtles in two sheds at the fish landing centre in Digha, which were subsequently confiscated by the forest department and released. In February again, some live turtles and dozens of dead turtles were seen at Digha, but it appeared that the main landing had moved inland along a creek. Despite all the enforcement, it is believed that 10,000 turtles were landed in 1982-83. Several thousand turtles were also shipped to Kolkata from Digha and other landing centres in 1983-84[58].

During the 1984-85 season, the CMFRI again carried out surveys in Odisha and West Bengal[59]. They reported that the large-scale take of the previous years had declined, but turtles were still being landed and sold in Digha (about ten to fifteen per day) and other locations in West Bengal.

S.K. Raut of the Department of Zoology, University of Calcutta, and N.C. Nandi of the Zoological Survey of India, also conducted a survey in 1983-84 on the West Bengal coast[60]. They recorded the capture of nearly 20,000 turtles from Digha to Jambudwip, about 8,000 of which were consumed locally and the rest transported to Kolkata. The turtles were transported through creeks to interior landing centres to avoid being caught by forest officials and thereon by 'rickshaw-van' and 'matador-van' to the markets.

ii The coast guard vessels, *Rajhansa* and *Rajtaranga*, patrolled the coast between Dhamra and Paradip during the 1982 season. During February 1983, two patrol vessels and a naval helicopter patrolled the area, and a shore liaison office was established at Gahirmatha; on 6 and 7 February 1983, the coast guard vessel *Rajhansa* was involved in a patrolling operation with the forest department and police, which resulted in the apprehension of ten country boats, three mechanized boats and about sixty poachers, all from West Bengal[61]. A total of 186 turtles were rescued, of which thirty turtles died in transit, but the rest were released (an event that Moll, Bhaskar and Vijaya witnessed)[62]. During 1983-84, the coast was patrolled by helicopter and low-flying planes of the coast guard[63].

FLAGGING SHIPS OF CONSERVATION

The 'world's largest sea turtle rookery'

I FIRST WENT TO Gahirmatha in 1997, on a field trip as part of a Northern Indian Ocean workshop of the IUCN's Marine Turtle Specialist Group (MTSG). This trip marked my return to the sea turtle world. After leaving the SSTCN and Chennai in 1991, I had spent a year teaching in a residential school, and then joined Indian Institute of Science, Bengaluru, where the twisting turns of academia led me to a Ph.D on small mammals in the Nilgiri mountains of the Western Ghats. Several years later, I heard about this workshop in Bhubaneswar and I decided to 'check out the sea turtle scene'. And I heard that Jack Frazier and a number of other international biologists were going to be at the meeting. I had read so much of their work and heard so much about them, I felt not terribly unlike a schoolboy about to meet his favourite sports heroes. It was not a disappointment. Jack gave some fantastic talks on sea turtle conservation, Colin Limpus and Frank Paladino lectured on aspects of sea turtle biology that I did not know existed, and I was enchanted once again by sea turtles.

Charles Tambaiah, who had helped the SSTCN get funds back in the early 1990s, was present, as were Peter Richardson and Thushan Kapurusinghe, who had helped start the Turtle Conservation Project, one of Sri Lanka's leading conservation groups. Pam Plotkin was then attempting to initiate a project on satellite telemetry on the olive ridleys of Odisha, following her work in Costa Rica, but the layers of bureaucracy eventually defeated her. Marydele Donnelly, the long-serving programme officer of the MTSG, Neca Marcovaldi, founder of Projeto Tamar, a pioneering sea turtle conservation effort in Brazil, and Douglas Hykle of the Convention on the Conservation of Migratory Species (CMS) were also there. And I also met B.C. Choudhury and Bivash Pandav for the first time.

After two days of brainstorming and lectures, the entire caravan travelled to Gahirmatha, first by bus to Rajnagar, and along a dust-laden track to Gupti. From there, we piled into a boat and chugged along the river for several hours before we reached the mouth of the Barunei. And then into smaller boats to reach the little islands off the river mouth. There are three islands: Inner Wheeler, Outer Wheeler, where the Defence Research and Development Organization (DRDO) has a missile testing range, and Long Wheeler Island or Babubali, where Bivash maintained his camp. Formerly a fishing camp, the semicircular thatched huts served as accommodation, while one large tent served as office, dining room, living room and so on. Two years later, when I camped there, and enjoyed and suffered the solitude, I was struck by the contrast with the DRDO base less than half a kilometre away, which had a tarred road, vehicles, a railway track and train, an officers' club, phones, whisky, normal food, beds, air conditioners and other assorted knick-knacks associated with civilization.

When the beach was discovered in the early 1970s, nesting occurred on a spit that was part of the mainland coast and extended across the Barunei river mouth. Researchers like Chandrasekhar Kar had used the Ekakula forest resthouse near the river mouth as their base. In 1989, the cyclone split the spit away from the mainland and the sand bar became an island that moved further away from the river mouth each year. Since then, nesting has occurred on the island, known as Ekakulanasi (nasi = island in Odiya). The island got further fragmented into two, Nasi 1 and Nasi 2, in the mid-1990s, and nesting occurred on both islands in 1999. Since then, one of the islands has become further fragmented, while the other has moved further and become contiguous with Outer Wheeler Island.

At night, we used the boats to get across to Nasi island, where mass nesting occurred. That year, the island was about 4 km long and a hundred metres or more wide. We saw no turtles that night, and there was multilingual cross-cultural disappointment at not having seen an arribada. Little did we know that there would be no arribadas in 1997 or 1998, setting the stage for large-scale concern about this population, and launching a major conservation initiative in Odisha.

The group left the next day, and I stayed on with Bivash for a few days, still hopeful of the arribada. There was little nesting, but we did see a few mating pairs out at sea. Not to mention the trawlers that were killing hundreds of turtles. The day before I left, we visited the navy camp at Ekakula on the Gahirmatha mainland beach. The coast guard and navy had been requested by the Odisha Forest Department to help monitor the offshore waters of Gahirmatha to prevent illegal fishing. A group of officers had been sent, but the forest department was supposed to provide the boat, which had not yet arrived. And now they were preparing to wind up the camp and leave. Disappointed that they had not achieved anything, they requested Bivash to take them out to sea, at least to see a few turtles.

The next day, we were all off together, and a little way out, we saw a couple of trawlers. The armed forces sprung into action, and we were off on a hot chase. We caught up with the trawler, and boarding it, the young army officer who was just back from gruelling times in Kashmir, wasted no time in quelling resistance. Pretty soon, a second trawler was apprehended. With both trawlers pitching and yawing as the army decided what to do with their conquests (the forest department guard had no idea either), the army officer who was the spiritual leader of this pack where no else had seen any real action, suddenly decided to be violently seasick, and collapsed in a heap. Realizing that we were outnumbered at sea, we used the psychological advantage to administer severe admonishments and reprimands and left. Not without some of their booty, a good catch of prawns, which livened up the next several meals.

It would be a couple of years before I returned to Odisha to study sea turtle genetics and subsequently became thoroughly immersed in the morass of turtle conservation in the state.

Shades of grey

The threat from trawling

The stranding of dead turtles on the beaches of Gahirmatha was first reported in the late 1970s and early 1980s. Many of these may have been discards from targeted take during the time, but by the early 1980s, it had become clear that the increasing number of shrimp trawlers were beginning to kill turtles as well. Chandrasekhar Kar, who started his work in Gahirmatha in the 1970s, says that a few hundred carcasses were found on the beach each year in the late 1970s. The significance of the threat from trawlers was recognized during the 1982-83 season, when Kar enumerated about 3,000 dead turtles on 10 km of beach in Gahirmatha[1]. Silas et al. estimated that there were 7,000–7,500 dead turtles along a 15 km stretch that year, based on the average mortality on the stretch they had encountered. They assumed that the stranding rate was consistent over the entire stretch and did not take the rest of the season into account. Nevertheless, there is an indication that thousands of turtles washed up at the shore that year. In their article, they devoted a section to the issue of incidental catch, calling it 'a very significant event'. It was here that they made their famous 'graveyard of ridleys' comment.

But before that, they talked about a turtle excluder device (TED) for shrimp trawling that had been developed in the US. Silas, ahead of things as always, started advocating the use of TEDs more than a decade before they became a current topic in Odisha. They also advocated the expansion of Bhitarkanika northwards to include Wheeler and Shorts Islands and southwards to include Hukitola Island and beaches near Paradip to protect the breeding aggregations and nesting beaches at Gahirmatha and the mouth of the Devi river.

Chief Minister J.B. Patnaik and Member of Parliament Jayanti Patnaik visited Gahirmatha on 31 December 1983 and 29-30 January 1984 to see the arribada[2]. Patnaik verbally agreed to the demands for a no-fishing zone in the offshore waters of Gahirmatha (up to a distance of 10 km) from September to April, from Palmyras Point (near Wheeler

Island) to False Point (near Hukitola Island), a distance of about 35 km. He would later announce this over the radio as well[3].

Over the next few years, hundreds of dead turtles – down from thousands – were recorded in Gahirmatha[4]. The reduced mortality was attributed to greater vigilance on the part of the government, perhaps Operation Geeturt and the coast guard. In the early 1990s, more than a thousand dead turtles were washing up on the beaches of Gahirmatha, and both gill-netters and trawlers were clearly recognized as threats, but attention was focused on a proposed fishing jetty at Talchua, within just a few kilometres of the mass-nesting site.

In 1993, WII student Bivash Pandav, started studying sea turtles in Odisha. Over the course of his research, the number of dead turtles recorded on the coast would increase from a few thousands in the early 1990s to more than 10,000 in the late 1990s; he and his team would count over 40,000 dead turtles in just a few years[i]. Given that the carcasses that wash up on the beaches are only a proportion of the animals that die at sea, the numbers of turtles killed have been monumental. The incidental mortality from trawlers incited action, namely Round 2 of turtle conservationists versus fishermen in Odisha.

When Pandav started reporting thousands of stranded carcasses in Gahirmatha and neighbouring beaches, it caught the attention of the media, but also of two conservationists who were interested in sea turtles. Belinda Wright had visited Gahirmatha with Shekar Dattatri and witnessed a really large arribada there. Following conversations with Banka Behary Das, a politician deeply concerned about the environment, Wright wrote about the little known threat from trawlers to highlight the issue.

Another conservationist in Odisha was also getting interested in olive ridley turtles. Biswajit Mohanty had become engaged with wildlife issues, and one that had slowly started to aggravate him was the unchecked mortality of sea turtles. Together, Wright and Mohanty conceived of and initiated Operation Kachhapa with support from the US-based Barbara Delano Foundation. In January, 1999, a meeting was held at Wright's house in New Delhi with the donor, with presentations by experts such as B.C. Choudhury. Having recently re-entered the world of sea turtles, I remained the proverbial fly on the wall. Eight months later, with more funds in the kitty, a key planning meeting was held at Wright's house again. This time, back from a full season of research in Odisha, I was more tuned in to the issues. Pandav, by now an outspoken biologist campaigning for the protection of sea turtles, was present at the meeting as well.

The meeting was largely constructive and involved planning the use of funds and strategies for fulfilling Operation Kachhapa's mandate. One of the stranger parts of the meeting, though, involved back-to-back discussions over the purchase of a firearm for Mohanty's protection and funds for food for compensating fishers when they were banned from fishing (only for traditional fisherfolk at mass-nesting beaches). The first discussion lasted all of five minutes and was easily approved by the American donor. The second took more than an hour and covered the gamut from the philosophy to the practicality of providing compensation. The amounts involved were identical. While we sat there, a cyclone was devastating the coast of Odisha. The compensation scheme, which was approved, was eventually used to provide support to communities at the Devi mouth region, which had been particularly hard hit by the cyclone.

Mohanty wrote about this eloquently in the first work report of Operation Kachhapa in its newsletter[5]. The cyclone had hit the Odisha coast on 29 October 1999, and most coastal villages were flooded and accessible only by boat for nearly two weeks. Many parts of Cuttack, the state capital, where Operation Kachhapa's office was situated, were underwater, and telephone lines were down through much of the region. S.K. Pani, Operation Kachhapa's project officer for many years, left for his village to look for his family and encountered harrowing scenes of death and devastation on his way. He reported that entire villages had been washed away and hundreds of bodies were washing up when the waters receded. The team attempted to visit the villages near Devi mouth, but the roads were washed away or blocked with fallen trees. In the third week of November, they were finally able to visit some of the villages to distribute relief material, including blankets and food to the families of fishermen and others of both villages.

This earned Operation Kachhapa some goodwill amongst those communities, at least in the beginning. The broad objectives of Operation Kachhapa were listed as follows[6]:

> ... to reduce turtle mortality and safeguard the future of the species by concentrating on three main activities: a) the prevention of turtle mortality by improving patrolling of no fishing zones and the protection of nesting sites; b) supporting legal action on turtle conservation issues and fishing law violations, and; c) building up public support and awareness of sea turtle conservation issues, including sensitizing the media, enforcement agencies and the judiciary about the large-scale turtle deaths.

Their planned activities under enforcement included providing a sea-going patrol vessel for the forest department and auxiliary support boats for the monitoring camps run by the department, providing equipment for mobile camps run by the forest department and researchers to protect and monitor nesting beaches and creating an incentive scheme for enforcement officers. They also aimed to assess enforcement action and penaltics incurred to present the results to the Odisha government's committee, follow relevant court cases, and pursue private proceedings against trawlers apprehended for illegal fishing and against poachers.

By this time, there were plenty of laws to protect olive ridley turtles in Odisha. In addition to sea turtles being listed on Schedule I of the Wildlife Act, the Gahirmatha Marine Sanctuary was declared in 1997 under this Act. Srimoy Kar's article[7] in 1997 reported on the issue, stating that the declaration of the sanctuary would 'probably bring an end to the age-old struggle between man and nature that has been going on in this region for years'. As he says, this was seen as a 'big victory for the conservationists' and 'widely welcomed', but he conceded that 'thousands of fishermen who have traditionally been dependent on this region for their livelihood, are very upset'. In closing,

Kar says, 'In its haste to protect the turtle, it cannot afford to throw its fishermen into the sea.'

The forest department had welcomed the notification as something they had long fought for and said that it would work closely with the coast guard to make sure that the law was implemented. Patrolling of nearshore waters, ever the problem, was to be done by a new boat for the forest department, which was repeated like a mantra each year, but never came about.

To help enforce these laws, Operation Kachhapa supported the forest department by hiring private trawlers to be used for patrolling the nearshore waters of Gahirmatha and paid legal expenses for prosecuting trawlers that were caught fishing illegally. In 1998-1999, a patrol vessel was hired for over three months along with a support boat and other equipment. Sixty-one trawlers and gill netters were seized for illegal fishing inside the Gahirmatha Marine Sanctuary, and their trial cases were contested by lawyers engaged under Operation Kachhapa on behalf of the forest department in the local courts. Apparently, trawler owners instructed their crews not to venture towards Gahirmatha and one of the indicators of decreased mechanized fishing was the steep fall in the sale of trawl nets, though this trend did not appear to last very long. But the boat was also rammed during repairs and put out of commission for a period. Though a patrol trawler was hired briefly in 1999-2000, state and NGO efforts were focused on cyclone relief.

In 2000-2001, a trawler was hired for nearly four months, resulting in the apprehending of thirty-one trawlers that year; the coast guard was also active, seizing over sixty fishing boats. The following year, patrol trawlers hired by Operation Kachhapa operated at Gahirmatha and Devi, with over a 100 trawlers and gill-netters apprehended in total. However, the conflict was increasing, and patrolling at Devi mouth was stopped after threats to the staff and crew by trawler fishermen. Patrolling continued in 2002-2003 with fisheries department boats and forest department staff[8].

During this period, Operation Kachhapa also raised awareness about turtle-related issues through the media. A turtle interpretation centre was established at Bhubaneswar to serve as a focal point for their education programmes. One of their more successful campaigns was the use of traditional wandering minstrels, who told the story of sea turtles and threats to their survival through a song in traditional style, written by Trilochan Dwivedi and set to music by Aditya Mohapatra[9]. The minstrels performed in over 100 villages along the coast using hand-painted backdrops, showing live nesting turtles, dead turtles with trawlers and other themes created for this purpose.

In addressing the issue of incidental mortality of sea turtles, however, conservation actions may have created more conflict than resolution. Both state and non-government agencies believed that enforcement of laws was the key. The three stated objectives of Operation Kachhapa (patrolling, legal support, and awareness) may have contributed directly to the polarization of fishing communities and conservationists. The tenor of many media reports (with titles such as '200 ridley turtles die on Orissa coast everyday'[10], 'Illegal trawlers mowing down Olive Ridleys'[11], 'No respite from trawling for Olive Ridleys[12]', 'Trawlers pose threat to Olive Ridleys: Report'[13], 'Flouting of Act leads to turtles' death[14]', 'Turtle death continues as fishing goes unabated'[15] and so on) certainly exacerbated the problem. The issue also received media coverage on a number of national and international television media including Star TV, BBC, National Geographic, Doordarshan, Zee TV and others.

The focus on trawlers diverted attention from a host of other threats that sea turtles face in Odisha. Light pollution and massive disorientation and mortality of hatchlings (particularly at Rushikulya), habitat degradation (mainly through the plantation of casuarina on vast expanses of nesting beach) and egg depredation by feral animals did not receive nearly as much attention. Not to mention the depletion of fisheries stocks, destruction of marine environments critical for fisheries and state support to poorly regulated fisheries export ventures at the cost of marginalized, small-scale fishers.

The most peculiar outcome was the use (or abuse) of the Orissa Marine Fisheries Regulation Act, 1981 (OMFRA), which stands out as a fine example of the perverse consequences of conservation action. The Act was passed in the early 1980s to regulate offshore mechanized fishing and to protect the rights of traditional fishermen[ii]. However, it was used by conservationists to promote the protection of turtles. Technically, there was nothing wrong with this, but it was promoted in a fashion that led to alienation of fishing communities (even non-mechanized ones) from conservation. Media attention on the OMFRA as a tool for turtle protection fuelled the perception that it was a bone of contention between traditional fishworkers and turtle conservationists. An Act that should have served to defend fishers was driving them out.

Operation Kachhapa reports clearly indicated that they were well aware that the laws should protect traditional fishers. One of their reports even stated[16]: 'Operation Kachhapa proposes to "level the playing field" and empower the local fishermen with the knowledge they need to stand up for their rights and protect their way of life.'

Even so, they were unable to prevent the spread of the idea that conservation actions were anti-people in general. This issue came to a head when a petition was filed with the Central Empowered Committee (CEC) of the Supreme Court of India, which was constituted in September 2002[17] to examine and ensure the implementation of legislation pertaining to forest and wildlife issues[iii]. In March 2003, the CEC set out interim orders with very specific recommendations largely pertaining to the better implementation of the existing legislation dealing with the mandatory use of TEDs and restrictions of fishing specified by OMFRA[18]. The document also summarized the views of the forest department, fisheries department and Trawler Owners' Association. The forest department, as always, bemoaned the lack of seaworthy vessels, manpower and funds. The CEC raised the issue of funds provided by Indian Oil Corporation (Rs 1 crore) towards sea turtle conservation, which was lying unutilized, pending governmental approval. The NGOs petitioning the CEC emphasized that 'three fast "sea going" boats' should be used for patrolling the three mass-nesting areas but also asked that casuarina plantations be stopped and the issue of artificial illumination addressed.

The fisheries department helpfully reiterated all the laws that had been passed, including the prohibition of fishing within 20 km of the Devi and Rushikulya mass-nesting beaches. Apparently 540 TEDs had been distributed to trawler owners free of cost and trials had been conducted. They also claimed that fisheries department vessels had been engaged in turtle conservation work and had seized twenty fishing vessels. The trawler owners claimed that they had been responsible for significant foreign exchange for the country and that conservation was depriving them of their livelihoods.

In passing its directions, the committee said it drew 'strength and guidance' from the 1998 judgement which was passed in the public interest litigation filed in the Odisha High Court by the Centre for Environmental Law, WWF[19]. It recommended that armed police personnel should be posted at each of the forest department camps, and that the coast guard intensify their patrolling and be authorized to impound boats. It also said that, once seized, these boats should be kept in 'well guarded locations' with armed guards. The emphasis on armed protection in this document is quite remarkable, and reflects a particular view of how wildlife conservation should be carried out. The order also instructed that all trawlers without TEDs should have their licences cancelled. Despite the (limited) cooperation from the fisheries department at that point, this was stretching belief a bit too far. That season, apparently two licences were cancelled for non-use of TEDs, out of the several hundreds of trawlers that were operating, needless to say, without using TEDs.

In 2001, Operation Kachhapa had documented a gill net that washed ashore near Gahirmatha with over 200 dead turtles entangled in it, suggesting that the threat from this fishery was also significant[20]. Though only a few large instances of such large-scale gill net mortality had been recorded, the CEC also recommended banning gill netters from operating within five kilometres of the coast at the turtle nesting sites, which certainly aggravated the conflict. Due to their lack of knowledge of laws or their own rights, local fishermen were often victimized by the forest department, whose enforcement sometimes lacked the discrimination it warranted. In addition, the state also made recommendations that the mass-nesting beaches at Devi and Rushikulya be declared as protected areas[21], which had the potential to further exclude people.

The CEC did, however, make a short field visit to Odisha in

February 2004 where it conducted meetings with fisher representatives (both traditional and mechanized) and conservationists and passed a further set of orders in 2004[22] with more nuanced fishing restrictions in Rushikulya and Devi.

TED talks

While much of the action revolved around zoning and time – area closures for various fishing groups – the subject of TEDs was never far from the surface. An apparently simple solution to the problem of incidental mortality in trawl nets is the turtle excluder device or TED. Many variations of soft and hard TEDs have been designed over the years, which are currently in use in the US, Central America, and Australia.

In the early 1980s, the US government introduced policy that required domestic shrimpers to use gear modifications such as TEDs to reduce the incidental mortality of sea turtles. In 1989, this policy was expanded to apply to all shrimp sold in their market. This meant that all nations importing shrimp to the US had to be certified by the US government. In response, the Association of Southeast Asian Nations (ASEAN), as well as India, Pakistan, Hong Kong, Korea, Australia, Mexico and Venezuela, protested to the World Trade Organization (WTO) that the US's implementation of this policy contradicted the rules of the organization. Following this, four countries – India, Malaysia, Pakistan, and Thailand – requested consultations where they argued against the US imposing domestic policy on foreign nations through import restrictions[23]. The US's influence came from its large market size – its shrimp imports totaled more than $1.2 billion per year[24]. Though India objected to the US's imposition of its domestic conservation standards upon shrimp importing nations, TEDs were eventually made mandatory on bottom trawl vessels in Odisha.

TEDs were first brought to Odisha during a workshop conducted by the National Marine Fisheries Service, US in 1996[25]. The workshop was locally organized by the Department of Fisheries and Project Swarajya, a local NGO, and attended by over 200 participants including trawler owners, operators and fishermen, and various national fisheries organizations[26]. Collectively, they came up with a large number of recommendations, including the mandatory use of TEDs, further research on incidental capture, and the consideration of financial incentives for the use of TEDs. Following a few years of campaigning on the issue and in response to a public interest litigation, the high court of Odisha delivered a judgement in May 1998 requiring all trawlers in the area to use TEDs to 'avoid entanglement of sea turtles'[iv].

The Central Institute of Fisheries Technology (CIFT), Kochi, conducted trials using the US designed super-shooter TED on the Andhra and Odisha coasts in 1995, demonstrating a catch loss of up to 30 per cent[27]. Apart from catch loss, this TED was considered unsatisfactory as Indian fishers targeted both shrimp and non-shrimp catch. Following this, Percy Dawson, M.R. Boopendranath and their colleagues at CIFT Kochi, developed an indigenous TED. This TED (christened CIFT-TED) was distributed free of cost by the Marine Products Export Development Authority on the east coast of India. The CIFT-TED even won a small award from the Southeast Asian Fisheries Development Corporation. In 2001-02, more than 500 trawlers in Odisha (about half the total number at the time) accepted these TEDs from the fisheries department of Odisha, but none of them installed the TEDs in their nets.

In general though, trawler owners in Odisha refused to accept the results or to use the TEDs. Though experiments with the CIFT-TED showed that catch loss was about 10 per cent, trawler owners believed that catch loss would be up to 50 per cent, and wanted an alternate device. They did not trust the results of the experiment, and firmly believed that the extra cod end (added to quantify catch loss) had been used to somehow deceive them. Like trawler owners in the US and Central America, these trawler owners argued that the loss of catch was too large, and that they had been subjected to unfair targeting as a principal cause of turtle decline, while equally significant threats such as habitat loss and beach lighting remained unregulated[28]. Trawler owners and captains also demanded a revision of marine protected areas and other areas closed to them.

A second workshop was organized at Paradip in October 2002 by Project Swarajya and Odisha fisheries department, with funds from the GoI-UNDP (Government of India-United Nations Development Programme) project[29]. There was widespread protest over turtle conservation measures by not just trawlers, but also gill-netters, who complained about: 'the absence of demarcation in the sea between the fishing and non-fishing zones, their treatment by the personnel of Coast Guard and Forest Department, increasing, multiple tax burdens ... the price of raw materials, declining catch, the lack of proper international or local markets ... the denial of diesel subsidies by the State Government.'

This resonated with the implementation of TEDs in the US in the 1980s which faced problems due to the 'economics of the fishery, perception of the "turtle problem", interagency coordination and uncertain regulations'[30]. The top down approach to implementation of TEDs in the US had also led to polarization that could eventually be resolved only by arbitration. In Odisha, a similar situation evolved. Neither the state nor the conservationists engaged the trawlers in dialogue. Rather, most media reports attacked the trawling community for being responsible for the turtle deaths. The fisheries department was not involved in the implementation of TEDs and enforcement of no-fishing zones, reflecting the lack of coordination between state agencies. The exclusion of the fishing community in the design of the device may also have been a major factor in their lack of faith in it[31]. Studies elsewhere have shown that the many TEDs that are currently in use are the ones where the trawling industry was involved in design and development[32]. After the 1999 cyclone, a local trawl operator developed a trawl guard, a mesh used at the mouth of the net, to prevent debris from entering it[33]. Chitta Behera and his organization, Project Swarajya, spent many years selling this as an ingenious and indigenous solution to the TED problem, but neither state agencies nor fishermen seemed interested.

In contrast to the situation in Odisha, some other state fisheries agencies seemed to encounter less resistance to the idea of TEDs, where there had been less confrontation and polarization on the issue of turtle conservation[v]. In the 2000s, some agencies continued to conduct research on TEDs, including the WWF, which attempted to engage trawler owners in dialogue[vi].

It was estimated that there would be an economic loss of over $20 million in India if TEDs were made mandatory, but the actual impact turned out to be quite small, probably because the export was mainly to the Japanese market. Officials responsible for exports in Odisha admitted that they certified all the shrimp shipments as aquaculture shrimp so that it could be exported. The local fishers believed, or at least stated, that these conservation ideas were Western or American notions and that we should not give in to their 'demands'. The word on the street was that this was a problem 'created' by conservationists and/or foreigners to harass the fishermen of Odisha.

Rashomon redux

By the mid-2000s, Operation Kachhapa had discontinued patrolling, but the conflict showed little sign of abating, thanks to the continued political and media campaigns. Not to be left out of the fun and games, the international environmental NGO, Greenpeace, embarked on a series of interventions along the Odisha coast as part of its international campaign titled 'Defending Our Oceans', starting 2005. The organization decided to undertake offshore patrolling with a view to 'making the nesting season of 2006 safer for sea turtles'[34]. They hired a trawler, and flying the Greenpeace flag, patrolled the offshore waters for a few months, which added to the impression that conservationists were against fishermen[35]. In January 2006, Greenpeace deployed buoys to demarcate the boundaries of the Gahirmatha Marine Sanctuary – a demand made by different categories of fishworkers and also a directive of the Supreme Court's Central Empowered Committee. This action was to meet with mixed reactions and also opposition from the representatives of fishworker groups, which expressed disappointment at being further restricted from fishing.

Two events aggravated the conflict. In February 2003, three forest guards were kidnapped by the crew of two gill-netters they had apprehended, and one was pushed overboard and drowned[36]. This was not the first attack on forest staff by fishermen, but it was the first fatality. In 2006, forest officers overreacted when they were approached by a fishing boat, and opened fire, killing one of the fishermen[37].

As in Akira Kurosawa's eternal classic, *Rashomon*, there was a dead body (several, actually) and an endless number of versions of what had actually happened. The trawlers had their version, the traditional fishermen their perspective, and the NGOs, state and biologists their unshakeable faith in each of their belief systems.

The conservationists and most biologists maintained that trawlers were principally to blame for turtle deaths. That trawler fishing was fundamentally bad for the environment, and therefore the community was negligent and irresponsible. And that ruthless enforcement of laws was the only way to resolve this problem. The state agreed with this assessment, with the caveat that it didn't have the resources to fix the problem and needed more money, more boats, more guns, more equipment and so on.

The trawler owners were not to be left behind. In desperation, they claimed that turtles died due to migration fatigue, pollution and (wait for it) labour pain. That sea turtles may have migrated to their breeding grounds for millions of years but were suddenly getting tired and dying en masse. At a meeting in Visakhapatnam, the president of the Odisha Trawler Owners' Association took me aside and gave me a detailed description of how painful his wife's labour was, and how much more pain a turtle must experience when giving birth to a hundred-plus eggs. 'Tears come from her eyes,' he said, referring to the salt excretion from their eyes. Of course, one hardly thinks that the trawler owners believed their own claims, but they were pursuing sound political strategy: countering rhetoric with rhetoric.

The traditional fishworkers, represented by the Odisha Traditional Fishworkers' Union (OTFWU)[vii], added their perspective. With the help of Aarthi Sridhar, who had just completed a study of the fisher-turtle conflict in Odisha, they wrote to the Central Empowered Committee of the Supreme Court that 'excessive turtle conservation measures have impacted their livelihoods'[38]. They complained that all responsibility was vested with the forest department with powers that were often abused at the village level, and that traditional fishermen needed to be made partners in conservation efforts at all levels. They suggested that joint management/ monitoring committees comprising traditional fishermen, officials and scientists should be formed for each of the turtle congregation areas, which would decide on the exact nature of restrictions. Critically, they voluntarily offered to ban three types of nets that trap turtles. They banned the stingray net (a large meshed multifilament net), the large meshed monofilament pomfret net, and the ring seine, of which the first two were known to cause turtle mortality. The OTFWU promoted this as its commitment to sea turtle conservation, but conflict within the community may have played some role in this decision.

In the very first issue of the *Indian Ocean Turtle Newsletter*, we attempted to arrange a dialogue between these warring groups by offering a forum for them to share and exchange views. In their statement, OTFWU reiterated their earlier points and demanded that fishermen who were displaced or prevented from fishing be adequately compensated[39]. OTFWU made three other equally critical points. First, they pointed out that they needed to be differentiated from mechanized fishing, a conflation that had created much conflict. Second, thcy stated that 'no traditional fishing gear should be banned without adequate and scientific study and data'. Finally, they offered suggestions for holistic marine conservation that included regulation of the mechanized fishing sector and of coastal industrial development.

The biologists (as represented by Bivash Pandav, Basudev Tripathy and myself) summarized the research on sea turtles in Odisha and (as expected) promoted more research and recommended using science for management[40]. For example, we said that the 'protection of the reproductive patches (rather than the entire marine sanctuary) is a more effective and efficient way of utilising the limited manpower resources of the state, and can involve local fishing communities'.

The trawler industry's position, summarized by Project Swarajya, stated that 'Indian trawlers should not be forced to use any prototypes or modified designs of TEDs, since these are American innovations ...'[41].

They even rejected the indigenously developed CIFT-TED, insisting that it resulted in too high a catch loss and suggested that both TEDs and fishing regulations be developed on the basis of dialogue between the government and the marine fishing industries. In closing, however, the trawler sector made the most politically correct noises possible, stating:

> The Government should make efforts to involve the trawling community in research and development programmes around turtle conservation, marine fisheries and biodiversity, and also in promoting eco-tourism along Orissa coast . . . The trawler associations are willing to extend their wholehearted cooperation to all turtle conservation measures at the ground level, provided these are adopted with their consent.

Most interestingly though, they suggested that 'the exact area and location of turtle congregation zones off the Odisha coast should be reviewed and redrawn from time to time in a collective exercise by both enforcement agencies and the trawling industry, in view of periodical changes found in the behaviour of migrating, congregating and nesting turtles in choosing new sites and abandoning old ones. While this may have been useful rhetoric, this reflected a fair knowledge of sea turtle biology, one that biologists, conservationists and the state forest department had ironically been unwilling to acknowledge.

The conflict between fishing communities and state continues till today. In January 2014, another fisherman was killed in an encounter with forest patrol staff. While the fishermen claimed that they had been fired on without provocation, the forest staff claimed that they had been attacked first by the fishermen. A forest officer was quoted as saying[42]:

> The joint forest patrol staff were forced to fire in self-defence because the crews of fishing trawlers had gone berserk and were firing at them. The rogue trawlers collided with a patrol vessel and sunk it. Outnumbered by the marauding fishermen, the patrol party had alerted the Coast Guard and retreated in a vessel leaving behind another, which was later found damaged.

The fishermen launched a major protest claiming that the forest department's story was fabricated and that their act was nothing short of murder. Neither side was short on rhetoric, and at the end of the day, one was once again reminded of *Rashomon*.

Consortium for conservation

In 2001, Roshni Kutty of Kalpavriksh had taken on a project to study community-based initiatives in sea turtle conservation in the country as part of the GoI-UNDP project[43]. She visited Rushikulya to explore the prospects for community-based tourism at this site. As part of her project, she conducted several meetings with the local communities[viii] and community groups at the villages near Rushikulya rookery. Her final meeting involved fishermen from all three villages, village heads, the wildlife and tourism departments, the United Artists' Association (a fishworker support group), OTFWU and Operation Kachhapa. The consensus at the meeting was to form a Uniform Protection Committee that would comprise the state (forest and fisheries departments and the police) as well as an equal representation from local committees.

This idea would develop independently a few years later. To explore common ground between turtles and fishermen, the Coastal and Marine Conservation Programme at the Ashoka Trust for Research in Ecology and the Environment (ATREE) created and facilitated a common platform for sea turtle conservation in Odisha[44]. In early 2004, Basudev Tripathy and I met Mangaraj Panda of the United Artists' Association in Ganjam, and Aleya, the secretary of OTFWU, to explore the possibility of working together. They were welcoming but guarded. Aarthi Sridhar, who joined ATREE in August of that year, and Tripathy conducted more small meetings, leading to a gathering in December 2004 in Bhubaneswar that was attended by fishworker organizations, local and large national NGOs and turtle conservationists. The group named itself the Orissa Marine Resources Conservation Consortium (OMRCC) or, in Oriya, Odisha Samudra Sampada Samrakshana Samiti. The stated mission of the OMRCC was 'to ensure sustainable use of marine living resources and livelihoods'[45].

The OMRCC's main objective was to reconcile conservation and livelihoods of traditional fishers[viii]. It aimed to demonstrate that collaborations between the state, fisherfolk, scientists and conservation groups could help sea turtle conservation in Odisha. It also intended to evaluate both the effectiveness of sea turtle conservation measures as well as their impact on local communities. From the outset, the OMRCC planned to campaign against large development projects such as the ports at the mouths of Dhamra and Jatadhar rivers.

Turtle conservationists

The negotiator

When Robert Bustard started his research programme on sea turtles, he made an attempt to mark all the turtles during an arribada in order to count them. One of the persons present at this marathon marking session was B.C. Choudhury, a young biologist. BC, as he is known to friends, says that they walked every night from dusk till dawn with a can of paint in one hand, and a brush in the other, marking every single turtle that came ashore. He says that this experience alone persuaded him to pursue a career on crocodiles rather than turtles. BC had been interning with newspapers after a journalism degree, though he had graduated in zoology and had developed some interest in conservation after reading Prime Minister Indira Gandhi's speech at the IUCN. Following the advice of senior colleagues at *The Statesman*, he decided to apply for a position on a crocodile project that had been advertised. BC remembers changing into jeans from his formal clothing when he realized that a white man was leading the interview. He then accompanied Bustard and the other young researchers (Sudhakar Kar and L.A.K. Singh) to Gahirmatha. He believes that he may have marked more than 18,000 turtles and decided that 'this was not my cup of tea'. But, more than the effort, it was the confinement to Gahirmatha that he dreaded. Wanting to travel the country, he decided to focus on crocodiles.

BC travelled to Tamil Nadu to get crocodile eggs to help establish a population at Simlipal in Odisha. He met and spent time with the Whitakers and went on turtle walks with Preston and Anne Ahimaz. Following some work on saltwater crocodiles in Andhra Pradesh, BC went on to work at the Central Crocodile Breeding and Management Training Institute in Hyderabad, established with the help of Bustard with support from the Food and Agricultural Organisation (FAO). Though his work was mainly with crocodiles, he also conducted sea turtle surveys along the Andhra Pradesh coast from the Godavari delta to the Odisha border between 1977 and 1984. He then moved to the newly formed Wildlife Institute of India in the early 1980s. Here he supervised major projects on freshwater turtles, herpetofauna, and later cranes and coral reefs.

From the town of Berhampur in southern Odisha, BC grew up unaware (like the rest of the world) that there was a large sea turtle rookery not far from his hometown. His father often told him, 'What is this nonsense and all, saving turtles, I think you know the government has nothing good to do and is doing these silly things?' But he would talk to him about the mouth of the Bahuda river, their ancestral home not far south of Rushikulya, and the turtles there, though not in the numbers one saw at the mass-nesting beaches. BC became involved with sea turtles again when his student, Bivash Pandav, decided to carry out a survey of nesting beaches in Odisha. This survey would lead to the discovery of the rookery in Rushikulya. BC gradually became more and more involved in the work on sea turtles through his supervision of Bivash's project. By the late 1990s, BC was often called on by both the Ministry of Environment and Forests, as well by NGOs, as an expert. One of the presentations to the Barbara Delano Foundation on behalf of the Wildlife Protection Society of India (WPSI) was made by B.C. Choudhury in 1998.

By the late 1990s, I had met BC at the Northern Indian Ocean workshop and was in regular communication with him. Though I had planned to do a post-doctoral project abroad on sea turtles, I had corresponded with BC about my ideas, and he was optimistic about getting funding for this research from WII. I put in a proposal, it was funded, and I arrived in WII in January 1999 to begin my post-doctoral research on sea turtle genetics.

During that year, the Ministry of Environment and Forests approached BC to help coordinate a national project on sea turtle conservation funded by the UNDP. BC took this project on and was responsible for many good outcomes, including promoting the involvement of small NGOs in sea turtle conservation. I worked closely with BC on this project for a couple of years. Apart from generating knowledge on sea turtle nesting and threats through a number of partner organizations, we produced several manuals for research and conservation and an edited volume, *Marine Turtles of the Indian Subcontinent*[46].

BC continued to be involved with sea turtles in Odisha over the next decade, supervising the dissertations of Master's students and coordinating projects in Odisha. One of the largest was funded by the Directorate General of Hydrocarbons (DGH) in India, with a view to examining offshore populations and migratory routes in the context of oil and gas exploration. This project tagged more than seventy ridleys with satellite transmitters, and obtained excellent datasets on their offshore movements and migrations.

BC made many contributions to conservation policy in India. In the early days, he recalls that Bustard and he used the former's access to Prime Minister Indira Gandhi's office to influence her to involve the Indian Navy in sea turtle conservation. BC was also trying to promote the farming of crocodiles, but says that the National Board for Wildlife was against it. In particular, he recalls that M. Krishnan, one of India's finest naturalists and spokespersons for wildlife conservation, was vehemently against the idea.

BC was also involved with the first drafting of the Convention on the Conservation of Migratory Species–United Nations Environment Programme (CMS-UNEP) and Indian Ocean and Southeast Asia (IOSEA) Memorandum of Understanding (MoU) on the conservation and management of sea turtles. He then frequently served as the Ministry of Environment and Forests representative at inter-governmental meetings of the MoU. He also advised WWF on their by-catch mitigation project in Odisha and conducted assessments of the Gulf of Mannar Marine National Park.

The balancing act

Passionate about ecology from a young age, Sejal Worah studied Life Sciences at St. Xavier's College in Mumbai, probably the first of many wildlife conservationists to do so. She went on to do a Master's programme in Wildlife Biology at Syracuse University in the US. She recalls her culture shock in Wildlife 101, when the instructor asked the class of over 100 students how many hunted. 'What a stupid question,' she thought, 'we are all conservationists!' and discovered to her shock that over 80 per cent replied in the affirmative. That was her first exposure to the 'Western' concept and philosophy of conservation. A greater understanding of the many worldviews of conservation would shape her future thinking and career trajectory.

Sejal returned to India determined to get involved in conservation, and started working for WWF in Mumbai, helping run their iconic nature camps that set so many of today's ecologists and conservationists on their way. She first encountered sea turtles through the WWF team in Chennai which had been involved in turtle conservation for many years. She left WWF to pursue her Ph.D which focused on the interface between communities and conservation, an area that was to become her main interest in the years to come.

After her Ph.D, Sejal moved to the UK, and then to Thailand to manage a regional capacity building project to promote linkages between conservation and sustainable development. During this period, she had the opportunity to work with marine conservation groups and sea turtle conservation projects in south-east Asia. We met in her office in WWF, New Delhi to talk about turtles. She said:

> And that's when my fascination and love for turtles sort of really came to the fore. We worked with groups in southern Thailand, and in the Turtle Islands in the Philippines, and in the Solomon Islands very specifically on turtle conservation projects. And once you've

seen a leatherback turtle nesting in the full moon, on an island, it's something that you never forget.

Sejal helped set up some community-based sea turtle conservation projects in south-east Asia and the Pacific and when she moved back to India, she brought her turtle interest back with her. While WWF was doing little work directly with sea turtles in the early 2000s, Sejal felt that the organization had a role. She said: 'I realized that there were a lot of people working on turtles. I think where we saw the value addition that WWF could bring is sort of this neutral broker-facilitator, bringing groups together, liaison with the government, possibly talking with the industry, and just, you know, process-related stuff, rather than on-the-ground stuff.'

Over the next few years, Sejal supported small projects by state offices on sea turtle conservation, but got more directly involved in a programme on by-catch reduction in Odisha in 2009, involving B.C. Choudhury and me as advisors. Over the next couple of years, the project conducted experimental TED trials and held stakeholder meetings with the trawl fishermen. She added: 'In my career I've lost count of the number of community meetings I've done, right? Hundreds, in so many different settings! And I have to say that the Paradip one was amongst the top three most difficult meetings that I have ever done. And it was not just the attitude, the abuse, the kind of antagonism and almost physical threat that you felt, but at the end of it, I had the sense that this was all an act.'

The very first question the fishermen asked was where they were from, and one of them said, 'If you were from Greenpeace, I would have thrown you in the water right now.' Fortunately, WWF had no presence in Odisha working with the trawler community. Ironically enough, the world's most recognizable conservation icon was unknown in Paradip, and this alone allowed them to get a foot in the door. Sejal felt that the situation had come about because of the polarization created by the conservationists. But her experience echoed what had gone before, both in India as well as in the US – that fishers could use the same powerful arguments that conservationists did. She said:

'And so the interesting thing to me was that people in Paradip knew the arguments, knew exactly what to say, had come prepared. They had their own papers, with their own statistics and their own figures. And they were sort of arguing with us and saying it's not by-catch, it's beach elimination, it's beach development, it's erosion.'

But, in the end, Sejal believes that it is important to weigh economic considerations. She pointed out a critical difference between south-east Asia and India: 'The hands-off, hands-on approach to conservation.' As she explained:

> In the Philippines, especially, turtles have a value and they have an economic value to people, and we need to figure out a way to use that economic value for the sustainable management of turtles. In India, it's very much hands-off – you can't touch the turtles, they're almost like state property, and therefore you cannot harvest the eggs and you definitely cannot harvest turtles.

Sejal had clearly gotten over her initial thoughts about conservation and use. With regard to why these policies exist in India, she said: 'I think it's us. I think it's urban conservationists who have to some extent reinforced this whole idea. We've sort of created this artificial thing that hunting is bad and no wildlife should ever be caught and killed.'

Sejal felt that while the cultural factors that have helped conserve a lot of the wildlife in India needed to be celebrated and strengthened, we also need to understand that this does not always mean strict protection. That the relationship between people and wildlife is not black or white – protect or consume – but much more nuanced. And that until we make the effort to understand this relationship, we will struggle with enforcing laws that are out of context – both with science and with society.

As a final word, Sejal emphasized the collaboration between different sectors, including the various strengths different kinds of NGOs bring, and working with communities to garner their support for conservation.

The new kid on the block

About a decade after Kar finished the major body of his research, Bivash Pandav, a wildlife student from the WII, would undertake a survey of sea turtles on the Odisha coast, and discover a third rookery at Rushikulya.

In January 1993, Bivash was conducting field research on monitor lizards in Bhitarkanika for his Master's dissertation, when he was invited to visit Gahirmatha by the Divisional Forest Officer (DFO). He had already heard about sea turtles from the herpetologist, Sushil Dutta, who had visited his college and given a lecture there. He travelled from Dangmal and saw his first mass nesting on Nasi Island, and like many others was greeted by a beach 'full of thumping sound'. He returned to Ekakula and shared a meal of hilsa fish with Chandrasekhar Kar. Over breakfast, Chandrasekhar, enthused by the young boy's enthusiasm, proposed that they should survey the coast of Odisha. Having discovered the second mass-nesting site a decade earlier, Chandrasekhar was convinced that there would be nesting, mass nesting even, at other sites along the coast. Bivash declined, as he had a Master's to complete, and returned to Dehradun a few months later.

Shortly afterwards, when his supervisor B.C. Choudhury asked him what he wanted to do, Bivash remembered Kar's suggestion and shared the idea with BC. Initially, BC was not supportive as Odisha was not an easy location to conduct research, but when the WII's research committee meeting came around, he asked Bivash to put a proposal together. In less than a day, Bivash wrote the proposal, but it had not been included in the agenda for the meeting. However, one thing led to another, and soon BC was discussing the proposal with the director, H.S. Panwar. On hearing that Bivash wanted less than Rs 20,000, mostly for a bicycle and his fellowship (less than Rs 1,000), he offered to fund the survey.

Thus Bivash landed in Odisha with a bicycle and food money in his pocket. He discovered serendipitously that the Department of Sports and Youth Services was organizing a trek from Konark to Gopalpur, a 250 km distance, as part of the Konark festival. Bivash spoke to the director of the department and joined the trek, covering the distance in six days. The trek was crucial in building his confidence to walk the coast. What followed, as he says, is this: 'And during the course of the six month survey, I walked the entire coast twice, and I bicycled it thrice with my Hero Ranger – a Rs 1,200 bicycle. I could manage to bicycle it thrice!'

Bivash had to overcome several challenges, including getting past the major river mouths, which he did with little or no planning. He would wait endlessly trying to wave down a fishing boat and eventually one would stop and take him across. He would stop at fishing camps and ask them for fish, which he would roast on casuarina leaves, and eat with onion and salt. And dream of drinking Coke or Thums Up at the next village.

Amongst other things, he found that a lot of turtles appeared to be nesting in Rushikulya. He had driven down on his father's old scooter with Chandrasekhar to visit the beach south of the Rushikulya river. As they sat at the fish landing centre, they saw two fishermen go past and called out to them to enquire if they had seen any turtles nesting nearby. The fishermen responded that they did indeed see a lot of nesting turtles at the site. Bivash and Chandrasekhar left them some self-addressed postcards, asking to be informed if there was significant nesting.

One of these fishermen was Damburu, who has since become the mainstay of the monitoring programme at Rushikulya, assisting all the researchers who have worked there in the past two decades. Damburu was awarded a president's award at the annual sea turtle symposium in Goa in 2010. In March 1993, Chandrasekhar received a postcard from Damburu that a lot of turtles were nesting near his village. It was several days before Bivash got the message, but he rushed down immediately and was escorted to the beach by Damburu, right in the middle of a wedding ceremony in his house. The beach was covered with predated nests and eggshells, and Bivash realized they had found another site. A couple of months later, they returned to witness mass hatching to confirm their finding.

Bivash did not often see mass nesting in Rushikulya. He was usually at Babubali island off the Gahirmatha coast, where he had set up camp. As soon as nesting started in Rushikulya, Damburu would travel all day and reach the Ekakula resthouse and flash a torch which was a general emergency signal to the team on the island. By the time Bivash reached Rushikulya, nesting was usually over. He believes that mass nesting lasted only two or three days during that period in the late 1990s, in contrast to more recent events that have lasted a week or longer.

After discovering Rushikulya, Bivash returned to WII full of excitement and enthusiasm. He attempted to convince the institute to fund him to tag turtles, but despite BC's support, he was told that putting tags on turtles would be like throwing a few lakhs of rupees into the sea. Bivash returned to Odisha with a small amount of funding to carry out surveys and surveyed the coast for nesting and mortality over the next couple of years. Without his own boat and field assistants, there was little more that he could do. But it was to bear fruit.

One day, while surveying the beach in Gahirmatha, Bivash met two women sunbathing on the beach. That, normally a stroke of fortune by itself for an isolated field biologist, turned out not to be the only lucky part of the story. The women, one from Norwegian Agency for Development Cooperation (NORAD) and the other from M.S. Swaminathan Research Foundation (MSSRF), were exploring field sites in India for the Norwegian funding agency to support projects in. Bivash, always ready to tell his tale of woe about tags, got their attention and was directed to meet one of their colleagues, Ian Bryson, in Bhubaneswar. Once Bivash met him, the deal was done. All that was required was that he collaborate with MSSRF as NORAD was already providing funding to their projects. Three months later, 35,000 tags appeared through a diplomatic bag at the US embassy in Delhi. They also provided funds to hire field assistants and a boat. As far as Bivash was concerned, Christmas had come early that year.

The following season, Bivash established five camps along the coast. He stationed himself at Gahirmatha, Basudev Tripathy at Rushikulya, and others at Agarnasi, Devi and Chilika. Bivash also had a number of dedicated field assistants at these camps. At Gahirmatha, there was

Madhu, the cook and major domo of the camp. The boatman (Subash) and the rest of the team (Kalia, Siria and Sahadev) were proficient 'turtle fishers'.

Bivash started his work on mating turtles as serendipitously as everything else. Riding down the Gahirmatha coast in 1996, he encountered several human bodies in the aftermath of a cyclone. Seeing something out at sea, and assuming it was another body, he took a look through his binoculars, and saw his first mating pair bobbing gently on the surface. He returned to Gupti where he had initiated a search for a boat, and hired Subash, his brother, and Kalia. Crossing the river mouth into the offshore waters of the Gahirmatha coast, Bivash was struck by the site of all the mating pairs in the water. Bivash wondered if it would be possible to catch these pairs, and by the next afternoon, Kalia had fashioned the famous triangular trap that would allow them to catch over 1,500 mating pairs over the next three years.

Many jokes were cracked about the karma of breaking up so many 'couples'. But very little can deter a male sea turtle. Bivash said: 'I used to catch a lot of triplets and quadruplets. So there were two males holding on to one female or three males holding on to one female. I remember I even once caught a quintuplet – four males holding on to one female, four males piled onto each other and one female.'

But his greatest excitement came a few months after starting the tagging programme. 'After finishing the exercise in '97, I came back to Dehradun, and BC gives me a letter from Sri Lanka,' he said. 'It's written in Tamil, and it says, 22 nautical miles off Kalmunai tag number so-and-so has been captured on 27th of April, 1997. So that was actually the first and most exciting information that came out of our tagging programme – first long-distance recovery.'

This turtle was tagged during the mass nesting in Devi river mouth on 13 March 1997, and captured forty-four days later about 2,000 km away. During that nesting event, a young schoolboy named Bichitrananda Biswal (Bichi) helped Bivash tag turtles and tried to impress upon him that he was passionate about sea turtles. Along with Tuku, Tulu and Bishnu, Bivash's older assistants, Bichi helped keep track of the dead turtles that were getting washed ashore on the Devi coast in large numbers. After working at his camp for three seasons, Bichi kept his interest in turtles alive, working for Operation Kachhapa and the forest department, and eventually started his own NGO.

Further south, a similar band of turtle followers were initiated into tagging and counting rituals at Rushikulya. Rabindranath Sahu was the leader of this group. After working for Bivash, some of the boys worked briefly for Operation Kachhapa. During the mid-2000s, they assisted Basudev Tripathy with his fieldwork for his Ph.D, and in turn, Basu helped them get a grant to build an interpretation centre and start their group, the Rushikulya Sea Turtle Protection Committee (RSTPC). Today, members of the RSTPC run the interpretation centre at Rushikulya, participate in state conservation activities, spread awareness in local schools and assist researchers who work on sea turtle biology at the site, in addition to their own ongoing monitoring.

The teams of field assistants were also instrumental in helping Bivash with his surveys of mortality. Over a seven year period, he and his field assistants physically counted 46,219 dead turtles, a number he reels off more than a decade later. Deeply concerned about this mortality, Bivash was encouraged by Chitta Behera of Project Swarajya, and the DFO for Bhitarkanika, Bitta Nath Naik, who he calls an 'unsung hero'. By the end of his research, Bivash became a very vocal champion for ridley turtles, campaigning strongly for their protection in various public forums.

Bivash inspired a number of young researchers who either volunteered with him or conducted their own research projects on sea turtles in Odisha. In 1999, when I conducted field research, Bivash was assisted by Dipani Sutaria, who would later carry out doctoral research on Irrawady dolphins in Chilika, and by Banugoppan, another ecologist in the making. The following year, Karthik Ram would follow up on Bivash's offshore distribution work during his Master's programme at Pondicherry University. And Basudev Tripathy, his research assistant at Rushikulya at the outset, would conduct research for his Master's, M.Phil. and Ph.D dissertations on various aspects of turtle biology at Rushikulya. Suresh Kumar, a junior from WII, also carried out his Ph.D on sea turtles at Rushikulya.

Having spent a few years working on tiger conservation in Nepal, Bivash continues now as faculty at the Wildlife Institute of India, and hopes to return one day to his project of tagging turtles in Odisha.

Flagships or gunboats?

Conservation has used a variety of approaches to achieve its goals. A set of popular concepts include using flagship, umbrella and keystone species. For example, protecting the habitats of umbrella species is supposed to protect a large number of smaller, less attractive species. However, many studies have shown that single species may not be particularly good surrogates for biodiversity, and protecting the habitat of, say, a large mammal, may not necessarily help protect frogs or insects. Keystone species are those that play important ecological roles or functions, whose removal would destabilize the ecosystem, hence justifying conservation attention to these species. Flagship species, perhaps the most commonly used approach, use charismatic, typically large species (mainly mammals such as tigers, rhinos, elephants, pandas or whales) to garner broad support for conservation.

Flagship approaches can sometimes drive the social change that is required to bring about positive conservation action and, at other times, create roadblocks to both environmental and social development. On the one hand, charismatic species can be 'flagships', when they gain support for conservation, but they can also become 'gunboats' where protection measures for these species alienate people and negatively impact conservation[47]. This usually happens when protection measures impact livelihoods or restrict access to traditionally used areas. In Odisha, one could argue that the very public image of olive ridleys may have driven a wedge between diverse communities of fishermen and conservationists.

In general, sea turtles have become widespread flagships for conservation, with champions amongst wildlife conservationists as well as local communities[48]. The mass-nesting olive ridleys of Odisha certainly served to attract attention to wildlife conservation issues in India. NGOs were very successful in raising the profile of the species both within and outside the state. In Odisha, Operation Kachhapa and Greenpeace attempted to mitigate mortality by patrolling nearshore waters and enforcing fishing bans. The strategy gained support from research which showed that sea turtles remain in small offshore congregations during the breeding season and are not diffusely distributed along the coast[49,50]. However, the location of these patches may vary over time. Thus, the creation of sanctuaries or protected areas with fixed boundaries may not be effective, but little thought was given to this nuance by the NGOs. And despite the investment of large amounts of effort and funds by the government and civil society groups to patrol nearshore waters, trawlers continued to fish illegally, causing continuing mortality of olive ridleys.

On the other hand, conservation measures created antagonism towards turtles in general, especially among local fishing communities who were the victims, or perceived themselves to be victims, of conservation action. Throughout this period, traditional fishers insisted (fairly accurately) that they were not killing turtles, and that they should be made partners in conservation. Indeed, strandings on the Rushikulya coast are much lower than elsewhere along the coast. Though conservationists agreed that 'traditional' fishing methods did not pose a problem for turtles, traditional fishermen were frequently victimized by the enforcement agencies. For example, in declaring Gahirmatha a marine sanctuary, the 'occupational interests of local fishermen' and 'rights of passage' were to be taken into account, as per the Indian Wildlife (Protection) Act, 1972, but in practice, this proved inadequate, thanks in, no small part, to the intentional or unintentional bureaucratic blockades of permits and licences. While nearshore fishing bans for mechanized vessels were instituted in many coastal states in India for the protection of traditional fishers, they were invoked in Odisha in the name of turtle protection, so that it became a 'people versus turtles' law. How did this come about?

Ironically, the process may not have been dissimilar from the threat it wanted to address. Like trawl fishing sweeps, all that comes before it without discrimination, the targets of media rhetoric can, in effect, be equally fuzzy. While most articles about the issue did mention trawlers, a general sense was conveyed that fishing was killing sea turtles. Many biologists and managers also disputed the notion of traditional or artisanal fishing and suggested that little if any fishing in Odisha could be considered truly artisanal. The origins of the marine fishing communities were also questioned. Thus fishing and fishing communities in general felt targeted by turtle conservation laws and measures.

A second mechanism was through what one might call the 'chain of reaction'. Media reports and constant badgering by the NGOs resulted in a stand-off with enforcement authorities such as the forest department. On one occasion, Greenpeace activists were arrested for bringing ridley carcasses to New Delhi as part of their protest. When the forest department was moved to action, instructions to enforce laws would pass down the hierarchy to the field officers, who rarely had the resources and authority to act against the politically strong trawlers. To demonstrate action, they would frequently harass traditional fishermen and since most fishers were unaware of the law, they had little recourse.

The lack of clarity in legislation about what constitutes a traditional method of fishing allowed mechanized fishers to garner the support of other fishing communities. Various laws and government orders tended to lump all boats with motors together, thus including small-scale fishermen along with trawlers and large gill netters[ix]. At the end of a workshop on TEDs in Odisha, the trawler owners submitted a memorandum which outlined their concerns[51]. The memorandum concluded by saying:

> In recognition of the spirit of self-sacrifice so shown by the trawling industry, and also in the interest of furthering the economic development of the fishing communities, necessary financial assistance should be provided to them on one hand, and their involvement in turtle conservation measures on the other.

Note how all fishing communities are lumped together, and the suggestion that all of them would be friends of sea turtles. As one of its first tasks, OMRCC conducted a meeting at Gundalba village near the

Devi rookery, when it became known that the forest department had wrongfully detained fishermen from that village and prevented them from fishing, even though the Supreme Court CEC's orders had by then explicitly recognized the rights of artisanal fishers. The fishing communities automatically assumed that turtle conservationists were to blame for this. Over the next few months, the team at ATREE worked closely with United Artists Association (UAA) and other OMRCC members to develop material to spread awareness about the laws. Large, visually explicit posters were developed for each of the mass-nesting sites, as were booklets in Oriya that outlined the laws related to fishing. These had some but limited impact.

On the whole, if fisheries' laws had been enforced for the reasons that they were originally instituted, namely to protect traditional fishermen and their livelihoods, it is likely that their implementation would have received wider support. And as a side effect, sea turtles would then have been protected from mechanized fishing.

From incidental catch to incidental conservation

Conservationists have made a significant contribution to ensure the continued survival of olive ridley turtles on the Odisha coast. However, most programmes failed to adequately address people related issues, probably because these are inherently more complex and take much longer to implement.

Most studies in the last couple of decades have focused on the biology of sea turtles in Odisha. Certainly, these are likely to provide some clues to their conservation and management. For example, understanding their distribution in offshore waters helps plan time-area closures (prohibition at certain times at fixed locations) for fisheries. However, several aspects of the conflict have not been studied, such as the origin and history of the communities and their engagement with state governance of fisheries and conservation. The diversity within the communities has also been overlooked. For example, no distinction was made between trawler owners, operators and workers. Many owners live in big cities like Bhubaneswar or as far away as Delhi and have little contact with the fisheries and little knowledge of or sensitivity to marine systems[x]. Operators are usually from within the community, as are the workers, but the latter may be from lower economic classes or may even have moved from traditional fishing when they could no longer support themselves. To club all these groups, and moreover, to assign to them the moral responsibility of the environmental degradation of bottom trawling, was hardly likely to result in a constructive solution.

There are differences within traditional fishing communities as well, keenly recognized by many Oriyas. In the north (in Balasore and Kendrapara districts), the marine fishing communities are mostly of Bengali origin who migrated to this region during Partition. Another wave of refugees arrived during the Bangladesh war in 1971-72. In their book on Bhitarkanika, Chadha and Kar said[52]: 'But the situation started changing very fast. With the immigration of refugees from East and West Bengal, large chunks of mangrove forests were reclaimed for conversion to paddy fields . . . what was wilderness to the inhabitants turned out to be bonanza for the new settlers.'

At various times, biologists, conservationists and managers stated that these fishermen were not Oriya, and had settled recently, and therefore did not have a right to fish in Odisha waters, or that they were not traditional because of the nature of their craft or gear, leading to the same conclusion. Marine fishing communities in central Odisha in Puri are Oriya in origin, but the communities in the south in Ganjam and Chilika districts are Telugu-speaking communities from neighbouring Andhra Pradesh[xi]. These communities have somehow received less criticism, probably because the southern coasts have had lower turtle mortality and not been as contentious as the northern ones.

In 2008, the Indian Institute of Science initiated a collaborative project on human-wildlife conflict with the Norwegian Institute of Nature Research, with funding from the Royal Norwegian Embassy and the Norwegian Research Council[53]. As part of this project, Aarthi Sridhar and I carried out a project on drivers of conflict between sea turtles and fishers in Odisha[54], in comparison to conflicts between green turtles and fishers in the Lakshadweep, a study carried out by Rohan Arthur of the Nature Conservation Foundation[55].

In order to understand the conundrum of turtle conflict in Odisha, we decided to look at the different actors in the theatre of conservation – the fishers, the NGOs, the scientists and the state – and examine their roles, perceptions and relations or interactions with one another. We found fundamental differences between conservationists and fishers with regard to the value of sea turtles. While many conservationists spoke of the turtles' inherent right to exist (with ecological and cultural reasons being secondary), the fishers mostly provided cultural reasons, and accorded livelihoods greater value. They also made a distinction between intentional take and incidental mortality, with less culpability attached to the latter, as something that just happened while they were going to fish. While conservationists and scientists believed that Odisha should take pride in its turtles and protect them, most local fishers saw them as a nuisance that had brought a range of unpleasant restrictions upon them, and negatively affected their livelihoods. Most stakeholders felt that they needed to be part of the movement, that much of the conflict came from a lack of transparency and alienation from decision-making processes.

Despite these small contributions, studies on social, cultural and economic aspects of the fishing community and the political ecology of conservation itself are mostly lacking. It is this understanding, coupled with engagement with communities, that has the potential to move sea turtle conservation in Odisha forward. At the time of writing, Madhuri Ramesh, a Ph.D student at ATREE, is pursuing research on the politics of sea turtle conservation in Odisha.

All the evidence suggests that conservation will not be achieved without the support of people, for which conservationists need to build a relationship with the other stakeholders based on mutual trust. The OMRCC is one such group; though it is a contentious group with many divergent objectives, it continues nevertheless to serve as a platform for argument, dissension, diatribe and occasional collaborative action.

Sea turtle conservation in Odisha is also served by local conservation organizations, particularly at Rushikulya. In the 2000s, nesting appeared to decline in Gahirmatha, but was stable or increasing in Rushikulya, at least during the latter half of the decade. Unlike Gahirmatha, Rushikulya is not a protected area. Its legal status (or lack thereof) has meant that it has been relatively accessible to NGOs, researchers, students, local communities and to the public. In the last five years, thousands of local tourists have visited (albeit in an uncontrolled fashion) to see mass-nesting or mass-hatching events. The local community at Rushikulya has been involved in turtle conservation through the projects of the Wildlife Institute of India, Operation Kachhapa and now through their own initiatives. Since 2007, Dakshin Foundation and Indian Institute of Science have conducted monitoring of nesting and offshore populations with local field assistants and the forest department[56]. M. Muralidharan, who carried out his Master's dissertation project at this site in 2009, has coordinated this project for the last few years, ably assisted by Nupur Kale, Chetan Rao, other researchers and a battery of assistants. The prospect of a locally managed sea turtle conservation programme has been mooted and if it could evolve with the support of multiple stakeholders, the ridleys may yet prove to be effective flagships in Odisha.

In summary, the globally significant Odisha ridley population, both in terms of the biological and cultural value of the arribada, has had very mixed results in generating support for conservation at the local level. Perhaps it has served as a flagship for conservation awareness amongst the urban elite in India, but has equally well become a gunboat for local fishing communities in Odisha. It would indeed be a pity to lose the mass-nesting populations of olive ridleys in Odisha, but it would be worse to lose the turtles as well as the sympathy of the people towards conservation.

Notes

i Pandav divided the 480 km coastline into sectors and systematically covered each sector once in two weeks. Dead turtles were marked with paint to prevent re-counting and were also measured and sexed. In the very first year of monitoring (1993/94), his team counted 5,282 dead turtles on the coast. By 1997/98, the number of dead turtles counted had gone up to a record high of 13,575, a total of 46,219 turtles over a five-year period[57]. Close to 90 per cent of these turtles were counted at Gahirmatha, Devi mouth and Paradip, which had the highest levels of shrimp trawling along the coast. In general, males predominated between November and January, and females between February and April.

In 1998-99, monitoring was carried out in collaboration with Operation Kachhapa, and over 13,000 dead turtles were counted along just 240 km of coast. Roughly the same stretch was monitored the following year with over 10,000 dead turtles. The dead turtle count peaked in 2000-2001 with 16,799 counted over that stretch of coast[58]. The numbers at Gahirmatha during this period are not known. Each year, since then, about 10,000 dead turtles have washed ashore.

The death of most of these turtles is due to drowning in trawl nets. Being air breathing vertebrates, sea turtles need to surface to breathe, roughly every half hour or so. The nets are operated for much longer, and the prolonged submergence and stress usually leads to their death. In 2001-2002, the Wildlife Institute of India carried out a study on incidental capture in trawlers by carrying out experimental trawling on the Odisha coast[59]. In all, they estimated that about 1,000 to 2,500 turtles were being captured in each of Gahirmatha, Paradip and Devi mouths. Not much can be inferred from this study, because 11,593 dead turtles were counted that year itself along just a fraction of the coast that the experimental trawling covered. The study did show, however, that most turtle captures were within 5 km of the coast, and at less than 10 fathoms of depth[60], consistent with earlier studies of the location of mating aggregations by Bivash Pandav and Karthik Ram.
ii The Odisha Marine Fishing Regulation Act (OMFRA) (1982) and Rules (1983) prohibits all mechanized fishing within 5 km of coast. As for the mass-nesting beaches, the Gahirmatha Marine Sanctuary was declared in 1997, and prohibited all mechanized fishing within 20 km of the Gahirmatha coast, which extends 35 km south from the mouth of the rivers Brahmani and Baitarani. Under OMFRA, orders were first issued in 1994 prohibiting fishing at Gahirmatha, and in 1997, prohibiting mechanized fishing within 20 km of coast around

both the Devi River mouth and Rushikulya from January to May. In 2005, the Odisha Fisheries Department announced a seasonal ban on fishing by motorized boats and trawlers within 10 km of Dhamra, Devi and Rushikulya under OMFRA[61].

iii In 1998, the Wildlife Society of Orissa had submitted a PIL petition to the high court in Odisha asking for the implementation of existing fishing regulations and to extend the ambit of the WWF case's judgement to Rushikulya and Devi river mouths as well[62]. The case was believed to have been slowed due to the political influence of the mechanized fishing lobby, but interim orders were passed in 2002. Another application was made in December 2002 with the CEC raising issues related to the conservation of turtles in Odisha[63]. The committee nominated Sanjeev Chadha, then assistant inspector general of Forests with the Ministry of Environment and Forests, and others to make a site visit in early 2003 to determine the facts. Sanjeev Chadha had served as wildlife warden of Gahirmatha during the early- to mid-1990s and later as divisional forest officer of the Puri range, which included the Devi mouth beaches. Chadha was an avid naturalist and sea turtle enthusiast. He produced a book on Bhitarkanika and remained active in discussions about sea turtle conservation.

iv TEDs were made mandatory through an order passed under OMFRA in 1998[64], followed by an amendment to the rules in 2001[65], and steady pressure was maintained on the state government by agencies such as the coast guard.

v TED demonstrations were carried out in West Bengal, Andhra Pradesh and Tamil Nadu. In Andhra Pradesh, the State Institute of Fisheries Technology at Kakinada was involved in promoting TEDs in the early 2000s, and received a fairly good response from trawlers there, including suggestions for improvement[66]. Several workshops, training programmes and awareness camps were conducted, and over 250 TEDs were distributed in Visakhapatnam and Kakinada.

vi In 2001-2002, the WII also carried out a research project on TEDs in Odisha[67]. Using data from experimental trawls, they estimated about 2,500 turtles drowned annually in Devi River mouth, about 1,200 in Gahirmatha zone and about 1,000 in Paradip. In 2009-2010, the WWF launched a by-catch reduction project in Odisha[68]. Their trials recorded a catch loss of about two to three per cent of the total value of the catch, mostly of low-value fish, but they acknowledged that, given low margins, this could be a much higher proportion of the profit. Based on exclusion rates of turtles, they estimated that approximately 7,000 turtles were being caught and killed by the trawlers operating out of Dhamra.

vii Though active in uniting the various marine fishworker groups across Odisha since 1989, the OTFWU was formed in 1995 with representation from fisher leaders in all six coastal districts. The OTFWU formed district level unions and linked up with Samudram, a federation of women fishworkers in southern Odisha.

viii In 2008, the OMRCC received funding support from the Ford Foundation, again facilitated by the team at ATREE, to promote local livelihoods. Coordinated by the UAA, the objectives of this project were to promote a sustainable conservation-enterprise model in fishery resource management, to enhance local livelihoods and to evolve collaborative fisheries management approaches. This included building bridges with trawler associations, value addition to fisheries products, promotion of co-management and responsible fishing practices, and production of environmental education and advocacy material (including a periodical newsletter in Oriya). One of the key ideas was to promote community-based tourism around turtles at sites such as Rushikulya. This remains a work in progress.

ix While OMFRA is clear with regard to sizes of mechanized vessels, the definitions of what constitutes a mechanized craft or mechanized fishing is absent. There is also no mention of regulations for motorized craft, especially those with outboard engines. These can vary in size and capacity, and many small-scale fishermen use these to reach their fishing grounds. Regulation of net types and sizes is also not adequately addressed. However, based on comments from numerous individuals and organizations, including OTFWU, the final report of the CEC in 2004[69] eventually did make a distinction between mechanized, motorized, and non-mechanized fishing vessels.

x In *Karate Kid 2*, when Miyagi san visits his village with Daniel, the karate kid, Daniel, asks why there is no fishing if it's a fishing village. Miyagi san responds that the 'villain' brought in a commercial trawler and in two years, all the fish were gone.

xi The fishing communities in southern Odisha are a mixture of Telugu (Andhra) and Oriya due to the political history of the region, prior to the formation of the Odisha state. The Telugu-speaking nolias are primarily marine fishermen, while the Oriya-speaking khandaras and kevtos are mainly inland fishers. Both these communities are generally more marginalized than other politically stronger communities, and do not have formal land deeds, though they have been occupying these areas for generations.

HYPE AND HYPOCRISY IN LAS TORTUGAS

Boys with toys

CIRCA 2005, RAPHAEL, MICHELANGELO, Donatello and Leonardo made an impromptu appearance in front of Bombay House, the head office of the Tata Group, in Mumbai. The four teenage mutant ninja turtles were on a new adventure, unauthorized, of course, by their creators. This was a crusade to save their brethren, the olive ridley turtles of Odisha, by preventing the Tatas from building a port at Dhamra, which would ostensibly spell doom for the olive ridleys that nest at Gahirmatha. Led by campaigners such as Sanjiv Gopal and Shailendra Yashwant, the turtles held up banners that said 'Don't say Tata to turtles'. The campaign led by Greenpeace India did not end here.

Greenpeace put out advertisements in newspapers in 2009 comparing the Nano small car manufactured by the Tatas (cheap) and the olive ridley (priceless). A while later, Greenpeace launched a Pacman-like online video game, called 'Turtle vs Tata', where little yellow (olive?) turtles had to eat white dots (=healthy food) while avoiding Ratty, Matty, Natty or Tinku (the bad guys). In response, the Tatas launched a trademark infringement and defamation lawsuit against Greenpeace India.

But this was hardly the first time that Greenpeace had courted trouble. When they started their campaign in Odisha in early 2004, there was already tension between fishers, conservationists and the forest department. Though some of my colleagues and I warned them that their entry might exacerbate the situation, Greenpeace decided that it should address the problem of boundary demarcation for the Gahirmatha National Park. When fishermen were caught fishing illegally inside the park, they often claimed that they didn't know where the boundaries lay, and the forest department claimed that this made enforcement difficult. Greenpeace decided that if they performed public service by demarcating the boundaries with buoys, both the fishermen and the forest department would be grateful.

Greenpeace consulted with the local forest department as well as fishers and received what they thought was support for this activity. Organized as a big media event, Greenpeace invited one of their key supporters, film actress Amala Akkineni, and sailed out to Gahirmatha from Paradip along with fishworkers and several observers. I met Amala at a sea turtle workshop in Visakhapatnam in April 2015, and we talked about the Gahirmatha trip and conservation in general. An avid diver and a turtle enthusiast from the time she rescued a sea turtle hatchling as a child, Amala had supported Greenpeace programmes for many years. She helped raise funds for the buoys and was later involved in a stray-dog sterilization programme along the central Odisha coast, aimed at reducing predation of turtle eggs.

Once they reached the park boundary, a group of dashing young activists threw their shirts off and themselves into the water. Amala recalled that it was really hot and they were all happy to be in the water for a while. Buoys were installed using GPS coordinates, and the crew returned, dare I say, buoyed by their success. But sadly, the fishermen were not in the least happy that they could no longer profess ignorance about the boundaries and mumbled their discontent. And the forest department, the supposed beneficiaries, pulled Greenpeace up for entering a protected area without permission.

As a final episode in this saga, Greenpeace campaigners ran afoul of the law again when they brought dead olive ridley turtles to the capital, New Delhi, to get the attention of Odisha Chief Minister Naveen Patnaik. They were then arrested on grounds of illegally handling and hunting a Schedule I animal, and spent two days in Tihar Jail. As absurd as the legal accusations against Greenpeace might have been, it is quite clear that the environmental NGO's intent was to create publicity around the issue. When I queried Amala about the fact that the issues that get the biggest publicity may not be the ones most important for conservation, she said: 'You hear the shout, but not the whisper.' There is little doubt that the individuals concerned are deeply passionate and committed to conservation. But the question is: does the shouting help conservation or does it just help the NGOs? At heart, are environmental organizations any different from the corporations they claim to battle?

Conservation versus development

As Nicholas Mrosovsky noted, hype has characterized many of the conservation narratives in the last three decades, sometimes to the detriment of conservation itself[1]. The early fishery drama in Odisha was no exception, as the exchanges between Davis et al.[2] and Frazier[3] had demonstrated. G.M. Oza, one of the actors in the original drama, was not done with that. A decade letter, he published a brief note titled 'Last chance to save olive ridley turtles in India'[4]. Oza started by referring to their 1977 article where they lamented the absence of nesting at Gahirmatha. Further down the article, he noted that 3,20,000 turtles were reported to have nested during 1992. Despite a decade-and-a-half of evidence that large numbers of ridleys were nesting at Gahirmatha, the title still painted a doomsday scenario. By now, however, the main cause for worry had changed. On the one hand, the incidental mortality in trawlers had become a concern. On the other, the talk of a port at Dhamra had begun. At the time, it was only a proposed expansion of the fishing port. Worse was to come.

Rockets and ridleys

As much as fishery related mortality dominated the headlines on turtle conservation, it was far from the only threat, and conservationists had to deal with a variety of problems, including development. The threats came from varied sources, including rockets, ports and oil exploration. But let's start with rockets. In 1986, Indraneil Das wrote about the establishment of a rocket testing range near Gahirmatha[5]. According to him, the government had decided to shift the missile testing range from Balasore to Satabhaya (near Gahirmatha) because of protests from local people. A national sea turtle specialist group which had been formed at the time recommended the creation of a marine national park at Gahirmatha. Shortly afterward, the government decided to establish the rocket testing range in Balasore to the north and not in Gahirmatha[i].

However, short-ranged missiles have a range of about 700 km, and could be fired eastward, but beyond this they would reach the Myanmar coast. Long-ranged missiles (with a range of more than 2,000 km) needed to be fired southward, which is why Baliapal block in Balasore had been chosen. After several years of dealing with local protests, the Defence Research and Development Organization (DRDO) gave up on this site and decided to establish their testing range at Wheeler Island at a considerable cost.

Strangely enough, the rhetoric died shortly thereafter, and the DRDO established the missile testing range on Outer Wheeler Island, not more than a few kilometres from the mass-nesting beaches at the Ekakula sandspit, with little protest from environmental groups. Eventually, the spit broke away from the mainland in 1989, and gradually moved across the river mouth. In the 1990s, Bivash Pandav's field research camp was located at Long Wheeler Island (Babubali island), and nesting occurred on the two fragments of the spit, Nasi 1 and Nasi 2, a short boat ride away. An equally short boat ride separated the two Wheeler Islands. On one was a modern missile testing range, a road and railway line that ran the length of the island, accommodation for the staff, an officers' club and so on. On the other, we lived in fishermen's huts and worked out of a single large tent.

Bivash Pandav and other turtle biologists and conservationists were then protesting about the lights that came from this island and likely misoriented the hatchlings. In 1997, at the inauguration of the

MTSG Northern Indian Ocean Workshop in Bhubaneswar, Dr A.P.J. Abdul Kalam, the future president of India (then the chief scientific adviser to the prime minister and the secretary of DRDO), gave an assurance that lights would be turned off during the nesting season. In response to petitions and pleas, the Wheeler Island management did sometimes turn the lights down during turtle nesting season, but the request usually had to be renewed each season.

But this was not the only threat that the defence establishment posed. While we were working on the mass-nesting beach in 1999, dogs had started to cross from Outer Wheeler Island to the beach during low tide. By the late 2000s, Nasi 2 island made contact with Wheeler Island, creating a variety of problems, including depredation of eggs by dogs, obstruction of turtles due to beach armouring and so on.

Port problems

Rockets were followed by ports. In the early 1990s, Gahirmatha was said to be threatened by a major fishing port at Talchua, about 10 km away. Again, there were several appeals and articles published in the *Marine Turtle Newsletter* (*MTN*) in 1993 and 1994. Harry Andrews of the Madras Crocodile Bank wrote about the threat from the large fish landing jetty and requested that 'all readers of the *Marine Turtle Newsletter* write letters, preferably on formal letterhead stationery, to the Government of India and the State Government of Orissa to motivate them to continue to protect Bhitarkanika Sanctuary'[6].

Basically, the only fishing harbour for mechanized boats at the time was at Dhamra, at the mouth of the Dhamra river. A new fishing complex was proposed at the south bank of the river at Talchua. However, unlike Dhamra, which was connected by road to the national highway, there was no road to Talchua. This meant that, in addition to the port, a long and circuitous road would have to be built through the sanctuary to Rajnagar and thereon to the national highway onwards to Kolkata.

A couple of issues later, under the title '"Urgent and Immediate

Action" needed to safeguard the world's largest aggregation of nesting sea turtles', Karen and Scott Eckert wrote in their editorial that 'the entire Gahirmatha population is seriously – very seriously – threatened'[7]. Their opening lines: 'These textbook statistics may be all that remain of the olive ridley sea turtle'. They bemoaned the disappearance of the mass-nesting beaches in Mexico, and counted Gahirmatha amongst the four remaining beaches worldwide (Devi river mouth had been discovered a decade earlier, but they did not seem aware of that, and Rushikulya would be discovered a year later). They then glorified Gahirmatha as the world's largest nesting site, and cited figures of about half a million nesting turtles during March 1991, numbers that are now known to be way overblown.

They were asked why we should be concerned when an 'obviously healthy number of sea turtles' were still nesting. With olive ridley turtles, the hype about endangerment has often been preceded by hype about their population size. Somewhat contradictory messages, but a recurring theme in the narrative about Gahirmatha. The Eckerts' answer was that the population was 'seriously, very seriously' threatened by the fishing port that Andrews had written about. According to them, 'there is no question but that the proposed development could literally spell the end of the unrivalled nesting beach'. They ended with the usual plea to write letters and make their voices heard.

Mrosovsky's article followed with the equally dramatic title 'World's largest aggregation of sea turtles to be jettisoned'[8]. However, Mrososvky, always the philosophical inquisitor, stated his case through a series of questions, and unlike the others, spoke to the use of the site both for tourism to benefit local communities as well as use of the turtle eggs as a source of protein. There are hints of hyperbole in his article, but also a case for approaches to conservation that go beyond the protectionism that the others advocate.

Karen Bjorndal, then chair of the IUCN-SSC Marine Turtle Specialist Group joined the bandwagon, writing that 'We must convince the government officials in India of the devastating effects of the proposed development' and 'alert the officials of the high level of international concern'[9].

An article from the *Indian Express* about the Talchua jetty was reprinted in this issue[10]. It noted that 'distinguished environmentalists from countries as far flung as Mexico, Ireland, Canada and United States' had written to the chief minister of Odisha to prevent what Belinda Wright called 'an environmental disaster of global proportions'. Wright, then leading a charge against the ports, and Shekar Dattatri, her friend and colleague in the wildlife film business, described the arribada phenomenon as 'one of the world's great natural treasures' and 'one of the few surviving naturalist spectacles' and suggested that it could be lost forever. The worldwide concern was also apparently reflected by the fact that 'this April, the 30,000 member Earth Island Institute wrote to Biju Patnaik [the Odisha chief minister]...' Almost as if each of those 30,000 had personally expressed grief. Interestingly, the state countered with an extension of the fishing ban on trawlers to a distance of 20 km from the Bhitarkanika coastline.

An article in *India Today* on the issue stated that the ecosystem had already been disrupted by encroachments by the 40,000-odd people who lived around the sanctuary. Belinda Wright was quoted in the article to say that the Talchua fishing complex would be the 'biggest single blow to the continued survival of the olive ridley'. Just as the rockets were before, and other ports would be after. Part of the continuing hype related to the disappearance of ridleys in Mexico, another recurrent theme not supported by fact. The article quoted Surya Narayan Patro, Odisha's minister of fisheries, as justifying the need for the jetty for the sake of the fishers, and protecting turtles by the creation of the 20 km limit for trawlers in the region.

A similar article was published in *Emirates News* in November 1993, and was reprinted in the January 1994 issue of *MTN*[11]. Following the reprinted news article, a small announcement followed, giving credit to the *MTN* appeals for the appearance of the piece in *Emirates News*, and expressing hope that the 5,00,000 Indian residents in UAE and the Indian embassy would generate additional pressure on the Indian government[12].

This spate of written appeals (and letters) may or may not have had an impact, but that is not the point of this detailing. What is striking is the degree of hyperbole that all the writers, including both the international and local conservationists, indulged in. Most of them did not hesitate to state that this population of olive ridleys *would* be extirpated if the plans for the fishing jetty were to go ahead. While there is little doubt that many scenarios would have resulted in negative impacts to the arribada beaches, that the population would be 'lost forever' was surely not a foregone conclusion. And much of the campaign was naïve at best, for example, the belief the Indian diaspora in the Middle East would give a damn about sea turtles in India, or that they would have any influence over the government.

Six months later, another announcement appeared in the *MTN*[13]. This time, the appeal came from Priyambada Mohanty-Hejmadi, a developmental biologist at Utkal University. Mohanty-Hejmadi was involved in several high level committees at the state and Central government on sea turtle conservation, and conducted research at Gahirmatha with support from the DRDO. An accomplished Odissi dancer, Mohanty-Hejmadi also served as vice chancellor of Sambalpur University and was an influential figure in sea turtle conservation in the 1990s. She wrote that construction of the Talchua jetty had, at first, collapsed, but the Government of Odisha had gone ahead and completed it. Roads were also being constructed within Bhitarkanika sanctuary, which was also a key concern. By this time, a PIL had been filed by the WWF's law division in April 1994 against the Government of Odisha in the Odisha High Court[ii]. The high court ordered a stay on the operation of the jetty and asked for an environmental assessment report. Simultaneously, the government was also removing sections of the sanctuary from protected status, which would mean that the jetty could be operated without permission from the Central government. Subsequently, Mohanty-Hejmadi wrote again that the court-appointed committee had submitted a favourable report recommending restricted use of the jetty, control of fishing, and suspension of the road construction[14]. She wrote: 'In the meantime, I and others have met with the Honourable Biju Patnaik, Chief Minister, Orissa, who is very much concerned about olive ridley sea turtles nesting in the Sanctuary. In response to his request, the United States Government has agreed to provide a technology transfer to enable the use of turtle excluder devices (TEDs) in offshore waters.'

It is not certain that the chief minister's request was directly responsible for the TED workshop that was conducted in Odisha the following year, but it was to start a new chapter of conflict over sea turtles in the state.

An editor's note followed in the *MTN*, highlighting the collection in the 1993 issue, and its role in motivating readers around the world to write letters, which had apparently led to the Odisha government's notification[15]. At last, a feel-good story. An occasional success in the conservation world. Coincidentally, this article was followed by two articles by Rene Marquez and his co-authors on the recovery of Kemp's ridleys in Rancho Nuevo, Tamaulipas[16] and olive ridleys at La Escobilla, Oaxaca[17], both beaches in Mexico.

Unfortunately, that was not the entire story. A few issues later, Banka Behary Das, the politician and environmentalist who inspired Belinda Wright, responded[18]. In the 1990s, Das became involved in environmental issues and was in touch with many international environmentalists, and also corresponded with Dr Abdul Kalam, then head of the Ministry of Defence. Das became a leading voice for environmental issues in the state, and formed the Orissa Krushak Mahasangh, an organization that sought to support both environmental causes as well as livelihoods for marginalized communities. In 1996, Das helped produce a comic book, called *The Real Story of Bhitar Kanika*, about the plight of sea turtles and the mangrove ecosystem of Bhitarkanika with support from the Mangrove Action Project and Sea Turtle Restoration Project in the US.

Das's article started with the lines: 'I was astonished to read in a recent issue of the *Marine Turtle Newsletter* that the Bhitara Kanika ecosystem in Orissa, India, has been "protected" by a court's decision.' Das wrote that the court had not yet passed judgement and hence Bhitarkanika was far from protected. While the court had indeed passed an interim order that the Talchua jetty should not be operational, work was continuing slowly on the roads through the sanctuary. He also pointed out that World Bank had 'not given a penny' for the so called eco-development of Bhitarkanika. Nor indeed had anyone been punished for the violation; instead the divisional forest officer in charge of the sanctuary had been transferred as he had been strict about enforcement. In his view, the petition filed by WWF was limited in scope. He pointed out that this was a 'double edged weapon': 'Failure in litigation encourages the adversary, while success, even if it is complete, will only solve the fringe of the problem.'

Das also talked about the efforts of the Orissa Krushak Mahasangh, who had engaged with the DRDO and, in their protest, 'asserted that this defence project will destroy the olive ridley nesting ground for all time to come'. Neither did the missile testing range 'scare away the turtles', nor did the DRDO make any particular concessions on behalf of the nesting turtles. Though he engaged in the same rhetoric about turtles as the other conservationists and NGOs, Das did not see the same solutions. He did not believe that litigation or adversarial action would provide long-term benefits. As a politician, he believed in mobilizing people and holding the state accountable.

In January 1998, Banka Behary Das gave an update on the status of Gahirmatha[19]. In 1997, no arribada had taken place at this nesting site. 'The silence,' Das says, 'was deafening'. Though records show that the arribadas have not occurred in many years, each time the arribada failed to happen, there was an outcry of impending disaster. In fact, no arribadas occurred in either 1997 or 1998 in Gahirmatha, but were recorded in Devi river mouth in 1997 and in Rushikulya in both years.

Das asserted that the failure of arribadas during the previous year was likely due to offshore trawling and accused the state of indiscriminately providing licences and, of course, of aggravating the situation by building the Talchua jetty. Das, like the conservationists, believed that the solution was simple. He said: 'The technology to save sea turtles from death in shrimp trawls is simple and inexpensive, the laws are already in place for time and area (fishing closures), and the groundwork has been laid for a marine sanctuary. All that is needed is the will.'

It turned out that the battle over TEDs was anything but simple. However, he was right about the marine sanctuary, which was reported shortly after. In 1998, the Odisha High Court finally passed judgement on the issue[20]. The judgement quoted extensively from the famous Chief Seattle speech in 1854. The lines, 'What is man without the beasts? If all the beasts were gone, man would die of great loneliness of spirit,' have been made famous in the last few decades. So much so that it has been taught in classrooms, and become gospel in the conservation world. It is ironic of course, that the chief did not actually utter these words; rather these were from a film script written in the 1970s. Of course, the 'fake' speech is far more famous than the story of its origins[iii]. No wonder then that a good part of it would show up in the high court judgement on development within the Bhitarkanika sanctuary.

The judgement, however, began with a summary of the arguments of the petitioners and the respondents. The petitioners had argued for the stoppage of construction of the jetty at Talchua and associated developments on Bhitarkanika. Opponents had, however, asserted that this was nothing but a publicity-seeking gimmick by 'environment lovers' and accused them of filing a 'publicity interest litigation'. Further, they said: 'These persons are busy bees, publicity hungry and make wild baseless allegations and present a distorted version far fetched from reality.'

Surely a sharp barb, considering that publicity seeking (or branding, as it is called internally) is indeed one of the chief objectives of many environmental NGOs. And, of course, the confident assertions of the immediate extirpation of sea turtle populations were a bit wild.

The opponents of the petition asserted that several lakhs of people would be affected and they would 'continue to live as nomads of prehistoric ages' and 'life would continue to be mere animal existence'. A bit harsh and not lacking in hyperbole, like the environmentalists. They were quoted as saying that the 'petitioner and so-called lovers of the environment shed crocodile tears for the Olive Ridley Sea turtles oblivious of [the] fate that would befall few lakhs of human beings'.

The judgement goes on to discuss environmental issues at length. From unnamed German philosophers to IUCN to the UN charter on nature, the judgement quotes widely to endorse the support of the environment. In the end, the judgement walks the line between supporting the environmental cause and local development. In its concluding paragraph, it says that these instructions were to be carried out by the Central and state governments 'with the hope that the olive ridley and the young fisher boy eking out his living by catching fish smile and not run away from each other'. The young fisher boy 'eking out his living', a very real image of fishing boys from traditional fishing communities in Odisha, indeed all along the east coast, was hardly the target of this judgement.

By this time, talk of the port at Dhamra had begun.

Dhamra drama

The Dhamra port first came into prominence in the late 1990s. The clearance to build a port was granted in 1997 taking advantage of the amendment to the Coastal Regulation Zone (CRZ) notification, which allowed the expansion of minor ports with clearance from the Ministry of Surface Transport rather than the Ministry of Environment and Forests. During that remarkable period, the power to clear ports was vested with the same government agency that built the ports. However, the power to clear such projects has since returned to the Ministry of Environment and Forests. The port was to be built by International Seaports Private Limited (ISPL) under a Build, Share, Transfer agreement with the Government of Odisha. A shoddy Environment Impact Assessment Report was prepared which was critiqued by several biologists[21].

Even at that point, however, it was becoming clear that this was only one of several projects to worry about. The Odisha government had signed numerous (apparently over thirty) MoUs with new companies as part of the state's development plan and had planned several major ports along the coast, which would potentially affect all the mass-nesting beaches in Odisha. One of the other ports that had been planned was a port at the Jatadhar river mouth, to be built by Posco, the Korean steel company.

In 2005, the Tatas entered into a partnership with L&T (one of the ISPL shareholders) to build the port to service a new steel plant they were planning in the region. Although the characteristics of their port proposal varied substantially from that of ISPL, the environmental clearance granted to the earlier proposal was used for this project. The opposition to this port from the angle of its impacts on turtles picked up sometime in 2005, with Greenpeace being its most visible critic[22]. Based on interactions with some conservationists and media reports, the Tatas then contacted several sea turtle biologists around the country and requested that we conduct studies (money for satellite telemetry was offered) to find out whether sea turtles would indeed be adversely affected by the port.

A group of us, including both biologists and conservationists attended a meeting with the Tatas in Mumbai to have a discussion on the subject. The Tatas expressed their keenness to get such studies going but did not agree to halt port construction. They stated that they would be willing to take required measures to mitigate impacts and would not rule out abandoning the project at a later stage if studies showed impacts on sea turtles. At that time, however, they were only interested in asking whether sea turtles were present in the offshore waters or not and wanted to dedicate only one season to fund such a study.

Based on discussions with colleagues, I advised them that any study should be collaborative and consultative and involve multiple stakeholders, otherwise the results would not be accepted. I also informed them that my team would be willing to conduct or help conduct these studies, but only if it had the endorsement of local groups and other NGOs. However, since neither the company nor conservationists held a consultation on this issue, there was never an opportunity to share perspectives and to even try to arrive at a consensus of any sort. We had also made it clear at that time that we were not interested in conducting a study with so narrow a scope that it only sought to ask whether turtles were present in those waters or not. We told both conservationists and the company that we should ask a larger question on whether the port would negatively impact sea turtles and their habitats in the region. Since this did not emerge, the coastal programme at the Ashoka Trust for Research in Ecology and the Environment (ATREE), with which I was then associated, decided not to get involved in any study at that point.

During this time, however, a grant was received by the WWF and returned when other NGOs protested about the narrow terms of reference of the study. The Bombay Natural History Society (BNHS) agreed to coordinate the project but also had to back down when petitioned by other NGOs. Many environmental NGOs demanded that the Tatas stop construction until results of the study were in. In an ideal world, this would indeed have been the appropriate sequence. However, considering that requisite clearances were already obtained by the Tatas and they insisted on carrying on construction, some conservationists felt that we should have continued to work together to engage with the company on various aspects. The NGOs, however, insisted that any attempt to work with the corporation would 'greenwash' the port and have long-term negative consequences for sea turtles, usually stating dramatically that it would lead to the extinction of olive ridleys from Odisha.

A few of us campaigned unsuccessfully with the NGOs that we should engage with the Tatas in participation and consultation with local conservation organizations. In particular, I engaged with a long-running battle with Greenpeace on what I considered a narrow approach to the issue. I suggested that there might be benefits to working with the Tatas, a company with a positive reputation for their contributions to education and other social causes. I argued that we needed to be working to counter the large-scale economic growth model that the government was proposing, not addressing each issue piecemeal, and recommended to the conservation groups (as well as later to the IUCN) that we needed to carry out a social and economic analysis to explore the value and viability of numerous ports on the Odisha coast. This was never done.

As an aside, around the same time, I also received a request from Reliance Industries Limited, which wanted to meet to talk about oil exploration plans off the southern Odisha coast. Concerns had been expressed not just about the exploration itself, but about the establishment of oil rigs and platforms, shipping and the pipes that would connect to a facility on the coast in the vicinity of the mass-nesting beaches. Reliance wanted to hire a scientist to conduct assessments and represent them in their negotiation with the government. I respectfully declined, and the project was taken up by the Behrampur University. Reliance continued to be interested in and associated with sea turtle conservation and quite eagerly supported the international symposium on sea turtle biology and conservation in Goa in 2010. Meanwhile, the Directorate General of Hydrocarbons was also pursuing plans for exploration off the northern coast in the vicinity of Gahirmatha. Given the concerns about sea turtle populations, the Wildlife Institute of India was asked to assess nesting along the coast and to conduct an extensive telemetry study to track sea turtles. The project, funded to the tune of Rs 1 crore, led to the tracking of over seventy olive ridley turtles, one of the largest tracking programmes in the world.

The IUCN–Tata deal

Over a period of a couple of years, Indian conservation NGOs declined to work with the Tatas and settled on an adversarial approach to the port at Dhamra, either forfeiting an opportunity or refusing to compromise, depending on which perspective you take. In response, the Tatas decided to go international. Tata/Dhamra Port Company Limited (DPCL) began a dialogue with IUCN, who, in turn, approached their specialist group, the Marine Turtle Specialist Group (MTSG), in 2006. At this time, I served as the regional vice chair of the MTSG. The correspondence is recorded in a series of emails between the co-chairs, Nicolas Pilcher and Rod Mast, and me. I provided a general background to the case, suggested that the Tatas had a good track record especially in terms of their support for education and that it might be possible to work with them. Asked for an informal opinion by the co-chairs over email, I suggested that IUCN not get involved in this project 'without consensus from local partners and groups'. I did not believe then that the entry of an international NGO and outside consultants unequivocally helps the cause of marine turtle conservation, especially when planned and implemented in a manner that was not transparent and excluded local partners. On several occasions and in numerous emails, the MTSG membership in India and I were assured that if and when the IUCN and MTSG did get involved, we would be given full information and asked for our advice and opinion. But at no time was I (or any of the other members in India) formally consulted by the IUCN or MTSG co-chairs[iv].

A critical point came during the first scoping mission, which the MTSG chairs had indicated that I would be invited to. In fact, in an email, one of the chairs indicated that the regional chair could easily replace him on the trip. But when the visit occurred, they were explicitly asked to exclude the vice chair (and therefore other Indian MTSG members), so that the review could be 'external'. This should have set alarm bells ringing, but obviously didn't. After the scoping mission to Dhamra, a project document was developed which outlined the terms of the contract between IUCN and Tata/DPCL[23,24]. Throughout this period, IUCN insisted that they were being advised and supported by the MTSG. However, the MTSG members who were aware of the situation in India strongly opposed the involvement of IUCN in this project.

Over the next year, several emails were exchanged between Pilcher and Mast and various turtle conservationists in India, including Greenpeace, and Romulus Whitaker[25]. Whitaker objected to the secrecy of the process and the exclusion of the Indian MTSG members in decision making, but Pilcher insisted that the members would be consulted when the time came. In an email, he said[26]:

> There is ample evidence around the world to show turtles and ports coexist, while there is no scientific evidence anywhere to show the Port is going to cause any sort of major catastrophe ... The petitioners have not a shred of evidence for anything they claim ... I fear these emotional petitions and cries for doomsday do little more than divert attention from the very positive things that are happening here, and from the potential example others could follow in actually engaging

in real, practical conservation work, hands on, rather than simply whining about it.

But all this only served to alienate the chair from the Indian members of MTSG. Other international MTSG members raised questions about the governance of MTSG and asked how dissenting opinions should be handled, advising that the issues should be discussed at an MTSG meeting, and determined by consensus, giving weight to the opinions of local/national experts. And that if agreement could not be reached, that the MTSG should abstain from offering advice or opinion. Despite all this, in October 2007, Pilcher sent a public email to the MTSG listserver requesting assistance in putting together a small team of experts to advise on the Dhamra port. Whitaker and other sea turtle biologists posted a letter on the CTURTLE listserver on 25 March 2008, objecting to the construction of the port and strongly opposing the involvement of the IUCN[27]. Following Pilcher's response, Whitaker wrote again in detail the following month leaving no doubt about the stand of the Indian MTSG members[28].

During that period, I exchanged several emails with the MTSG leadership. In the beginning I believed that I could make them see my point of view. I wrote about the IUCN that 'on the whole, if you can dissuade them from getting involved, great. I don't think they can help at this point, and locally, they may do damage.' But in the end, given the lack of transparency, and the complete disregard of the local membership, I had no option but to resign[v].

Meanwhile, IUCN had presented the project to their membership in India in August, 2007. WWF India and Wildlife Protection Society of India protested strongly, while other members raised concerns. Though these concerns were noted, IUCN signed the contract with DPCL in November 2007. Members also raised concerns and protested IUCN's involvement at a meeting in February 2008, but were construed as a minority. It is not clear if even a single member was strongly in favour of this engagement. IUCN also ignored the protests of many conservation NGOs that were completely against their involvement with Tata and DPCL. Finally, several members of IUCN in India and most MTSG members sent a joint letter to IUCN on 7 May 2008, outlining these issues[29].

The letter from the sea turtle conservationists included most of the members of the MTSG[30] and the IUCN national committee[31] in India. It reiterated the points made by the conservationists that the environmental impact assessment was inadequate, and that wider social and environmental concerns (pollution from shipping and associated industries, bilge water disposal and invasives, etc.) simply had not been taken into consideration. More importantly, there had been no consultation with local communities, and the local membership of MTSG and IUCN were being ignored. The conservationists urged the IUCN to withdraw support to the port, insist on a holistic EIA, and engage with the membership.

The IUCN formally responded in a letter where they defended their position by suggesting that the port was a foregone conclusion (since all clearances pre-dated their involvement) and they were only able to offer scientific advice for mitigation, specifically in terms of turtle friendly dredgers and lighting[32]. They also claimed that they had been guided at all times by scientists of the MTSG. But this was misleading; they were guided by their international experts, specifically one of the co-chairs, but ignored by all the Indian MTSG experts.

Whether from naïveté or otherwise, the IUCN claimed to be 'designing programmes to generate awareness amongst local communities, especially children, about the turtles and the need to conserve them' and to be 'working with fishermen to encourage them to use turtle excluder devices (TEDs) to reduce the incidence of by-catch.' Based on the history of conflict in turtle conservation, only someone completely ignorant would write this. The IUCN also claimed in their letter that they were developing a lighting ordinance with the Government of Odisha, but a decade later, no such thing is even in the pipeline.

After a great deal of debate and controversy, aired in meetings, letters and online fora, the editors of *MTN* and *IOTN* (Indian Ocean Turtle Newsletter) decided to jointly publish a collection of views on the issue[vi]. A number of Western and Indian scientists, conservationists,

IUCN and MTSG chairs, and a representative of DPCL contributed to this collection. Janaki Lenin and Rom Whitaker added a short and scathing article aptly titled, 'Membership Excluder Devices', where they documented the illusion the MTSG created of consultation[33]. There were several strong critiques of the IUCN and MTSG, in particular that they were greenwashing the port. Lenin and Whitaker also raised the issue of conflict of interest over 'using the credibility of the combined membership of the MTSG to raise funds for one's own organization or personal benefit', and asked how an overseas co-chair of the group got appointed as a consultant.

Nicolas Pilcher, as the chair of MTSG, then set out to defend his position[34]. He wrote that 'the bottom line is that the group exists to do what it can to conserve marine turtles'. He believed that 'When I first went to Dhamra, we walked into a Port under development, and were asked to help where we could – and there was plenty of scope for that.' Pilcher claimed in his article, as well as in several emails, that the Indian MTSG members had been consulted, including the vice chair (me) over several months. Though we exchanged numerous emails with both co-chairs, we were urging them not to get involved with the Tatas and DPCL. So, we were consulted, but our advice, offered in increasingly strong tenor, was repeatedly ignored. It seems absurd that such communication can be considered as consultation, when the views of the majority were so thoroughly dismissed. Pilcher said: 'The very reason for establishing a regional Vice-Chair network was so that there would be an avenue of communication and correspondence between members and the Co-Chairs, and in this case it worked particularly well.'

If it had indeed worked well, why would Indian MTSG members have been so disillusioned, leading to the resignation of the vice-chair? Pilcher, who no doubt had the best intentions towards turtle conservation, stated repeatedly that he believed in collaborations and partnerships, and was saddened by the turn of events in Dhamra, but never appeared to understand why this had come about. In the end, he said. 'We save turtles, it's what we do all over the world, and what we continue to do in Dhamra.' Several years later though, a similar controversy arose over a port in Brazil, and Pilcher decided to step down as MTSG chair.

Hawks and doves

An unexpected outcome of the MTSG's involvement in Odisha was triggered by an article written by one of their consultants. Eric Hawk, a scientist with US National Marine Fisheries Service, was invited to provide advice on minimizing sea turtle and dredging interactions during the construction of the port. Having been to Odisha for a few days, Hawk wrote an article titled 'The continuing shame of Orissa' on the horrors of the sea turtle situation there, and what the government needed to do[35]. In his opinion, the large-scale mortality in trawl fishing was the overwhelmingly large problem and everything else paled into insignificance. Decrying the scientists' tendency to remain aloof, he wrote: 'Sometimes it's good—even necessary—to get mad, indignant, horrified, outraged. Certainly the carnage at Gahirmatha merits that reaction. It's time to blow the horn loudly for Orissa's sea turtles, but let's focus our anger and concern at the main threats, not the red herrings.'

The solution was simply better enforcement. He wrote: 'Somcone in authority (i.e., State and Federal Government ministries) has got to get some courage and step to the plate. Maybe the United States and other signatories to international sea turtle conservation agreements should get together and apply diplomatic pressure.' And then he asked: 'Where is public outcry? Surely, the situation is egregious, outrageous, and shameful. It smells, literally, of rotting sea turtles.'

Of course, this had all been done. Public outcry, anger, indignation, outrage and horror. International pressure. National rage. In a remarkable response, a group of over fifteen authors – including scientists, wildlife biologists, government researchers, local conservationists, national conservation NGOs and activists – who had never agreed on anything before – came together to write a scathing critique of the article[36]. Nothing unites people like a common enemy. IUCN and MTSG had succeeded in doing just that.

The article began by 'acknowledging that Hawk's opinion piece, though remarkably ill-informed, limited and naïve in its understanding of the issue of turtle conservation in Odisha, appears well intentioned'. Providing a historical and political background, the article countered the points that Hawk had made, and pointed out that IUCN and MTSG had in fact undermined the very processes that he was recommending. Remarkably, this group, which included a wide diversity of approaches and philosophies, agreed sufficiently to say: 'Thus, we are likely best served by examining why past conservation measures have failed by critically analyzing them, rather than repeating these mistakes. One step forward is to analyse and acknowledge history.' In conclusion, the authors said:

> Unfortunately, the IUCN and the MTSG have chosen to ignore this vast constituency, and done more to undermine ridley conservation than any good they may have done by saving a handful of turtles from a few port related threats such as lights and dredgers. This only demonstrates the importance of understanding history, and socio-political contexts, in order to be successful. With consultation and participatory decision making, not just with each other, but with local communities, conservation organisations (international ones in particular) can achieve a great deal more towards long term conservation.

At some point, a huge swathe of conservationists of different hues[vii] decided that they were getting nowhere and decided to deal directly with the port promoters. The collective wrote letters, one from a large group of international and Indian biologists[37], and held meetings between October 2008 and February 2009, including one at the construction site at Dhamra. The discussion revolved, as before around a fresh environmental impact assessment, suspension of construction, and exploration of an alternate site. It is interesting that the Tatas decided to consult directly with the groups despite having the IUCN on board as a consultant. Clearly, they were concerned about the adverse publicity but that wasn't enough to stop construction.

After nearly two years of entrenched positions, the conservationists proposed a compromise in February 2009, and requested the company to suspend dredging at least during the turtle nesting season. But now, the firm rejected this on the basis of the advice they had received from 'experts', apparently their IUCN advisors. The company attempted to engage the National Geographic Channel in making a film about their involvement in sea turtle conservation, but the NGOs wrote to the latter and protested against this[38].

Around this time, the IUCN conducted a one-day technical workshop in Bhubaneswar. The agenda was largely dominated by IUCN and MTSG presentations, and came under fire from the conservation groups. Eventually, most organizational members of IUCN as well as MTSG members in India refused to attend, because the organizers simply did not include their main concerns in the agenda. A formal letter was written to the organizers pointing out that the general consensus amongst MTSG members and sea turtle conservation colleagues in India was that 'the IUCN should never have been involved in the Dhamra project without first fully consulting its Indian members' and that 'the process of the IUCN's involvement has been neither transparent nor democratic'[39]. Objecting to the invitation of a select group while ignoring various concerns that had been raised, the letter said that the invitation to the workshop 'indicates IUCN's disrespect of our concerns'. They called on the IUCN to host a meeting with its 'disaffected members in India and the others named above to find common ground' but said that it 'would be futile and premature' to attend the technical workshop unless those concerns were addressed.

This did not, however, stop the IUCN from passing the workshop off as a part of the consultative process to the media and to the global conservation community. The conservation alliance then turned its attention to getting documents related to the consultancy, such as the terms of reference, scope and contract[40]. Repeated requests elicited no response, and when documents were provided or shared on the website, crucial financial details were said to be confidential. All of this, the conservation alliance said, 'contradicts the lofty rhetoric on the IUCN website' about transparency and partnership. In summary, they characterized the IUCN in India as 'naked, but not transparent'[41].

Over the years, alliances and allegiances shifted in the landscape. At no time was there the obvious 'us' and 'them' where all the 'good' environmentalists were on one side, battling the evil corporation. Were these alliances formed on the basis of either common principles or mutual trust or were they just marriages of convenience at a given point in time?

Campaigning for conservation

Wright again

Belinda Wright is best known today for her work on tiger conservation and regulating trade in shahtoosh and other wildlife products through her organization, the Wildlife Protection Society of India. But, like many others, turtles had touched her life in a significant way. Belinda had seen turtles and eggs in the markets in Kolkata, and heard about the nesting beaches at Gahirmatha from her mother, Anne Wright. Belinda started visiting Odisha in the early 1980s and realized the increasing problems being created by trawl fishing for the turtles.

Her contact there was Banka Behary Das, a rare politician who cared deeply about the environment. But what may have turned her attention irrevocably was a visit in the early 1990s with Shekar Dattatri. He was there to do some filming for a British TV company.

Shekar had already been to Gahirmatha a couple of times before. In the early 1980s, he had visited Kolkata and gone to Bhitarkanika with Anne Wright and Indraneil Das. They reached Dangmal in the mangroves, but due to some logistical problems, were not being able to proceed further to the nesting beach, and came back without seeing any nesting. Shekar returned during the season of 1983-84 to do a photo shoot for *Sanctuary* magazine. He recalls taking the Coromandel Express to Bhadrak, a bus to Chandbali, then a motor launch, and then after many hours waiting for the right tide, being dropped at a mud flat some distance from the beach. Shekar recalls:

> I got off the boat and started wading towards the beach, but very soon started sinking in the black mangrove mud. In a few minutes I was trapped up to my waist in the gooey substance and couldn't move. Luckily, a local fisherman and his canoe came to the rescue and I managed to extricate myself and reach the beach. After cleaning myself in the sea, I started walking towards Chandrasekhar's camp, which was six or seven kilometres away. It was dark now, and I began seeing turtles emerging to nest. Every wave brought more females to the shore, and I started photographing the nesting process. Towards 3.00 or 4.00 in the morning I saw a few lights on the beach coming towards me. It was Chandrasekhar and his team, carrying kerosene lanterns and counting nesting turtles.

He returned to camp and slept till lunch the next day, with lots of prawns and crabs on the menu and Jack Frazier and Rom Whitaker tucking away into the feast.

Shekar next visited Gahirmatha in 1992 after he returned from a nine-month stint with Oxford Scientific Films in the UK. He was on an assignment to help a British film crew document the arribada. Unfortunately, the producers had not heeded Shekar's advice, and the timing was off. They had also ignored advice about the water off the coast being too murky for underwater filming and spent many futile days sitting on the beach waiting for the nesting or diving in the offshore waters in the hope of filming mating and swimming turtles. Given that this shoot was a washout, Shekar offered to come back during the hatching (which was more predictable) and shoot a sequence for the producers. So he returned several weeks after the arribada had taken place, with a few rolls of film. This time too, he was accompanied by Belinda. This visit turned out to be spectacular, as not only was there a mass hatching of millions of baby turtles, but also nesting by thousands of adult females in a second, smaller arribada. As the hatchlings raced towards the sea, the females were clambering onto the beach to lay more eggs. 'It was an unforgettable spectacle,' recalls Shekar. He would return in 2002 and 2003, spending six months in Odisha on each occasion, to research and shoot his documentary, *The*

Ridley's Last Stand, largely a self-financed effort but with some logistic support from Belinda Wright and Biswajit Mohanty.

During their trip together, Belinda took some memorable pictures of Shekar filming turtles, nesting adults with hatchlings crawling over, under and around them. She wrote[42]:

> Experiencing the arribada was like adding a whole new dimension to my life . . . You did not see the classical beauty of nature there as you would in, say, a flower or a tiger gliding through the bamboo. It was elemental, powerful, violent even.

Her experience persuaded her that she needed to do something to save Bhitarkanika. After returning, she put together a document detailing the importance of Bhitarkanika and the mass-nesting beaches at Gahirmatha, with the help of herpetologists at the Madras Crocodile Bank. Belinda's photographs and the document were circulated widely nationally and internationally, numerous articles were published, and letters sent to the government. During this period, she continued to interact with Banka Behary Das. When she returned to Bhubaneshwar to talk to government officials and conservationists, she remembers being followed by plainclothes police and denied permission to visit Bhitarkanika.

In the late 1990s, Belinda decided to initiate a project for the conservation of sea turtles in now with Odisha. She brought in a major donor and teamed up with Biswajit Mohanty (the Wildlife Society of Orissa) to launch Operation Kachhapa. Together, they formed a formidable team to promote enforcement and legislation in favour of sea turtle conservation.

When I met Belinda with her mother, Anne, at their residence in New Delhi, she said:

> How quickly this trawler thing became a problem. It went from a handful to a hundred thousand turtles dying, in no time at all. And the reason that it couldn't be stopped was purely political. Because these boats were owned by politicians, bureaucrats, businessmen,

> you know, by people who thought they were above the law. And the fact that they were all making so much money from trawling and prawn fishing. I think it's a great sadness – none of these things benefitted the local people. It all stayed in the hands of a few influential people.

Belinda helped support Operation Kachhapa for many years and co-founded the newsletter *Kachhapa*, which would eventually morph into the *Indian Ocean Turtle Newsletter*. She remains actively involved in campaigns for sea turtle protection in Odisha, and is a significant voice for the conservation of endangered species in India.

Reminiscing about their trips to Odisha, Anne asked what it was that made turtles so attractive, why so many people were fascinated by them. Belinda suggested that it might be to do with the fact that they're such mysterious animals, which appear so briefly on land and people see so little of them, along with the many mysteries about where the turtles go and what they do. It might also have a lot to do with the romantic idea of walking on remote beaches on moonlit nights, wide open spaces and the wild weather that usually accompanies a mass nesting …

We talked a little bit about the fact that she thinks that if she hadn't become involved with turtles she probably would not have become a wildlife activist. She said that as long as she was working in Madhya Pradesh, she was fairly shy and reclusive, and inclined to keep to herself and do her own thing. By her reckoning, it was the 'powerful and strange experience' with the turtles that led her towards activism. When she realized that her efforts to spread information about the turtles' plight was actually making a difference and motivating others, it brought her out in the world … to do all the things that she's done since, both with turtles as well as with tigers. She said:

> The idea was to try and get Orissa to be proud of the fact that it had this extraordinary phenomenon on its coast, before it was too late … It was like lighting a flame and, after watching it splutter and sputter, seeing it suddenly take hold and burn brightly … I hadn't solved the

problem but I had succeeded in motivating people to stand up and say, 'We must save the sea turtles!'

Accounting for conservation

Operation Kachhapa was headed by Biswajit Mohanty, a towering hulk of an individual. Biswajit was professionally an accountant and became a wildlife enthusiast through visits to sanctuaries and parks with friends. I had interacted with Biswajit on issues related to sea turtle conservation for over a decade and met him in his house in Cuttack. Biswajit said that he was not even aware of sea turtles till the late 1980s and became interested in wildlife conservation only in the mid 1990s. He said: 'Then we got the reports of Banka Behary Das, the grandfather of conservation in Odisha, who is no more with us. So we read about the issue of turtle deaths on the Odisha coast. Then we got in touch with Belinda ...'

Biswajit was introduced to Belinda Wright by the forest officer, Sanjeev Chadha, who had worked as divisional forest officer of Rajnagar forest division, which included Gahirmatha beach. Biswajit and Sanjeev Chadha had conducted a sting operation on the leopard skin trade, which is how Belinda initially got involved with wildlife issues in Odisha. Biswajit and Belinda launched Operation Kachhapa in 1998, jointly with Project Swarajya and Wildlife Society of Orissa and Wildlife Protection Society of India, and other local groups and individuals who were interested in saving the turtles.

Biswajit Mohanty became a major figure in sea turtle conservation in Odisha in the late 1990s and early 2000s. Running Operation Kachhapa's day-to-day activities more or less single-handedly, which ranged from patrolling to legal cases to awareness campaigns, he was a frequent figure at field sites during mass nesting and hatching and an active force in policy campaigns. He said:

> Well, I used to believe that if the state government used to plead its helplessness because they said that we don't have money for boats, we don't have boats, and we don't have legal support also ... basically

> it's a question of lack of money for a lot of things. So we sincerely and honestly thought that if we would be able to raise funds for that and provide them that gap in the support services, then things would improve. That enforcement would be better and all that. But unfortunately it didn't happen because then we realized that it was not actually the lack of money, it was actually lack of will and lack of commitment by the government as well as the political leadership.

Biswajit has long been a critic of the government, believing firmly that top-down approaches can bring about changes quickly if there is will. He does not believe that the problems are confined to the forest department. He said:

> I think that is the general problem with most government departments in our state. Most of them are unaccountable and most of them do not carry out their mandated duties sincerely or properly. And we seem to catch on to the forest department because we are focusing on its work and we look into their activities almost like with a microscope. So that is why we're finding so many weaknesses and faults and deficiencies. But I think it's more or less the same situation with almost all the government departments in the state. It is not something to be surprised with because that is the way the government functions in Orissa.

But clearly, it was the lack of accountability in the forest department that vexed him the most. He lamented: 'Thousands of turtles die every year. Not a single forest officer is accountable – why they're killed, what were you doing when they were in your territory?'

Biswajit had long been taking the forest department to task over the funds they received from the Indian Oil Corporation in 2000 – about Rs 100 lakhs for turtle conservation. He said that, instead of using the money to buy high-speed patrol boats, they had spent the money on buying SUVs for the officers to travel. Apparently, they had been unable to find a suitable boat supplier even after four years of getting the funds. He wondered how they could patrol the seas with hired fishing trawlers which were dilapidated and much slower than the trawlers they were supposed to catch. He said: 'In fact, the insouciant senior forest officers sitting in Bhubaneswar never think before ordering the unarmed and untrained and underequipped patrol staff to prevent turtle deaths by using these trawlers as patrol boats.'

Biswajit felt that their focus on marine conservation made them realize 'how useless they [the forest department] are in protection as far as waterways are concerned or marine areas are concerned.' But he did not feel they were significantly better in other areas. He added:

> I mean, they do a slightly better job in terrestrial areas but still I would say that they're not good because there's also the problem with elephants, tigers and other animals in the state and we've got huge poaching rackets going on ... Wildlife crime is a raging issue in the state and none of the tiger reserves or sanctuaries are safe for endangered species.

The Greenpeace gang

Sanjiv 'Sanju' Gopal, Areeba Hamid and Ashish Fernandes were Greenpeace's oceans campaigners. Of these, Sanju was the most active in the sea turtle programme in Odisha. Sanju had started out, like many others, participating in turtle walks on the Chennai coast. By the late 1990s, he had become one of the senior student members of the students' group and was one of its leaders. Sanju assisted me when I visited to collect blood samples from the hatchlings in their hatchery for my turtle genetics project. As he admitted later, for a conservation-oriented youngster, the research seemed a little arcane, not entirely relevant and perhaps even a little counterproductive in that I was subjecting the hatchlings to 'unnecessary cruelty'.

In the early 2000s, Sanju joined Greenpeace and, along with Shailendra Yashwant, coordinated the visit of Greenpeace's campaign ship, *Rainbow Warrior*, to Chennai and Odisha in 2004[43]. Shortly afterwards, in 2006, he helped Greenpeace launch its sea turtle conservation programme in Odisha with a 'turtle witness camp' at the Devi river mouth. They hired a trawler, the *Sugayatri*, painted it in rainbow colours, and installed a wooden turtle on the front[44]. The trawler was used for patrolling the waters off the nesting beaches. Later that season, Sanju and his colleagues staged a protest outside the chief wildlife warden's office, recreating a 'turtle graveyard' there with wooden crosses. They also carried out a protest with dead turtle carcasses at the chief minister's residence in New Delhi[45]. This created a furore, and three of them (Ashish Fernandes, Bidhan Chandra Singh and Imran Khan) were arrested and spent a couple of days in jail, where they were treated by the inmates with a combination of curiosity and wonder that they were willing to go so far for their cause. Bidhan even sang people's movement songs for or with the inmates.

Ashish Fernandes, who had previously worked at *Sanctuary Asia* magazine, was quoted as saying:

> It's time the Chief Minister woke up to the urgency of the situation. He can no longer evade his responsibility for the annual turtle genocide in Orissa. The evidence is before us – the state's failure to protect this endangered species could well result in the total collapse of the turtle population. He needs to take action, and do so now!

Ashish was another conservationist who had graduated from St. Xavier's College, Mumbai, but he studied economics there. While he was initially motivated by biodiversity and wildlife, he quickly realized that it was not sufficient to focus on species or even ecosystems while ignoring the larger drivers of environmental destruction. Aside from running the editorial side of *Sanctuary Asia* and *The Ecologist* magazines, he was involved in conservation and development issues, including large dams for the next few years. Ashish put his work on forest issues aside for a while when he joined Greenpeace in 2006 to work on their oceans campaign.

While the camp and patrolling were operational only for a season, Sanju and Ashish continued to be involved with sea turtle conservation in Odisha for many years. Sanju pursued the fisheries issue by working closely with fishworker unions, and also by interacting with politicians (members of the legislative assembly) from the constituencies of Kendrapara and Dhamra, where many of the fishermen were from. He also interacted frequently with the OMRCC, believing that this consortium held promise for taking turtle conservation in Odisha forward.

Ashish and Sanju were also at the forefront of the campaign against the port at Dhamra. They led the charge against the Tatas, as well as against IUCN and MTSG, when these groups agreed to work with DPCL. While there were some conservationists and scientists who contemplated negotiations with the company, they remained steadfastly opposed to the port and any negotiations, unless the company stopped all construction till a fresh environmental impact assessment and studies on turtles had been carried out. Starting in 2005, they staged several protests in front of Tata House in Mumbai, with the activists dressed as turtles on many instances, creating considerable adverse publicity for the company.

In 2005, returning to Odisha after his previous visit with the *Rainbow Warrior*, Sanju had written[46]: 'My journey back to Orissa was to check on whether any change for the better had happened on ground, in lieu of promises and commitments we received from politicians, governmental departments and supposedly concerned corporations on my last visit there, during the ship tour.'

Over the next few years, he would discover that promises are easier made than kept, that change is not simple, neither politically nor practically. But Sanju persevered for the few years he worked in Odisha, determined to make a difference to the conservation of sea turtles. Many of us may have disagreed with their methods, but as many activists have said, 'Their job is to make the world take notice.' Sanju and Ashish and the team at Greenpeace certainly succeeded on that count, whether or not it eventually made a difference for turtles.

The conservation circus

Hype hurts

In 2001, Nicholas Mrosovsky, the conscience of the conservation community in the sea turtle world, exercised his emeritus editorship in *MTN* to write a scathing piece on the impact of hype[47]. Irked by the constant doomsday predictions in conservation propaganda, which he felt blocked many legitimate approaches to management, Mrosovsky believed that biologists needed to be more objective in their assessments of populations. He also criticized the IUCN red-listing system, which, he felt, forced biologists to impose higher categories of threat because of conservation concerns rather than hard data. He pointed out examples of long-term monitoring of sea turtles where the data indicate upward trends. The 'endangered' olive ridleys (now classified as vulnerable) were nesting in the hundreds of thousands in Mexico, Costa Rica and India. How could a species with such great numbers be considered on the verge of extinction?

Mrososvsky correctly pointed out that exaggerated claims were likely to lose their effectiveness once people found out that that they were not true. But what was worse was for such hype to become accepted practice in the scientific community. Mrosovsky remarked:

> When their mentors, supervisors, and seniors are putting such statements in their grant proposals and pamphlets to the public, it becomes the accepted practice. Hype perniciously downgrades precisely what one should wish to encourage in scientists: an overriding respect for the truth. Hype corrupts.

Mrosovsky wondered about the imposition of cultural convictions in science: 'If people place greater moral disapproval on killing a turtle than killing a fish, that is their prerogative. But the scientific aspects of conservation if they are to be truly scientific must necessarily stand outside personal values.'

Georgina Mace and Russell Lande, influential figures in the

formulation of the IUCN Red List assessment criteria, stated that the biological assessments should be objective and scientific, while priority setting can be based on other considerations, including economic, political and scientific factors[48]. To this list, Mrosovsky said 'one might add the charisma or huggability of a species, personal preferences and ethics, and religious beliefs'. In conclusion, he wrote: 'When can we stop pretending? Most sea turtle species are not on the brink of extinction, but some populations are. Let's start with the facts, and base our actions accordingly.'

Mrosovsky's point, made in different forms over the years by Robert Bustard, Peter Pritchard and Matthew Godfrey, resonates strongly with the situation in Odisha. If not for all the breast-beating about the demise of this population, perhaps stronger alliances could have been built with local fishing communities. Perhaps some degree of utilization could have provided local incentives for conservation. On the other hand, activists firmly believe that an extreme approach is sometimes necessary to balance the push for development and 'to make the rest of us look reasonable'. They often succeed.

BINGOs: world's noblest profession or oldest?

Large international NGOs such as Worldwide Fund for Nature (WWF) with its iconic panda and acronym[viii] and Greenpeace have been at the forefront of environmental conservation for the last half century. They have also become increasingly prominent as a global force in the last couple of decades, with organizations like Conservation International (CI), The Nature Conservancy (TNC) and Wildlife Conservation Society (WCS) operating in numerous countries and exerting considerable influence on both policy and action. Agencies like the IUCN (originally the International Union for Conservation of Nature, now World Conservation Organization-IUCN) have also been tremendously influential. IUCN in particular is Hydra-like, functioning simultaneously as a trans-governmental collective (both countries and NGOs are members of IUCN), as an informal consortium of experts (its various commissions), and a large multinational NGO, which receives grants for projects and competes with other international and local NGOs.

Founded in 1948, the world's first global environmental organization has more than 1,200 member organizations including more than 200 government and 900 non-government organizations[49]. Over 10,000 voluntary scientists and experts from 160-odd countries work as part of the six commissions, including education, policy, environmental law, ecosystem management, protected areas and species. The IUCN Species Survival Commission, the largest one, consists of over 7,500 members belonging to over a hundred specialist groups. The specialist groups are responsible, amongst other things, for assessing the threat status of species under the IUCN Red List[ix], which is one of the most ubiquitous and powerful symbols in the conservation world. Most accounts of large vertebrates, whether academic or popular, often begin by citing the threat status, with endangered and critically endangered species garnering most of the attention and money. So much so that conservationists are paranoid about 'downlisting' species even when they are doing well, because they believe this might have negative consequences.

IUCN projects itself as 'a neutral forum for governments, NGOs, scientists, business and local communities to find practical solutions to conservation and development challenges'[50]. But they also employ over 1,000 staff in more than forty offices around the world, and have, by their own estimation, 'thousands of field projects and activities around the world'. And they are funded by a variety of governments, donor agencies, and foundations. The IUCN even has official observer status at the United Nations General Assembly.

All this sounds wonderful, so what is the problem? The 'secretariat' which operates projects uses the information derived from this enormous body of voluntary experts to buff up its resume. In essence, the large 'voluntary' arm provides an enormous body of pro-bono expertise and knowledge, which the 'corporate' arm can then use to leverage funds for projects. In Odisha, for example, even though some MTSG members were paid for their contributions (which is reasonable) IUCN was invited and funded not on the basis of any

UMEED MISTRY

A diver watches a juvenile green turtle resting on the reef at Kadmat in the Lakshadweep Islands. Green turtles are herbivorous and feed primarily on sea grass and algae. The lagoons of Lakshadweep have large sea grass meadows and serve as excellent foraging grounds for both adult and juvenile green turtles.

SUMER VERMA

A male hawksbill turtle swims away from the camera at Manta Point near Bangaram in the Lakshadweep Islands. Hawksbills are closely associated with coral reefs and are amongst the very few animals that feed on sponges.

ROBERT BUSTARD

Arribadas were discovered in Gahirmatha in the early 1970s by Robert Bustard, with over 1,50,000 turtles recorded during this particular event. Bustard's research team would walk the beach each night counting turtles, painstakingly marking each one with paint to avoid counting them again.

In the late 1970s and early 1980s, tens of thousands of olive ridley turtles were captured off the coast of Odisha and shipped to Kolkata. Vijaya's pictures of this illegal take created a furore that eventually led to stringent conservation measures.

J VIJAYA

ZAHIDA WHITAKER

Vijaya was a young and dynamic herpetologist whose surveys of Odisha led to conservation actions for olive ridleys. She also redicovered the forest cane turtle, which is now named after her. Vijaya died tragically in 1987.

In 1992, Shekar Dattatri and Belinda Wright visited Gahirmatha together to film mass hatching. Here, Shekar is surrounded by thousands of hatchlings scrambling towards the sea.

BELINDA WRIGHT

SUMER VERMA

Both male and female olive ridley turtles migrate to the offshore waters of Odisha for breeding. Mating usually occurs between November and February within 5 km of the coast in small aggregrations that are about 50 sq km in size.

Bivash Pandav carried out his doctoral reseach on the biology of olive ridley turtles on the Odisha coast. He captured mating pairs in offshore waters and tagged them for the first time.

BASUDEV TRIPATHY

C. S. Kar started his research on olive ridley turtles in Gahirmatha with Robert Bustard in the 1970s. He worked with the forest department till 2014, when he passed away.

KALYAN VARMA

Rushikulya was discovered in 1994 by Bivash Pandav and C. S. Kar. Large arribadas have been recorded there in recent years with over 1,50,000 turtles nesting in March 2013 (below) and March 2015 (above). Dakshin Foundation and Indian Institute of Science have been monitoring this rookery since 2007.

KARTIK SHANKER

PRESTON AHIMAZ

Ann Joseph (now Ahimaz) was one of the pioneers of the turtle walk movement on the Chennai coast in the 1970s. Here, she is seen in one of the first hatcheries, sponsored by Grindlays Bank, with a young Shekar Dattatri.

The SSTCN established their first hatchery during the 1988-1989 season. To the far left is Ramachandran, the watchman; in the front row are Maya and Tara Thiagarajan and KS; sitting behind are Tito Chandy (left) and Arif Razack (right), the founders of the group.

Satish Bhaskar conducted sea turtle surveys along much of the mainland coast, Andaman and Nicobar Islands and Lakshadweep Islands. Here, he is seen with a juvenile green turtle at the Madras Crocodile Bank.

ROM WHITAKER

Dhrubajyoti Basu, Satish Bhaskar, Rom Whitaker and Allen Vaughan show off their haul after a lobster hunting expedition off the coast of Mahabalipuram. Using socks for protection on their hands, they pulled up the lobsters, while Basu floated on a car tube and collected them in a snakebag.

KARTIK SHANKER

Hundreds of leatherback turtles would nest each season at the Galathea beach on Great Nicobar Island. The beach was destroyed during the December 2004 tsunami, but has re-formed on the other side of the river mouth.

ADHITH SWAMINATHAN

West Bay and South Bay, Little Andaman Islands, are today the main nesting beaches for leatherback turtles in the Andaman group of islands. Dakshin Foundation and Indian Institute of Science have conducted monitoring at South Bay since 2008 and at West Bay since 2011.

KARTIK SHANKER

In 2007, Uncle Pa-Aung (on his last field trip) and Saw Agu helped conduct surveys of West and South Bay beaches with KS and Manish Chandi. Uncle was the boatman for innumerable surveys with ANET for over two decades. Agu (right), who helped establish the leatherback monitoring camps at Galathea on Great Nicobar Island and on Little Andaman Island, continues to be an invaluable field assistant.

expertise of its own staff, but on the strength of the knowledge of one of its specialist groups.

The large NGO paradigm first received a jolt when Mac Chapin wrote a critique in the Worldwatch Institute's magazine[51]. Chapin traced the growth of the three largest conservation NGOs – WWF, TNC and CI – into multi-million dollar entities by the mid-1990s. In fact, by the end of that decade, it was suggested that conservation funding had declined in general, but funding to these organizations had increased in both relative and absolute terms. Large NGOs spend inordinate amounts of money; one of CI's annual dinners is believed to have cost $750,000. Christine McDonald wrote a damning expose of the organization she worked for in her book, *Green Inc.*, about CI[52]. The dominant paradigm of conservation in these organizations had become exclusionary where people were not seen as part of nature. In this framework, indigenous or traditional people had no real role in conservation. While the paradigms of sustainable development, community resource management and community-based conservation evolved in parallel, and many NGOs jumped on this bandwagon, very rarely did such programmes originate from or truly empower those communities.

Often, these agencies were funded or backed by governments and corporate donors, whom they were unable to oppose when they were involved in environmentally destructive activities. Chapin pointed out that this combination of a 'conflict of interest' in their engagement with those most likely to harm the environment, and their approach to local communities, was resulting in a model of conservation that was deeply troubling from the point of view of social and environmental justice.

A few years later, Mark Dowie wrote a scathing critique of these NGOs in *Orion Magazine*, where he questioned the entire conceptualization of conservation 'when protecting nature means kicking people out'[53]. Dowie pointed to a massively growing class of citizens around the world who had been displaced by conservation activities, mainly the creation of parks and reserves. Globally, the number of parks increased from about a thousand a half century ago to over a hundred thousand in the 2000s. Backed by largely a protectionist and exclusionary philosophy, which is based on the notion of wilderness, or what Daniel Brockington calls 'fortress conservation'[54], many of these protected areas resulted (or are resulting) in the relocation of people, now numbering millions. Dowie pointed out that at an international forum for indigenous mapping in Canada, 200 delegates signed a declaration that 'activities of conservation organizations now represent the single biggest threat to the integrity of indigenous lands'.

Conservation was being identified as the 'new and biggest' enemy to indigenous peoples and cultures. And big conservation NGOs were leading the charge. Dowie argued that large NGOs – by now BINGOs – were perpetrating acts of grave social injustice either directly or indirectly across the world. Since then, many anthropologists and sociologists have criticized this approach to conservation[x]. Brockington, James Igoe and others have characterized this as neoliberal conservation, where money rather than principles drive decisions or action, with no concern for social or environmental justice[55]. Brockington and his student, Katie Schofield, also came up with a taxonomy of NGOs including such WANGOs (Wonderful Animal Focused NGOs), PONGOs (Wonderful People Focused NGOs), MANGOs (Memorial NGOs), DINGOS (Dabbling In Conservation NGOs) and so on[56].

While many have questioned the absence of a social ethic in these organizations, perhaps the question is the wrong one. After all, why should one expect such organizations to behave any differently from other corporations? In the end, are they about social change and education, once called the world's noblest profession, or are they about selling themselves?

What about BINGOs in Odisha then? When Greenpeace decided to start a marine conservation campaign, one of the sites and projects they chose was sea turtles in Odisha. Perhaps this was influenced by the fact that their programme officer, Sanjiv Gopal, and their director, G. Ananthapadmanabhan, were former turtle walkers from Chennai. Sanjiv met us to discuss their plans, which focused on patrolling, strengthening enforcement and media campaigns. Aarthi Sridhar and I pointed out that Operation Kachhapa had tried all these approaches, with far greater resources and effort, and failed. Greenpeace was unmoved. After all, they were Greenpeace and the *Rainbow Warrior* sailed before them.

We pointed out that local communities and groups (and the state) could resent these actions, especially by external groups, and asked if Greenpeace would be willing to attempt the same activities, but without any publicity. Would a large NGO carry out its activities in a self-effacing manner? Well, not in this instance! Though there may have been instances where Greenpeace acted quietly and behind the scenes, they remained convinced that greater publicity would yield better results in this situation. The camp that was set up at Devi mouth, and the trawler that was hired to patrol the offshore waters, all flew the Greenpeace flag proudly. This inevitably created resentment amongst all parties as we predicted it would.

WWF was no better. At Rushikulya, they established themselves in Gokharkuda village, since Purunabandha was already 'occupied' by Operation Kachhapa, and first proceeded to put up posters and murals that displayed their logo prominently. They also made a small film taking credit for the efforts of local groups, which they had funded in a small way. Without a doubt though, IUCN was the worst offender in Odisha. They acted with complete disregard for local opinion and expertise and, in the end, did no favours for sea turtle conservation in the state.

My way or the highway

Throughout the 1980s and 1990s, numerous articles appeared in the media stating that half a million or more turtles nested at Gahirmatha. The figures were repeated from one publication to another, often getting changed along the way[57]. Regardless of the veracity of those numbers, there is no doubt that this is an important population of sea turtles, which merits conservation for a number of reasons, including its spectacular arribadas. Clearly, it is under threat due to a variety of factors. However, it has also become clear that the population will not go extinct this year or the next as many newspaper reports and conservation activists have repeatedly suggested. In the penultimate paragraph of an article in *India Today*, having wound their way through argument and counter argument, the authors say[58]: 'Not much is known about them [turtles], though one fact has been confirmed: when faced with a threat, turtle populations collapse.'

The hype perpetrated by conservationists could not have resulted in a statement more insulting to what we know about turtles, or indeed these creatures themselves, which have survived millions of years. Credibility is critical in the global movement to save the environment. There is no point in making alarmist claims that will be shortly proved wrong or false claims about the importance of a species or habitat for short-term gains.

Unfortunately, in Odisha, both the truth and turtles have been the casualty of the war between conservation activists and trawler owners. A telling point is that twenty years ago, discussions related to sea turtles in Odisha revolved around exactly the same problems and solutions[59]. As a consequence of the hype that has been generated over this issue, conservation actions have mostly been a hurried reaction to a particular threat, resulting in short-term remedial measures that have created more heat than light. Although some of these actions were necessary, conservationists also needed to pursue long-term solutions, which embraced the needs of the local people.

Where conservationists mirror corporations

Greenpeace, WWF and IUCN are all large international NGOs. Tata is a large multinational corporation. Generally, corporations act out of self-interest, and conservationists are self-righteous in their rhetoric. But are they fundamentally different; and did they act differently with regard to sea turtle conservation in Odisha?

In the first round of the battle over Dhamra, both the corporation and the NGOs largely stuck to their agendas. Tata and DPCL refused to acknowledge problems with the environment impact assessment and clearly intended to proceed with port construction while exploring mitigation strategies, rather than re-examine the port project itself.

Additionally, in numerous negotiations with NGOs, the Tatas repeatedly assured NGOs of their commitment to environmental conservation and sea turtles. At a meeting, one executive said something like, 'The Tatas do not want to be responsible for the extinction of olive ridleys in Orissa'. But their promises of a long-term commitment to sea turtles in Odisha proved empty as they have sold their share and moved out, and are no longer 'responsible'.

At the time, it was clear that some number of ports would be built on the Odisha coast with no environmental safeguards for turtles or for conservation. On the one hand, it seemed that the NGOs and individuals against the port did not consider that Tata's willingness to accept some environmental safeguards may have been an opportunity to mainstream some of these as regulations in port and coastal development. In the long-term, this may have had net positive impacts for the coast and marine turtles. On the other hand, many NGOs such as Greenpeace felt the best outcome for marine conservation was to take on the most visible and developed of the port proposals and lead a high publicity campaign against it. Whether the fuss over Dhamra slowed down the other port proposals, or made it easier to fight them, remains arguable. That the Tatas have since pulled out of the port, and the Adani Group, not particularly well known for their environmental track record, now owns the major share, does little to support either argument.

Both conservationists and corporations were remarkably similar in their singular approach to meet their mandates. If anything, the corporation seemed more ready to negotiate at that point. In their obsession with marine turtles, the big conservation NGOs seemed to ignore a range of other issues such as impact on social development, fisheries, introduction of invasives through bilge water disposal, and so on. Many environmental NGOs do argue, however, that communicating complex issues to a large, general audience needs a hook or a symbol and the turtles served that role in Dhamra. However, the disadvantage was that other issues were more easily forgotten. Thus, if studies had shown that sea turtles were not found in the port area, or would not be adversely affected, all opposition to the port project could have collapsed. Also, once the port was completed, and sea turtle populations did not decline, the current positions taken by groups such as Greenpeace would seem alarmist in the extreme, leading to loss of credibility, as had happened before.

Thus, it would seem that IUCN's involvement was seeking this important middle ground. However, in securing the contract, they left aside many principles that they claimed to abide by, namely the precautionary principle, transparency and democracy. For many conservationists in India, the objection was not that IUCN or MTSG engaged with a company, but that the process was not transparent or democratic, and completely dismissive of the entire local membership.

In my article in *IOTN* 8[60], I had pointed out that Dhamra port might not even be the biggest problem in Odisha for marine turtles. The state government was already planning many other ports, and other major constructions (Posco at Jatadhar) and expansions (Paradip port) were in the pipeline. We needed to be working to counter the large-scale uncontrolled economic growth model – with little concern about environmental and social impacts – that the government was proposing. Certainly, the subsequent trajectory of development in Odisha has shown this to be true. Conservation in Odisha has been driven more by rhetoric than action. Conservation organizations have a lot more in common with corporations than they would like to believe, in the way they function and make decisions, but particularly in the way they like to use information selectively. If nothing has indeed changed in the thirty-odd years of sea turtle conservation in Odisha, could it be possibly because all the players (the state, conservationists, corporations, academics, fishers) intentionally or institutionally continue to pursue agendas and strategies that are geared to help mainly themselves, regardless of whether it helps sea turtles in the long run or not?

Notes

i A British firing range was established in Chandipur more than 150 years ago. The weapons were originally sent to England for testing, but because of the long delays, the then port officer of Kolkata recommended Chandipur. Given its low tidal volume, where the tide recedes up to 4-5 km, it was possible to recover the shells after firing. In 1894, a Proof and Experimental Establishment was established (originally known as PEE, but now as PXE) for weapons testing. In 1958, the PXE came under the newly established Defence Research and Development Organisation (DRDO) and become their Interim Test Range for missile testing. Now known as the Integrated Test Range, this is still used for testing short-range missiles.

ii The WWF case was heard in court in August 1995, and the recommendations of the committee were largely accepted, especially that there should be no violations of the Wildlife (Protection) Act, 1972[61]. In October 1995, the court ruled that the interim stay would continue and that the Central government should initiate legal proceedings against the officers who were responsible for constructing the jetty in violation of the Coastal Regulation Zone Notification under the Environment (Protection) Act, 1986. The general tenor of the article is reflected in its title, 'Bhitarkanika ecosystem protected by court's decision'. It suggested that a sanctuary would be funded by the Ministry of Environment and Forests, and that the World Bank had sanctioned a fund for 'holistic eco-development of the area' and stopped aid for commercial aquaculture in the region.

iii The Indian chief's speech was made in response to an offer to buy their land. Chief Seattle (Seathl, to be accurate) made his speech to the Commissioner for Indian Affairs in his native language, and it was translated by Dr Henry Smith, a physician. Smith visited the chief many times to get the translation right and published it thirty-odd years later in the *Seattle Sunday Star*. Eighty years later, Prof. William Arrowsmith of the University of Texas discovered the speech, edited it to a more 'authentic' style and published it in 1969. When Arrowsmith read the speech out to a large student gathering on the first Earth Day in 1970, a young film professor, Ted Perry, was in the audience. Perry adapted the speech to write a script about pollution and the state of the planet, borrowing the spirit of the speech and applying to contemporary issues. But the speech acquired a life of its own. A couple of years later,

Environmental Action magazine published it as a letter from Chief Seattle to President Franklin Pierce. The speech became (further) immortalized in a children's story which sold hundreds of thousands of copies. Eventually the story of the fabrication of the speech emerged in the 1990s. Many people still do not believe that the speech is fake and think that Perry was trying to take credit for monetary benefit. But what is probably more ironic is that the fake speech is far more famous than the story of its origins. The story of the speech and an interview with Ted Perry is published in *The New Earth Reader: The Best of Terra Nova*, 1999[62].

iv In an email to the MTSG co-chairs in September 2007, I wrote[63]: 'As the chair, I think you were obliged to consult with the regional vice chair, who would if asked, have been obliged to consult with the members of that region (at least). I did object then (and have the same objections now) to the MTSG sidestepping its regional leadership and membership. To clarify, I knew that the visit was taking place and very little else. I was not asked to participate or provide inputs. I did provide you with a detailed background to this issue on good faith, but I am afraid I have to agree with Rom [Whitaker] and others here who have raised objections to IUCN's involvement.'

v In an email to the MTSG listserver, I wrote[64]: 'I understand that the MTSG is not a democracy, but I remain somewhat foolishly wedded to democratic and participatory processes. The company, Tata, had contacted me to undertake this project more than two years before they contacted IUCN. As a scientist, I would have been happy to work with them … However, I did not do so, simply because this did not have the support of the sea turtle biologist/conservationist community in Orissa (or India). I believed that a unilateral approach would eventually have more negative consequences for ridleys in Orissa. This is the exact advice that I provided to the MTSG and IUCN. However, when they decided to proceed, I felt that I could no longer be party to this process of decision-making within the group.'

In a subsequent email, I further clarified[65]: 'Almost every single individual who has protested the IUCN's involvement in this project has been involved with sea turtle conservation in Orissa for between one and three decades, and the amount of effort they have spent is extraordinary ... With regard to this project, we were assured on several occasions that if and when the MTSG did get involved, we would be given full information and asked for our advice and opinion. We were not ... The main objection therefore is not that IUCN or MTSG engaged with a company, but that the process was not transparent and in many ways, insulting to the entire local membership.'

vi The collection was published in *Marine Turtle Newsletter 121*, and *Indian Ocean Turtle Newsletter* 8. The editors, Matthew Godfrey (*MTN*) and Chloe Schauble (*IOTN*), sought contributions from various parties, while I recused myself as an editor as I had been involved in the debate and wanted to contribute a perspective piece[66]. Douglas Hykle's editorial provides a good overview of a number of the issues, and ends with a plea for representativeness and transparency in the engagement with Dhamra[67]. This was followed by two letters, the one jointly authored by a number of sea turtle conservationists/NGOs in India to the IUCN, and the response from IUCN.

In his article, Mrosovsky pointed out that several decades of national and international campaigning (of which *MTN* had been a part) had perhaps been effective but in the short term, and the threats resurfaced[68]. He suggested that this might in part be due to the fact the campaigns had served as alerts but did not provide any solutions. Mrosovsky returned to a pet topic, suggesting that consumptive use had not been given a fair shake in the region, wherein communities could have benefitted from the resource and would have supported conservation, despite all the biological arguments in favour of it. Whether community support for conservation would have helped sea turtles is another matter. Intense protests from land owners and fishing communities marked the site of another port further south at the Jatadhar river mouth by the Korean multi-national company, Posco. The story of the development of this port and the protests surrounding it, from local communities to national groups, is a saga in itself.

Ashish Fernandes reiterated Greenpeace's objections to IUCN's involvement with DPCL, suggesting that the environmental organization was greenwashing Tata's construction of the port[69].

Fernandes suggested that if the IUCN had backed the Indian NGOs, then the company may have had no choice but to capitulate to the combined pressure from national and international agencies. Instead, the support they received from the international NGO strengthened their hand in their conflict with the local conservationists. Sudarshan Rodriguez and Aarthi Sridhar provided a comparison of the projects as originally proposed and cleared (through environmental channels) and finally planned by the Tata consortium[70]. The differences, they pointed out, were sufficient to justify demands (from NGOs) for an entirely fresh Environmental Impact Assessment.

Holly Dublin, then chair of the Species Survival Commission (SSC), said that 'Science and objectivity form the Holy Grail of the SSC' but pointed out that there cannot always be agreement between different constituencies within the IUCN or SSC[71]. Critically, though, she acknowledged that they 'may have fallen short of bringing all concerned parties and individuals along with us in this role.'

Amlan Dutta, environment manager for the Dhamra Port Company Limited, also presented the company's point of view[72]. Quite rightly, he started by saying that the Dhamra port had 'a history of mistrust, misconception and hysteria'. He continued by saying that the Tata's 'association with IUCN is looked at with derision and cynicism, once again regardless of the fact that this is the first such association of conservation science and Indian industry'. Dutta asked why conservationists and corporations could not work together to bring about overall positive benefits for conservation.

vii Ashish Fernandes (Greenpeace India), Debi Goenka (Conservation Action Trust), Mitali and Prahlad Kakkar (Reefwatch Marine Conservation), N.D. Koli (National Fishworkers' Forum), Janaki Lenin (as regional chair of the IUCN's Crocodile Specialist Group), Biswajit Mohanty (Wildlife Society of Orissa), Divya Raghunandan (Greenpeace India), Bittu Sahgal (Sanctuary Asia), Ravi Singh (WWF India), and Belinda Wright (Wildlife Protection Society of India).

viii In fact, WWF is so possessive about its acronym that, in 2000, it sued World Wrestling Federation (the premier global sporting event of predetermined results) when they started using the same acronym.

The latter capitulated and changed their name to World Wrestling Entertainment and ran an ad campaign which announced that they were WWE and 'Get the "F" out'.

ix The IUCN Red List (originally the Red Data Book) provides a conservation status for plants and animals that have been globally assessed using a standard set of criteria (see www.iucnredlist.org). Based on the assessments which use population size or decline, and/or information on distribution, species for which adequate data is available, are classified as 'Least Concern', 'Near Threatened' or 'Threatened', which is further divided into 'Vulnerable', 'Endangered' and 'Critically Endangered'. There are numerous critiques of the red-listing process, practice and philosophy.

x Brockington, Igoe and their colleague Sian Sullivan organized a coalition of the concerned in Washington DC in May 2008, which Chapin and several of us from around the world attended. The group shared experiences and deliberated the role of large NGOs in conservation and displacement, and discussed possible ways forward. The meeting resulted in the publication of a special issue of *Current Conservation* on the theme of conservation and displacement[73], and the setting up of an online forum on Facebook and the Internet called Conservation Watch. In recent years, many of the major conservation NGOs, including CI and TNC, have undergone something of a make-over and are professing to focus on human development as well as conservation.

RIDLEYS IN THE BIG IDLY

The beach by the city

WE ARE SCRAPING THE sand with our hands, gently poking a short stick here and there. Suddenly, the stick plunges in. I stop, heart thumping. Dig in frantically, but it's a crab hole. More scraping of sand, stick poking, and discussions of where the up-track ends and the down-track begins. The stick plunges in again, and I follow its path. Here the sand is soft and loose. I think I've found it, I say. A few seconds later, my fingers swim into a sea of eggs. They are warm and wet, and that feeling is one of the high points of a life where success is finding a nest and relocating it to the hatchery.

Some decades earlier, my late grandfather sat on Besant Nagar beach with a friend and watched desultorily as a turtle crawled up to them and laid her eggs in the sand. So little did he think of this event that he did not even report this to the rest of the family till several days later. Sea turtles were barely visible on the horizon for the public of Chennai in the early 1970s. In contrast to my grandfather's remarkable lack of interest in these animals, a group of dedicated wildlife enthusiasts and nature lovers started walking the beaches of Chennai to document the status and threats to olive ridley turtles. Amongst these were Rom Whitaker, Satish Bhaskar, S. Valliapan and many others.

Fast forward three decades. Every January, the newspapers are full of the olive ridleys which migrate to the beaches of Chennai to nest.

They lament the death of the turtles, most likely drowned in trawl fishing nets, and eulogize the many groups working towards their conservation. This media attention is no accident.

In northern Tamil Nadu, nesting occurs primarily along a 50 km stretch from the mouth of the Adyar river, Chennai, to Kalpakkam to the south. Eggs were collected and eaten and sold by people from nearby villages as well as nomadic tribal communities. More recently, threats from fishery-related mortality have emerged, as on much of the Indian coast, with about 20–30 dead ridleys washed ashore every season. There is increasing depredation by feral dogs, and worse, as residential, middle class colonies spread along the coast, beachfront lighting and subsequent misorientation of hatchlings has become an increasingly widespread problem.

The stretch of coast within the limits of the rapidly expanding Chennai city, down to Muttukadu estuary in the south and beyond to Kovalam and Mahabalipuram has received more attention than any other beach in the country barring those in Odisha. Relatively few turtles may nest here, but the beaches have been monitored, surveyed, walked on, slept on, written about, and talked about for decades. By NGOs, by the government, by researchers, by students, by people not associated with turtles in any way before or after. None of this should be surprising. After all, Chennai is in many ways the birthplace of sea turtle conservation in India.

Turtle walks forever

The beginning

Chennai, on the Coromandel Coast in the southern part of the Indian peninsula, is one of the major metropolises in India. In the nineteenth century, the city became the capital of Madras Presidency, the southern division of British imperial India. After Independence in 1947, it became the capital of Madras state, which was renamed Tamil Nadu. More recently, the city was renamed Chennai.

By the early 1970s, the Madras Snake Park was well established and receiving nearly a million visitors a year. The snake park was associated with all things reptilian, so it seemed like the logical place to bring a sea turtle if you found one on the beach. For reasons buried in history, a fisherman brought an olive ridley turtle to the snake park one day. Of course, this caused much excitement, as the nesting of sea turtles on this coast was not known at the time. Despite the decades of capture and egg collection in Odisha, and centuries of capture in the Gulf of Mannar[1], this knowledge had been kept safe from the prying eyes of the scientific community. Romulus Whitaker and colleagues released the ridley into the sea, but decided to start monitoring the beaches. At the time, they had no idea that turtles nested elsewhere, that arribadas occurred in Gahirmatha, and that turtle trouble was brewing in West Bengal and Odisha.

Around the same time, they started getting reports of turtle nesting from the artists' village at Cholamandal, about 20 km south of Madras. They decided to investigate and started surveying the coast down to Kalpakkam, 50 km south of the city. The pioneers included Rom Whitaker himself, Valliapan, an employee of the Central Leather Research Institute, and a few others.

Valliapan and Whitaker wrote of their first survey from December 1973 to March 1974, covering about 10–15 km each night on the 50 km stretch[2]. The fishermen were familiar with sea turtles, including leatherback turtles, but they rarely nested by this time. Whitaker and Valliapan observed that most fishermen did not collect eggs, and even when they did, 'left some eggs in the nest to "keep the race going"'[3]. Apparently, white people were interested in buying turtle eggs for consumption. Many years later, I was told that the late Sivaji Ganesan, one of the leading film stars of Tamil cinema between the 1950s and 1970s, paid for several hundred sea turtle eggs each season. Non-fishing communities like the Irulas and Kurvikars (nomadic hunting tribes) did collect sea turtle eggs regularly. The Irulas called the ridley the 'egg giving sea turtle' and even had a myth about a turtle called the 'sea goddess' which laid two enormous eggs, which no human could see.

Dead stranded turtles and egg depredation by dogs were common even at that time. On one walk, they encountered forty nests that had been depredated, mostly by dogs but a few by humans. Soon they had a hatchery on the beach, the first in India, in the backyard of a friend's house on the beach. The friends, Jean and Janine Delouche, supported their efforts for several years. Jean Delouche is today a historian with the French Institute in Pondicherry, who amongst other things has written on historical trade routes in south India, when tortoiseshell was surely part of the cargo. During the first season, they transplanted just ten nests and released 925 hatchlings, with a hatching success of 73 per cent, which is pretty good for a first attempt. The Delouches supplied 'innumerable cups of tea' and Mrs Delouche said that 'every time a hatchling emerged, she heard a mild buzz and "pop"'[4]. Plans were already afoot to start the 'Croc Bank', then conceived as a project of the Madras Snake Park. They had already acquired 2.5 acres of land along the coast, and planned to start a turtle research centre at the site as well.

Once the turtle walks were introduced, they quickly became popular. A number of bored corporate employees, friends, volunteers at the snake park and later, Madras Crocodile Bank, and students from various colleges would join in. While these were initially focused entirely on protecting turtles and collection of eggs for the hatchery, involving the public became an integral part of the exercise. One of the first was a group of school teachers from Neelbagh, brought along by Amukta Mahapatra, a teacher at a local school and a regular on turtle walks.

The snake park team collected ten, eighteen and forty-two nests in their first three seasons from 1974–1976, using the Delouche backyard as their hatchery[5]. Together with the Central Marine Fisheries Research Institute (CMFRI), they collected 125 nests in 1977; CMFRI paid professional egg collectors for the eggs, and 9,000 hatchlings were released[6]. Totally, about 13,000 hatchlings were released during this period, with a hatching success of about 60 per cent. The snake park also tagged about fifty nesting females during this period; several ridleys, a couple of green turtles, and a hawksbill were also reared and subsequently released by the snake park. One of the green turtles was from Lakshadweep while the other was from the Coromandel coast;

the other turtles were a hatchling ridley collected in the Lakshadweep and a hawksbill collected in the Indian Ocean by the Indian Navy[7]. The hatchling hawksbill was presented to Vice Admiral Manohar Awati while he was in the Seychelles in late 1977; about 10 cm long (carapace) when it arrived at the Madras Croc Bank after transiting in a Mumbai Aquarium, it grew to a length of 36 cm in about two years before it died[8].

Captive sea turtles were also reared by the CMFRI and the forest department. In the case of the latter, a few ridleys were maintained in the 1980s at Point Calimere and the zoological park in Chennai. S. Paulraj, assistant director of the park, and one of the supporters of the Students' Sea Turtle Conservation Network's (SSTCN) brief zoo volunteer programme, wrote of their experience in rearing three ridleys at the zoo[9]. The animals died after about a year in captivity.

The panda and other animals

In the early 1970s, Ann Joseph, then working at the snake park, became involved with the turtle walks. A few years later, she went to work at the Officers' Training Academy at Chennai, under Brigadier E.A. Thyagaraj. The brigadier was interested in wildlife conservation and in reviving the Tamil Nadu chapter of the WWF. The WWF office had long functioned out of snake park with Rom Whitaker as its regional representative. His wife-to-be, Zahida 'Zai' Futehally, had helped her father, Zafar, and sister, Shama, get WWF off the ground in Mumbai, through the Bombay Natural History Society. The snake park was a natural choice for a southern WWF office, but it was not very active as everyone was occupied with the snake park and crocodile bank's activities.

Ann had helped with WWF activities while at the snake park and was enthusiastic about getting it going again. The trustees included stalwarts such as M. Krishnan, one of India's finest naturalists. Ann was already continuing the turtle walks with other enthusiastic students and volunteers. As part of WWF's activities, Ann started mailing WWF subscribers and inviting them to join the walks. Slowly, the enthusiasm grew and a number of college students and others began participating in the walks. Brigadier Thyagaraj also had friends in Grindlays Bank with a house by the beach. Grindlays Bank sponsored a hatchery that the volunteers could relocate their eggs to. During the 1982 season, Ann Joseph and her volunteers relocated forty-four nests to this hatchery, but there is no data for other years[10].

Preston Ahimaz, the WWF coordinator, and Bhanumathi, continued these walks for many years. In addition, Bhanu started a series of nature clubs in schools. Soon, a number of other smaller, local groups became involved. Often, the groups would collect eggs during their walks and relocate them at one of the forest department's hatcheries. Some of these 'walkers' were in the prime of their youthful enthusiasm (or foolishness) when the forest department decided to close down their hatcheries in 1988.

The two faces of the state

The Central Marine Fisheries Research Institute maintained a hatchery for research at Kovalam from 1977 to 1982. E.G. Silas, director of CMFRI, and M. Rajagopalan were at the forefront of this effort, and several small research projects were carried out during this period and published as a collection[11]. These included studies on yolk utilization[12], food intake[13], and growth and health of ridley hatchling and juveniles in captivity[14]. CMFRI even attempted to tag hatchlings, fitting fifty 'green button-type plastic tags'[15], unaware perhaps of the futility of tagging hatchlings. After all, only one in a thousand survive, and a palm sized animal grows into a 50 kg ridley or 500 kg leatherback or anything in between. Two decades later, genetics would provide answers that tagging could not.

The CMFRI team covered about 20 km of coast and collected between 100 and 300 nests each year between 1977 and 1983, with a maximum of 309 nests in 1979[16]. At the beginning of their programme, they overlapped with the Madras Snake Park, and by the end, the WWF and the Tamil Nadu forest department were active on the Chennai coast. While most of the hatcheries on the Chennai coast from Whitaker's early efforts to the SSTCN were primarily aimed at conservation, the CMFRI hatchery was distinct in that it was established primarily for research. In the early 1980s, they were replaced by another state agency.

The forest department likes to grow things, which can perhaps be traced to their origins as a forestry department. In keeping with this philosophy, they had started setting up hatcheries in various states including Gujarat and Karnataka. From 1982 to 1988, the forest department set up several hatcheries along the Tamil Nadu coast, two or three near Chennai and three near Nagapattinam[17]. Along the Chennai coast, the forest department collected hundreds of nests each year. Unfortunately, there are few detailed records of their collections at the hatcheries. During this period, WWF and other local NGOs also conducted turtle walks and deposited the eggs they collected at the forest department hatcheries.

In the late 1980s, the forest department decided to shut down its hatcheries, though it is not entirely clear why. However, one senior forest department officer is supposed to have said that they had 'released so many hatchlings over the past few years that the Bay of Bengal had reached its carrying capacity for sea turtles'. Even if it could be estimated, that number is likely far in excess of current adult populations, and given hatchling mortality and other threats, it is unlikely that the forest department's programme was contributing to an increase in the population, let alone approaching carrying capacity.

Expansive statements apart, the decision had been taken. The groups that ran turtle walks were dismayed. Where would they take the eggs? What would be the focus of the turtle walk programme? As always, crisis brought forth enterprise. Two ambitious youngsters had the craziest idea ever. Why not run the hatchery programme through a students' group? Sea turtle conservation had begun as a volunteer programme, but the CMFRI and forest department paid for eggs, or had their staff carry out collections. It was time to return to the roots, to the spirit of volunteerism in Chennai.

Madness in the method

The Students Sea Turtle Conservation Network (SSTCN) was formally (if such a word can be applied to it) started in October 1988. In December of that year, the group established its first hatchery. The beach was patrolled every night by two or three core members and volunteers, and public 'turtle walks' were conducted on weekends[18]. More than seventy nests were relocated in the hatchery[19] and despite financial and other troubles during the season, it was considered a fair success. Despite various ups and downs, the SSTCN has continued its core activities over the next two-and-a-half decades and continues to do so[20]. This included beach monitoring, maintenance of a hatchery, protection of clutches left on the beach ('*in situ* nests'), and education and awareness programmes. Since 1988, the group has patrolled the same 7 km sector of beach every night from end-December through mid-March. In recent years, the patrolling has extended northward as well. Due to high levels of egg predation by feral dogs and people, most nests along this stretch have had to be relocated.

Each season, a hatchery would be established in late December or early January, often using materials left over from the previous year. The entire infrastructure consisted of a fenced hatchery (with a chain link fence about four feet high), and a thatch hut to store equipment and for volunteers to rest in. Some years, the hut had a single steel cupboard to store materials and equipment. During one of the early years, two freshwater turtles (*Melanochelys trijuga*) served as companions at the hatchery and lived in a little stone tank. Tweedledum and Tweedledee were generally well behaved, but one of them was quite the escape artist. How he got out of a tank where the lip was well out of his reach from the food platform, we never did find out. However, we did find him outside his tank, outside the hatchery, nearly 100 metres away, gamely marching on.

Till 2004, the hatchery was established at Neelankarai, about 8 km south of the Adyar estuary. After the tsunami, the hatchery was established near the estuary, as this was closer to the city, and volunteers could find their way home after the patrols. Once the hatchery was up, the volunteers would start the night patrols. These would typically begin at 11 p.m. or later, and the volunteers covered the beach looking for sea turtle tracks.

Encountering a track was always exciting, and more so, when the return track was absent, meaning that the turtle was still nesting. If the turtle had finished nesting and returned, finding the nests could be tricky. Professional egg collectors or 'poachers' find the nest easily by inserting a long probe in the sand; the probe sinks easily where the sand is loosely packed which usually marks the nest. But egg collectors use long probes, which enter the nest cavity and sometimes break a few eggs. Conservationists typically exercise more care, and use shorter probes, but this method is more uncertain, and demands digging to confirm the location. Nests can take between a few minutes to hours to locate. Even though ridley nest sites are small, relative to leatherbacks which leave vast craters on the beach, nests could still sometimes prove frustratingly elusive. At the hatchery, nests were placed a few feet apart and marked. A few days prior to the expected emergence of hatchlings, they were covered with plastic or thatch baskets, to prevent the hatchlings from crawling on to the beach. Hatchlings were generally released at the edge of the sea the same night of emergence.

On weekends, the group would convene near a pre-determined spot where they would meet with members of the public who would join them on the turtle walk, thus carrying on a tradition started from the very first days of sea turtle conservation in Chennai. During the first few seasons, the SSTCN also dabbled with other programmes including a volunteer programme at the Chennai zoo, and marine education camps at the Theosophical Society[21]. However, most years, the activities were confined to seasonal turtle walks and the hatchery programme.

For more than a decade, the SSTCN was organized and operated by students between the ages of fifteen and twenty-five, many of them students of undergraduate and postgraduate courses at various colleges in Chennai. Several non-students also participated in running the organization, and would advise, assist on turtle walks, and help raise funds. These included corporate employees, lawyers, professional biologists, doctors, veterinarians and so on. But the main leadership remained in the hands of students. This was not planned. When the students finished their courses, they would often leave, having contributed to or led the organization for two or three years. Inevitably, one of the younger members would step up to lead the group forward, despite concerns over whether the group would survive. The first decade-and-a-half was marked by not just constant turnover of members but leadership; not merely new leadership, but young, inexperienced and perhaps even naive leadership. The structure and administration was often disorganized to the point of being chaotic, but the core activities of SSTCN were always carried out. That the group survived and made such an impact is testimony to the determination of all those youngsters, and to the power of an idea once it has taken root.

Often, individuals would be replaced by groups, who found strength in collective leadership. Till about 2000, the group resisted, perhaps not consciously or actively, efforts by older or more experienced individuals to manage the operations. In the early 2000s, however, V. Arun, a school teacher at the imaginatively named 'The School' in Chennai, began to play a more active role. For over a decade, he provided some stability (if not structure) to the leadership of SSTCN while students continued to lead the on-ground activities.

Much of the strength of the SSTCN owes to its oral tradition. Not just within the SSTCN years, but to those who came before. Satish Bhaskar, between surveys in Papua and the Andamans, was in Chennai during the late 1980s. He was a significant mentor to the founders of SSTCN and spent a lot of time on the beach with us. Rom Whitaker and Harry Andrews were active at the Madras Crocodile Bank and provided guidance and support.

The objectives of the group ranged from species protection to habitat protection to environmental education, depending on the leaders of each generation. In the very first report, the stated programme objectives included conservation of sea turtles, generation of scientific information and 'environment and conservation education programmes among the coastal residents, fisherfolk, student community and general public'[22]. Since 1988, these programmes have included the participation of the public and students during weekend 'turtle walks' and during the release of hatchlings. In our newsletter, we wrote that the SSTCN 'continued to focus on the student and the younger section of the populace in the belief that they will one day grow into responsible, aware citizens of the 21st century'[23]. Of the fishermen, we remarked: 'Their way of life is under threat … The SSTCN has established an excellent rapport with the fishermen and we hope that they will imbibe the spirit of conservation and actively support the cause in the years to come.'[24]

In the mid-1990s, an attempt was made to start in-situ conservation, where a certain proportion of nests were left in the beach where they were found, and subsequently monitored. Though the temptation to remove the eggs to the hatchery where they would be 'safe' ran strong within the group, there was also a sense that perhaps it was not sufficient for conservation. They wrote in their 1993 report[25]: 'Though the ridley trail continued to occupy a major portion of the season's activities, it was with a difference – beach management was introduced for the first time on the Madras coast wherein the turtle walks were conducted merely to locate and monitor wild nests …'

The hatchery was obviously an effective short-term measure, but the notion that a wild species could only be saved by constant and permanent human intervention did not perhaps sit well. Alternate (and perhaps more appropriate) models of management needed to be part of the conservation programme. This was continued for a couple of years[26,27], but the programme was discontinued due to the loss of eggs, and the amount of additional effort required for monitoring in-situ nests. In any case, the hatchery had a range of uses. It provided a physical focus for the group's activities. Fishing communities, students and the public who were taken on turtle walks associated the structure with the SSTCN, giving the group identity. It was functional, providing shelter for the group at night and for storage of equipment. It served as a meeting place, and centre for hatchling display and release activities.

In the early years, SSTCN also grappled with issues like temperature sex determination and hatching success in the hatchery. While increased spacing of nests in the hatchery had led to better hatching success[i], there was still the worry that relocation might be altering sex ratios. The Madras Crocodile Bank had sporadically been involved in turtle walks and nest protection, particularly along the stretch of the beach adjacent to the park. Urged by Whitaker and Andrews to monitor temperatures, the SSTCN borrowed thermistors from them, and spent considerable effort collecting temperature data for one or two seasons. Fifteen years later, the data still sits in a box somewhere and the SSTCN still worries mildly over the sex of its offspring. Hatcheries, contrary to some opinions in the 1980s, have not been banned or closed down. It is recognized that as long as natural conditions are maintained, they will likely produce near natural sex ratios.

After nearly four decades of surveys, one would imagine that one could answer that burning question: is the number turtles in Chennai declining? It turns out that these animals guard their secrets well. Since sea turtles are slow growing, late maturing, long-lived species, impacts on their populations take a long time to show up. There is another problem; not every turtle nests every year. Some nest once in two years, others once in three years, and each olive ridley turtle may lay one or two or three nests per year. So the number of nests is a poor index of population size or health.

Moreover, much of the data for the Chennai coast comes from hatchery collections. During these years, though some students and researchers at SSTCN enrolled in Master's programmes and used their patrols or surveys for their dissertations, only a few volunteers were biologists, and research was not necessarily a core activity. Hence, there was often lack of information on effort (number of nights covered, length of beach covered, and so on) which would have been necessary to compare data across years.

In the 1970s, Whitaker and colleagues had estimated nesting to be about 100 nests per kilometre per season. However, this may well have been on particular stretches, such as near river mouths, which are known to get much higher densities than other parts of the coast. Over the last two decades, average densities on the beach ranged from 10–15 nests per km, and SSTCN has collected between fifty and 200 clutches per year[28].

Periodically, nesting has declined to the point that the population was thought to be on the verge of extirpation, but it has recovered in subsequent seasons. For example, only fifteen nests were collected during the 1997 and 1998 seasons, but the numbers went back up to between fifty and 100 over the next few years. Similarly, after some unspectacular years, 2011 was a bonanza year with ninety nests on the regular stretch of beach, and an additional ninety nests on Marina beach, a crowded beach right in the middle of the city, where little if any nesting had been known to occur previously[29]. In what is surely a record, the group collected seventy-two nests in a ten-day period in late February, thirty-six on either side of the estuary. However, it is fashionable to say that these turtles have declined dramatically. The SSTCN continues its monitoring and outreach with the same enthusiasm, and spectacular 2011 and 2013 seasons probably gave an even bigger fillip to the programme and to its volunteers. Arun notes that over 1,500 people joined the groups on turtle walks in 2011, and that over 25,000 people have joined the SSTCN walks over the years[30].

Crusade against casuarina

The SSTCN has remained active and successful through the 2000s, and has become involved in some environmental campaigns through its older members. The history of casuarina planting along primarily the east (but also west) coast of India as a vegetation barrier has a long history, dating back to plantations in Odisha (Anne Wright opposed it during her years of involvement there) and in Tamil Nadu in the 1960s[31]. Despite little evidence that these barriers are effective – the 1999 super cyclone in Odisha caused substantial devastation in areas with casuarina[32] – they have been planted by all the states on the east coast.

In different decades, a variety of reasons have been presented to justify the plantations, from coastal protection (from storm surges and cyclones) to social forestry (fuelwood, and so on)[33]. Many of these efforts have been supported by the Central government as well as international development agencies (such as World Bank and the United Nations), with scant attention to data on their impact. Recent reviews show that the science supporting the positive effect of vegetation as coastal barriers is equivocal[34]. But science, if it was used at all, was used selectively to support these plantation efforts at various times. Given the amount of money available for these initiatives, it is no surprise that they have been pursued despite the apparent costs and less apparent benefits.

More than a third of the Tamil Nadu coast was planted with casuarina after the tsunami under the Emergency Tsunami Reconstruction Project funded by the World Bank. However, many of these plantations were in areas where no vegetation had existed before. Even more strangely, they were not planted in front of villages – where they could potentially have mitigated storm surges – because the fishers did not want to lose access to the beachfront where they kept their boats[35]. These so-called bioshields for coastal protection were planted in vast tracts between villages, where they could not possibly have attenuated any wave action.

In many areas, the plantation extended up to the high-tide line, and potentially impeded turtles from nesting. In 2010, Swapnil Chaudhari, from Pondicherry University, worked on the impact of these casuarina plantations along the coast on sea turtle nesting beaches[36]. His work provides some evidence that they negatively impact nesting beaches especially when planted very close to the high-tide line. Many ecologists pointed out that the plantations were adversely affecting both natural sand dunes and were interfering with the natural transport of sand, which was essential for beach dynamics[37].

Akila Balu and Arun of SSTCN joined forces with Shekar Dattatri and campaigned for the removal of the exotic casuarina from the coast of Tamil Nadu, which had been planted as a coastal bioshield against storm surges after the 2004 tsunami[38]. They asked for the removal of casuarina upto fifty metres from the high tide line. In 2007, they sent a letter to the president of the World Bank, copied to a number of other environmental organizations. This resulted in World Bank officials meeting with the campaigners and assuring them of action, but nothing came of it initially. However, their campaign over a year and a half, combined with the efforts of other groups such as the TREE (Trust for Environmental Education) Foundation, eventually resulted in a government order from the state ordering the removal of casuarina upto a certain distance from the high tide line in areas that were critical for turtle nesting. While the extent of removal of casuarina is uncertain, the order did lead to the clearing of casuarina from many areas and the fact that their efforts influenced government policy is remarkable[39].

The group also had some success in getting floodlights on some beaches switched off during the nesting season[40]. In the mid-2000s, the city corporation installed large floodlights on several public beaches in Chennai. After many years of campaigning against the lights, in 2010, the chief wildlife warden arranged for SSTCN to make a presentation to the secretary of the Ministry for Environment and Forests, and other key officials. The secretary was sympathetic and directed the corporation to switch the lights off during the nesting season. The following year, the chief wildlife warden was successful in getting a government order passed to switch these lights off between 11 p.m. and 5 a.m. between January and April each year over a 5 km stretch. The group also helped campaign against a 10 km-long elevated expressway that would pass over the main nesting beaches in Chennai. Apart from these occasional forays into environmental activism, the core of the group remains its turtle walks and hatchery, a tradition started four decades ago, and carried on still.

Down memory lane

Snakeman turtleman crocman

Romulus Earl Whitaker has been a profound influence on the conservation of reptiles in India. Best known for his work on snakes – including setting up the Madras Snake Park and the Irula Cooperative – he also established the Madras Crocodile Bank Trust and field stations in the Andamans and at Agumbe in the Western Ghats. Late twentieth century herpetology in India benefited because Rom's mother married a filmmaker and moved to Mumbai in 1951. After his schooling at the Kodaikanal International School, where he divided his time between herping, flirting and musical appreciation (Tom Lehrer, mostly), Rom returned to the US to attend college (aborted after a year or so). Drafted by the army in 1965, Rom served in Texas and Japan. While posted at El Paso, Rom used the opportunity to travel and explore Texas and Arizona, looking for rattlers and other snakes.

Before he went into the army, however, Rom worked in Florida, helping at the Reptile World Serpentarium, as well as in snake collections. Under the not so cruel tutelage of Bill Haast (who after all only injected himself with snake venom for immunization), Rom learnt many of the tricks of the trade. Haast was a legend and his blood was used as anti-venom on more than two dozen occasions. Rom spent a considerable part of his time on snake hunts with Heyward Clamp, Schubert Lee and others.

In 1967, Rom decided to find his way back to India. With the Suez Canal closed, it took two months for the Greek freighter to get from New York to 'Bombay'. After setting up a venom extraction facility near Mumbai for a couple of years, Rom moved to Chennai, a homecoming of sorts, after his schooling in the region. A couple of years of snake collection for venom extraction passed before his idea of setting up a snake park became a reality. The Madras Snake Park was established in 1969 at a village near Tambaram with the help of his brother Neel and sister Nina.

They found a half-built house with the foundations of the walls that provided perfect snake pits. In 1971, the government provided land adjacent to the Guindy National Park and the snake park moved to its current location. Very soon, they were getting half a million visitors a year. A trust was established to oversee the snake park, and though differences between Rom and the trust led to a separation, the park had been established as one of Chennai's attractions. Crucially also, Rom and others impressed upon Prime Minister Indira Gandhi, during her visit to the park in 1977, the importance of the adjacent scrub forest with its resident blackbuck population, and this was shortly afterwards declared as Guindy National Park.

In 1974, Rom married Zai and they launched many herp projects together. The Madras Crocodile Bank Trust was started by them in 1975, and sea turtle surveys were initiated around the country by the one-man show that was Satish Bhaskar. Rom also initiated crocodile surveys around the country; with a small grant from WWF and with his Irula field assistant, Rajamani, he surveyed the Ganges, Chambal and Mahanadi Rivers for gharials. Muggers were also surveyed in these rivers and in other parts of the country. Over the years, the Croc Bank carried out many crocodile surveys, and their gharial and freshwater turtle projects in Uttar Pradesh, Rajasthan and Madhya Pradesh are still ongoing. Of course, the snake work continued as before. Zai's book, *Snakeman: The Story of a Naturalist*, provides an account of this period, imbued with the passion and spirit of adventure that herps inspired in Rom and all those around him.[41]

The crocodile work took Rom and Zai to other countries as well, including Sri Lanka, and Papua New Guinea, where they spent a couple of years. The work in the Andamans was also initiated around this time with surveys for the king cobra and saltwater crocodile. Fascinated with the islands, Rom bought land there which allowed them to eventually start a field base in Wandoor, 30 km from Port Blair. In 1975, Rom became an Indian citizen so that he could work in the Islands. The Andaman and Nicobar Islands Environmental Team (ANET) was started in 1989 and later developed into a fully functional field base.

More recently, Rom has started the Agumbe Rainforest Research Station (ARRS) in the Western Ghats. A long-term initiative at this site is the king cobra rescue and monitoring project, and hundreds of these huge snakes have been rescued from houses and fields in the area, where people are remarkably tolerant of them. Several animals were also tagged with radio transmitters to study their movement patterns. Both ANET and ARRS now partner with many national research institutions and NGOs to strengthen research at these sites, to initiate long-term monitoring programmes, and to provide a base for students and researchers working on a range of taxa and topics for their Master's and Ph.D dissertations and other research projects.

Rom's contribution, however, goes beyond setting up these institutions and field bases. More important than the places he set up were the people he set on their way. Amongst the first persons that Rom inspired to a career in wildlife was Satish Bhaskar. Satish, never really committed to his course at the Indian Institute of Technology in Chennai, was dividing his time between the beach and the snake park. When I interviewed Rom at his home in Chengalpattu, he remembered telling Bhaskar:

> If you just concentrate on sea turtles, you'll become 'Mr Sea Turtle'. Because there's nobody else doing it! You know, there's Archie Carr, and there's George Hughes, and all these people (and he was probably wondering what the hell's this guy ranting about). But he eventually read all these reprints that we'd started collecting and getting feedback from all these fantastic, wonderful people from all over the world ... sending their reprints, folded carefully, and their notes all the way to India ... and that was the thing that got us excited about continuing, because there were other people doing things much more seriously than we were.

Rom gives honourable mention to the girls (his sister, Nina, Satish's wife Brenda, Ann Joseph, Wendy Bland) at the typewriters sending letter after letter to various famous sea turtle biologists around the world – Archie Carr in Florida, George Hughes in South Africa, George Balazs in Hawaii, Jack Frazier, and others. Many of Chennai's conservationists started their careers at the snake park or on a turtle walk. Others dedicated a few years of their life to turtles and went on with their day jobs. Satish may be the best known for his work on sea turtles, but there are many others who cut their teeth as volunteers at the snake park, to whom Rom may have said: 'Hey man, how do you feel about cleaning out a snake pit?'

Early enthusiasts

Valliapan was the heart of the group in the early years. He walked frequently, collected data, and published some of the earliest notes on the sea turtles of Chennai[42]. One of these was co-authored with Solomon Pushparaj, from the forest department, who died early of alcoholism. They were part of a cohort that joined the snake park in its earliest days, along with 'Motorcycle Mani', all enthralled by snakes and turtles and such. Mani was the first assistant at the Madras Snake Park and went on to become assistant curator; he acquired his tag when he inherited Rom's 1967 Java motorcycle. Rom remembers hiring them as teenagers and having many adventures together, including a near drowning with Solomon.

Amukta 'Amu' Mahapatra, then a teacher at the Krishnamurthi Foundation school in Chennai was a regular at the walks. Introduced to the world of wildlife by her father and his friends, and a frequent visitor to various sanctuaries in south India, Amu decided to volunteer at the snake park to get over her fear of snakes and reptiles. She and her friend, Aarthi, also from the school, would go to the snake park with Siddhartha Buch, a well-known Chennai naturalist. After snake handling, (which was awful, but she got over her fear), she joined a group of volunteers on the turtle walks.

Amu and Valli wrote an endearing article in the newsletter of the Indian Institute of Technology, Chennai. After describing her first encounter with a ridley turtle, Amu urged youngsters and students to do something[43]. A call to arms that was taken up then, but with even greater vigour a decade later.

Amu went on to start Abacus, a well-known Montessori school in Chennai in the 1980s, and spent several years conducting Montessori training programmes in Tamil Nadu and Karnataka. Though she was never involved with sea turtles directly, she remembers her turtle nights vividly. And strangely enough, turtles did not relinquish their connection with her. Her home was my base in Bhubaneswar when I started research on ridley turtles in Odisha. Her husband, Aditya Mahapatra, composed the music for the turtle song for the 'minstrels' of Operation Kachhapa and translated my children's book *Turtle Story* into Odiya.

Many others drifted in and out of this circle. Dhrubajyoti Basu, who went on to become the doyen of gharial research and conservation, was part of the group then. Basu, as he was popularly known, died tragically in 2011, while on a field trip in Assam. Priya Davidar, daughter of E.R.C. Davidar, one of India's early naturalists, and a well-known ecologist at Pondicherrry, joined the walks frequently too.

SuperwomAnn

One of the earliest organizers of turtle walks was Ann Joseph. Ann started as a volunteer at the snake park in the early 1970s. Like many others, Ann's fascination with snakes started with some degree of fear. Ever so often, after her shorthand and typing class, she would go and hang about the snake park. Rom eventually noticed and offered her work there, much to her mother's chagrin. Ann negotiated that she would work at the snake park just till she finished her course and then she would get a *regular* job.

Her first assignment was putting up signs in the enclosures for the visitors. Slowly, she got used to the snakes crawling around her, cobras included, and started doing the hourly demonstrations with non-venomous snakes. Ann worked as Rom's secretary for a while at the snake park, and then, in keeping her promise, took a job at the Officers' Training Academy.

After she started working for the academy in 1976, Ann was still very much the organizer for turtle walks. Along with her boss, Brigadier Thyagaraj, Ann helped establish and promote WWF in Chennai, and started conducting the walks as a WWF activity. Ann remembered her very first turtle walks with the WWF subscribers: 'When I first sent out an invitation for the turtle walk to the WWF subscribers, and they all came, and I gave them a little talk about the different types of turtles and how you go about this turtle walk. And they said, we thought turtle walk means walking with the turtles.'

Shekar Dattatri recalled Ann's transformation from office goer to wildlife woman: 'She was the real catalyst. And I can clearly remember how Ann used to come to the beach straight from the office, impeccably decked up in a sari. She'd disappear behind a parked boat with her bag and emerge a few minutes later in jeans and t-shirt! It was like Superman going into a phone booth kind of thing ...'

Ann was also instrumental in starting the WWF nature camps in parks such as Mudumalai and Anamalai. Like the snake park and the turtle walks, the nature camps made a considerable contribution to awareness of wildlife ecology and conservation in the country. A remarkable number of biologists got their first taste of wildlife at these camps. One of these was Preston Ahimaz. Preston also started volunteering at the snake park and met Ann at one of the camps. A few years later, they were married, making them one of the first turtle-walk couples.

Preston also started out as a volunteer in the snake park, another of Rom's recruits. Despite much advice to the contrary, including from persons in WWF itself, Preston decided to take this up as a career (with a degree in commerce, he could have easily pursued chartered accountancy or banking). Preston was the first coordinator of the WWF office in Chennai, starting with a tiny office and a table and a chair in 1982. He built their nature camp programme, training many youngsters who went on to become wildlife biologists and conservationists. The WWF turtle walks lasted through the 1980s under the stewardship of Bhanu (who joined later) and Preston.

Movers and Shekars

Yet another career – this one in films – was born through Rom's encouragement of students who needed a place to hide away from class. Shekar Dattatri was only thirteen when he started skipping school as often as he could to be at the snake park. He joined in 1976 as a volunteer and became involved with childlike enthusiasm in everything that was possible: 'I was like a mongoose. I didn't care whether it was turtles, or snakes, or whatever. I was out all the time. I was hardly in school.'

One of Shekar's close friends was Viji; the two of them would go on snake catching expeditions with the Irulas, and on turtle walks, where they would mask the tracks rather than collect eggs for the hatchery.

Between 1979 and 1983, Shekar regularly visited Sri Lanka, where his father was on a four-year assignment with the UN. He backpacked around the country and spent many nights at the Kosgoda nesting beach about 60 km south of Colombo. Unlike beaches on the Indian mainland or in most places, this beach was visited frequently by four different species of sea turtles – greens, hawksbills, leatherbacks and ridleys, and maybe even the occasional loggerhead. Although Shekar himself never saw all species on one night, some of his Sri Lankan naturalist friends showed him photographs that they had taken of at least three species on a single night. At Kosgoda, he established a friendship with a local fisherman, Similias Abrew, who ran a small hatchery with his own money and without any fanfare. In the following years, the establishment of turtle hatcheries along the southern coast picked up, initially started by a few naturalist-hoteliers. These soon became tourist attractions, and eventually attracted criticism from scientists and conservationists. But at the time, it was innovative and well-meaning.

Since little was known about the sea turtles of Sri Lanka, Shekar sourced a small grant of $400 from the Sea Turtle Rescue Fund in 1982, and convinced a friend with a motorcycle and an interest in natural history to join him on a survey. They surveyed much of the coast, both east and west, and even visited Jaffna, which the Sinhalese rarely visited, even though this was before the civil war started[44]. They visited Kandakuliya, on the seaward side of Puttalam lagoon in north-western Sri Lanka. Shekar told me:

> At that very remote spot, we found the biggest graveyard of sea turtles I have seen, before I went to Orissa. These were not fresh carcasses because it wasn't the season. But everywhere in the sand, you just had to scuff your foot and you came up with a piece of a plastron or a shell or whatever. It was predominantly ridleys, but there was other stuff also, green – carapaces, plastrons, just pieces, everywhere. Obviously, it was just a *huge* massacre happening there every year at a certain time of the year.

Shekar finished his zoology degree from Loyola College, Chennai, in 1983, though it was a couple of years before he could graduate, given that he still had some courses to pass. Around this time though, Shekar got drawn into the world of film-making. Over the next couple of decades, Shekar worked first with Rom and then with international television broadcasters and production houses to make films on snakes, rats, sea turtles and a variety of other wildlife subjects.

Student soap opera

The SSTCN was started in 1988 by Tito Chandy and Arif Razack, a passionate but volatile pair. They worked on several projects together, mostly related to wildlife tourism. Having been on turtle walks the previous season, though never having seen a turtle, I joined enthusiastically, along with a few others. Our most exciting new recruit was Tharani Selvam, from a fishing village near Ennore, then doing his Ph.D in Chennai; Tharani passed away tragically a few years ago. We also inducted Tara Thiagarajan, who saved the SSTCN from bankruptcy several times in the first few years. Tara organized a signature campaign at a couple of well-known city schools, which brought in a then astonishing Rs 10,000 and kept the organization afloat. The following year, while studying at Brandeis University, she visited the Centre for Marine Conservation (CMC) in Washington DC (now the Ocean Conservancy) and obtained a small grant for the group. Charles Tambaiah, a Sri Lankan turtle conservationist, and Marydele Donnelly, long-time programme officer of the IUCN Marine Turtle Specialist Group, both then at CMC, helped obtain these grants. Marydele said recently[45]:

> I loved that Small Grant program at CMC – real bang for the buck, so to speak. Over time my bosses cut and then eliminated the funding despite my protests and even offers to take a pay cut to support it. Being able to help eager and dedicated young people was not only rewarding for turtles but for me personally as it hugely leveraged what I could do alone.

The SSTCN's contact with the global sea turtle community was established that year. Tito attended the Smithsonian field course in Malaysia, where he met Jack Frazier, already a veteran of sea turtle conservation, and an Indophile. Yohan Thiruchelvam, a young volunteer with the SSTCN straight out of school, spent a summer in Malaysia working on the beach as a volunteer with Jeanne Mortimer, one of Archie Carr's students.

During the first season, we also hired Ramachandran, a fisherman from the nearby village of Chinna Neelankarai. Ramachandran had had a stroke that had left a part of his body paralyzed, and this prevented him from going out to fish. Had he owned a motorized boat, this may have been less of a problem, but the physical coordination required to handle a catamaran was quite intense, as we discovered during our rides with his son, Raja. Ramachandran stayed with the SSTCN for many years and his meagre eager salary was one of the few non-negotiable items on the annual budget, which rarely exceeded Rs 20,000 and was often far less.

In 1990, in only our second season, Tito Chandy, Yohan and I were the only regular turtle walkers. We covered a 15 km stretch every night without fail, and came up with fifty-five nests. This was depressing to the point where we questioned whether we should continue at all. From the beginning, it was never very clear how the mantle of leadership would pass on in the SSTCN. But once Yohan and Tito left, it fell on me to carry the SSTCN on. Just two seasons old, there was no certainty that the group would (or should) survive. As with earlier efforts, the SSTCN risked an early demise with the departure of its founders. With a little persuasion, however, a group of marginal members decided that they would join the effort. In transitioning from one leadership to another, the group became aware that this was possible. Tara managed to get us another grant from the Marine Conservation Society, and we built a spanking new thatch hut, and installed a steel cupboard, that somewhat ironically belonged to my grandfather who had thought so little of a nesting turtle.

Several individuals became very involved, and our meetings often had eight to ten participants[ii]. One of them, John Mathew would (before his two Ph.Ds on butterfly phylogenetics and history of science at Harvard) go on to write *Olive*, a musical on the life of a young female turtle and her journey from hatchling to adulthood. The plans for this were hatched on late nights during our turtle walks, and the show was produced a few times in Chennai in the 1990s. It was a cohesive group who really felt a common ownership of the SSTCN, and appropriately enough, it was an inspiring season that brought 206 nests. Fittingly, a local television company made a film about the SSTCN, which was shown on Doordarshan.

Over the years, many volunteers continued to work on sea turtles as part of their Master's dissertations and became ecologists of one hue or another. Abi Vanak, now a carnivore biologist, joined as a thirteen- or fourteen-year-old in 1988 and was for long the youngest member of the young organization. Subramaniam, who also joined at about the same age and would continue to be involved off and on with the SSTCN over the next fifteen years, remains one of its most loyal members. Arjun Sivasundar, part of the group in the mid-1990s, conducted the first study of nesting leatherback turtles in Little Andaman Island. In the late 1990s, Karthik Ram ran the organization for a couple of years, and later carried out his Master's dissertation study on the offshore distribution of olive ridleys in Gahirmatha. Abi Vanak worked with Operation Kachhapa in Odisha that year. Sanjiv 'Sanju' Gopal, a leader of the group in the early 2000s, became a campaigner for Greenpeace and worked for its oceans programme for several years, leading campaigns in Odisha. Adhith Swaminathan, who started turtle walks as a ten-year-old, has spent the last several seasons as a researcher at our sea turtle monitoring projects, first with olive ridleys in Odisha, and then with leatherbacks on Little Andaman Island.

In the late 1990s, Arun joined the group and has stayed involved with them since. Arun would become the pillar around whom the group has revolved for nearly two decades now. Akila Balu, who went on turtle walks in her college days, joined Arun in coordinating the turtle walks in 2007 and has become another core member of the group. In 2009, the group was finally registered as a formal trust, with Arun, Akila, Adhith and Sanju as its founding trustees. In 2011, Arun wrote a touching tribute to the volunteers of SSTCN[46]; that season was, fittingly, marked by remarkable numbers of nests. The season served to build momentum for the group as it celebrated its twenty-fifth year.

Hot and cold eggs

Sea turtles have proved enigmatic to study and understand, often to the dismay and sometimes great embarrassment of biologists working on them. In the 1970s, the US government decided to intervene to save the endangered Kemp's ridley turtle in Mexico. Tens of thousands of eggs were transported by air to be incubated in laboratories in Texas[47]. The eggs were incubated in Styrofoam boxes and the hatchlings were released, sometimes after rearing them to a particular age (known as headstarting), so that they would avoid the initial high mortality that hatchlings face. The idea was that if these hatchlings were imprinted to the sand in Padre Island, they might return there and establish a new nesting colony. The 'headstarting' itself was controversial but there was another problem. Around this time, the work of Nicholas Mrosovsky and others demonstrated that in turtles, like crocodiles and some other reptiles, sex was determined by incubation temperature[48]. There was a fairly narrow threshold or pivotal temperature, where both males and females were produced, but above this threshold, only females were produced, and below this threshold, only males.

However, the squeaky clean Styrofoam boxes, effective as they were in producing high percentages of hatchlings, were maintained at temperatures much lower than the pivotal temperature. Thus, for many years, the US conservation programme probably produced and released only male Kemp's ridleys. In fact, eggs incubated in Styrofoam boxes in numerous conservation programmes around the world probably suffered the same fate[49]. Subsequent research has shown that, in nature, clutches rarely produce both males and females equally; in fact many beaches produce female biased sex ratios[50]. Hatchling sex ratios vary between seasons and years and on different kinds of beaches, hot dark sand beaches or cool white sand beaches.

The evolutionary significance of temperature-dependent sex

determination (TSD) has been widely debated; does it confer any advantage to organisms? One hypothesis that has some support is that TSD may be adaptive when the incubation temperature differentially affects the fitness of males and females[51]. While higher temperatures produce females and lower temperatures produce males in sea turtles, the pattern is exactly the opposite for crocodiles. While the causes for this pattern are not known, it may have something to do with the nature of sexual dimorphism in the two groups of reptiles. In both groups, the larger sex is produced at higher temperatures, which suggests that the pattern may be linked to body size.

In recent years, TSD has come under greater scrutiny because of the potential impacts of climate change[52]. Many beaches already produce female biased hatchling sex ratios, and with rising temperatures, the sex ratios could become even more female biased which could eventually affect the health or stability of these populations. Sea turtle research and conservation programmes around the world are today monitoring temperatures and estimating sex ratios to establish baselines, and estimate and predict the impact of climate change.

Another emerging paradigm in the 1970s was that of natal homing. Scientists were starting to believe that sea turtles returned to the beaches where they were born to nest as adults. Questions were beginning to be asked about how they did this, and it was believed that some imprinting might occur when they left the beaches as hatchlings. Was there a critical period when this imprinting occurred and was this affected by releasing them as one- or two-year-olds? Headstarting the turtles thus became a questionable conservation strategy. A few of the headstarted Kemp's ridleys have returned to nest in Texas in the last few years[53], but whether it increased their survival is still unknown. All this shows that conservation does need some biological knowledge.

There are, in fact, many concerns about in-situ conservation. Is the sex ratio of the population affected? Do we create feeding stations for the fish by releasing hatchlings at a regular time or place? Does the decrease in hatching success of hatchery clutches affect the population? Does developing in the hatchery environment decrease the fitness of the hatchlings? And, by protecting a particular kind of clutch, that is those below the high-tide line or in vegetation, are we in fact distorting the gene pool of the species? Almost all hatcheries show a reduction in hatching rates (probably due to translocation and micro-environment in the hatchery), but often translocated clutches are those that have no chance of survival in the wild.

Nicholas Mrosovsky, in his classic book, raised a number of these and related questions about sea turtle conservation approaches as far back as 1983[54]. Though there are no definitive answers yet to these important questions, hatcheries continue to play an important role in sea turtle conservation programmes around the world. After all, hatcheries also serve another purpose for conservation, they provide a platform for education and awareness which has fairly far reaching impacts.

The accidental icon

In previous chapters, I reflected on the use of sea turtles as flagships, and wondered if they had really served more as gunboats. What has four decades of sea turtle conservation in Chennai achieved? Numerous people have expended considerable effort, sometimes over years, to keep the programme going. The sum total of this effort in terms of direct conservation action is the release of a few thousand hatchlings each year. Given the low survival rates of sea turtles from hatchling to adulthood, this represents just over a hundred adults in a couple of decades. Considering the simultaneous threats to adults from fisheries and other sources, it does not seem likely that the hatchery programme does that much to conserve ridleys on the coast of Chennai.

On the other hand, over the years, thousands of people in the Chennai area have been on a turtle walk, some have had the good fortune of seeing a nesting turtle, helped relocate nests, seen hatchlings, held and released them, and felt a connection to nature, to oceans, to the distant seas that these turtles come from. More importantly, many students have gone on to careers in ecology, wildlife biology, environmental activism, conservation and eco-tourism. During the 1980s and 1990s, almost every wildlife biologist in the country had got a start at a WWF camp (these were held in sanctuaries and national parks) or at least attended and enjoyed one. The number of active ecologists, environmentalists and conservationists who have been on a turtle walk or participated in running SSTCN for a season or two, is similarly impressive. These turtles have helped motivate and inspire young ecologists and conservationists to work with turtles or other species of wildlife elsewhere. In addition, a number of other NGOs have sprung up around the country, many inspired or motivated by the example set by the sea turtle conservation movement in Chennai.

Perhaps the one criticism that can be directed at the SSTCN is that they have not been greatly successful in involving the local fishing communities. As students of urban schools and colleges, the target remained an urban, English-speaking audience in the early years. Even though Tharani Selvam, one of the founders, was from a fishing community himself, the engagement with fishers was not extensive or sustained. That role was taken on by another citizens' organization, TREE Foundation, which works with youth from the fishing villages along the Chennai coast. The youth patrol and monitor their beaches, protect nests both in-situ and through relocation, and spread awareness about the need for sea turtle conservation.

Given the threat from coastal development and fisheries to the nesting beaches and turtles, the long-term prospects for the olive ridleys of this coast are far from certain. Yet the students and other volunteers are happy to go out year after year, and work themselves raw, often combining night walks with day jobs, or classes, to keep the programme going. Surely, the mystery and charm of marine turtles has also something to do with the numbers that are attracted to the endeavour. In conventional (largely terrestrial) conservation prioritization, small populations that have little chance of survival (or small forest patches) are often discounted, as they are believed to have little value. But the olive ridleys of Chennai have made an enormous contribution by inspiring support for conservation, even though they might not themselves survive. In that sense, they may have been the best flagships of all.

Notes

i After a dismal second season, the third season (1990-91) was a veritable bonanza. More than 200 nests were collected, partly because there was an increase in nesting but also because additional coastline was covered[55]. However, the downside was extremely low hatching success (less than 50 per cent). In general, hatching success is lower in hatcheries, but such low survival rates were particularly discouraging. We noticed that hatching success was particularly low in large nests and towards the end of the season. As students of zoology, we attempted to logically work out what the problem might be, but the lack of access to literature (not that there was much work at that point), and the lack of formal training in research were constraints. Even so, we hypothesized that high late season temperatures and crowding in the hatchery may have contributed to this mortality. That year, we experimented with shading nests and with splitting large nests, which was somewhat successful. After discussions within the group, it was decided that the spacing between nests would be increased, and a particularly large hatchery was constructed the following year. In 1991-92, a large number of nests (about 195) were collected again, but this time, spacing was doubled[56]. Hatching success went up from 40 per cent the previous season to over 80 per cent. Something was working. Of course, in the absence of controlled experiments, it is hard to say whether spacing by itself was responsible, but recent studies have shown the effects of crowding on hatching through oxygen deprivation, carbon dioxide accumulation and spread of microbes and pathogens[57].

SMALL BEACHES, BIG BATTLES

Submarine standoff

THE DECOMMISSIONED NAVAL SUBMARINE, *Kursura*, sat on the sand like a beached whale. We waited with a largish crowd to get entry into this tourist attraction. My companions, Pradeep Nath of the Visakha Society for the Prevention of Cruelty to Animals (VSPCA), and his colleague warned me not to call them by their names once we were inside the submarine. Odd as this request seemed, there was a reason. Pradeep and Swati had protested long and hard against the plan to set up this decommissioned submarine for tourism on the Visakhapatnam beach. They had claimed that it would disrupt the nesting of olive ridleys on the beach. While they sought support at the international level, they were disappointed that many sea turtle biologists such as I had refused to back them. I had argued, in email correspondence and in articles, that this was not an important beach, given the much larger nesting beaches on the east of India, particularly in Odisha. What's more, the northern Andhra coast itself had many more important nesting beaches, and this one was right in the middle of the city, with considerable disturbance from vehicular traffic, lighting from beach-front apartments, sewage and tourism at Ramakrishna beach.

Tempers flared, and one of their avid supporters, Lily Venizelos, a veteran of sea turtle conservation campaigns in Greece, accused us of being deskbound bureaucrats with no knowledge of on-ground reality.

Time passed, Lily and I met at many sea turtle symposiums, and while agreeing to disagree on this issue, became good friends.

Meanwhile, the submarine project went ahead, Pradeep continued protesting, and became embroiled in a series of controversies. He was harassed by various authorities, and attained a certain notoriety. Little wonder then that he did not want to be recognized in the closed confines of the submarine. Once you get into the tube, it's a fairly long narrow passage, and with people ahead and behind, the opportunities for escape are limited. I thought he was being a bit dramatic, but since I had persuaded him to make the trip to humour me, I was willing to let this pass. I was in Visakhapatnam for a sea turtle workshop which the Wildlife Institute of India (where I was then working) had organized. Pradeep attended the workshop and we were able to share our philosophical approaches to conservation without coming to blows. We enjoyed an uneventful visit and bonded sufficiently that we agreed to keep in touch with regard to our work on turtles.

Pradeep's campaign against the submarine is one of the many campaigns that small sea turtle conservation groups around the country have launched, some successful and others not. What is more interesting than the campaigns or their success is the diversity of actions and approaches. Some, like VSPCA, are strongly driven by animal welfare motives, others by species centrism, and a few by community conservation. In this chapter, I explore the history of these groups and examine how a variety of groups and individuals, with their varying philosophies and ideals, approach conservation.

Arabian nights

Sea turtles and whale sharks in Gujarat

Gujarat has one of the longest coastlines in India (roughly 1,600 km) with two bays, the Gulf of Kutch and the Gulf of Khambhat (Cambay). Sea turtles in Gujarat were first surveyed by Satish Bhaskar between 1978 and 1981[i]. On his first visit in June–July 1978, Bhaskar visited Bhaidar Island, one of thirteen uninhabited islands in the Gulf of

Kutch, and noted the nesting of olive ridley turtles[1]. At a nearby creek, fishermen caught and released some green turtles; however, they were often killed for fat to be used for caulking in fishing boats (as in Lakshadweep), and occasionally flippers would be hacked off to make shoes[2]. There was egg collection by the some communities on both the Saurashtra and Kutch coasts. Though turtle meat was not widely consumed, there was a small market at Mithapur, where turtles caught accidentally in trawl nets were butchered and sold, mainly for their apparent medicinal value.

In August 1981, Bhaskar returned to Bhaidar Island, and as usual, the trip involved some adventure[3]. He hired a three-man crew and set out in a fishing sailboat from Beyt Dwarka Island, a journey of six hours to Bhaidar. He carried rations for five days including two large cans of water, since the island had no potable water. Arriving at low tide, he had to carry his rations more than a kilometre, wading through knee deep mud. And though his rations ran out early, his sailboat crew arrived and they left, leaving in his words, 'Bhaidar to its rightful owners – the turtles, birds and crabs.'

In May 1990, a green turtle with a tag was captured by a fisherman off Bhaidar Island. The information was communicated through a researcher in Saurashtra University to Fehmida Firdous, who managed the sea turtle project in Pakistan. Firdous reported that she had tagged the turtle in September 1989 at Hawke's Bay beach in Karachi[4]. This programme had tagged over 1,000 green and olive ridley turtles in the 1980s, and went on to tag many more, and has now monitored the beach for three decades. The turtle had travelled over 300 km to the Gulf of Kutch to feed, demonstrating that sea turtles have little respect for political boundaries, even contentious ones.

In 1987, Jack Frazier visited Gujarat for a few days; Dr and Mrs Frazier proceeded from Sasan to Mangrol by public transport and visited the beaches there[5]. South of the Mangrol port, there was a sandy beach with a forest department planted casuarina plantation. On both beaches, there was evidence of green and olive ridley turtle nesting. He also visited the hatcheries, one at the forest range office and another inland of one of the beaches in a coconut plantation.

The hatcheries were constructed with metal fencing and were heavily shaded. Frazier likely thought of the impact of sex ratios when he saw this and commented on it in his recommendations. He then reported in detail on two dead green turtles that he encountered at Okha and Beyt Dwarka, noting the predominantly algal contents in the stomach of one of them[6].

On this trip, Frazier also visited the Amateur Rangers Nature Club, Mithapur, having heard of their programme to monitor nesting sea turtles[7]. Satish Trivedi, founder of the club, escorted Frazier on a visit of Tata Chemicals. Notable naturalist Lavkumar Khachar had produced a report on the conservation potential of the land owned by the company, which included a coastal stretch with green turtle nesting. Frazier produced a report recommending the establishment of a bird reserve and a coastal reserve, including monitoring and management of marine turtles. He noted that while the 'Tata family of companies is famous for its humanitarian contributions' and their list of educational facilities and communities was impressive, they were 'curiously' not known for their advances in environmental conservation and awareness.

A few years later, in 1991, I visited the Gir Forest National Park and Wildlife Sanctuary, and heard about the hatcheries in Mangrol, not far from Junagadh. I undertook a brief journey in an old ambassador taxi with nine other passengers, all wedged in somehow. The enclosure was still there, some distance from the beach, and not particularly well maintained. The range officer who was in charge did not seem to know too much about the sea turtles that nested on that coast, but the staff seemed sincere. Frazier's coaching a few years earlier appeared to have failed to transcend the language barrier. Or perhaps the staff had changed.

In the 1990s, information on sea turtles in the state came mainly from researchers and environmental educators. One of these, Pranav Trivedi, a graduate of the Wildlife Institute of India, worked with WWF-Gujarat for close to a decade. Trivedi ran education camps at a number of sites on the Gujarat coast, including Beyt Dwarka and other islands, including Pirotan, and recalls seeing sea turtle nests and carapaces. E.K. Nareshwar, who earlier worked with one of the many fly-by-night groups on the Chennai coast, was working with the Centre for Environment Education in Ahmedabad, and conducted education and awareness programmes at Beyt Dwarka. Naresh reports, in 1998, that the island had nesting of both green turtles and olive ridleys, and that eggs were depredated by jackals and wild boar[8].

In the early 2000s, Wesley Sunderraj and Justus Joshua of the Gujarat Institute of Desert Ecology, in Bhuj, undertook a systematic survey of sea turtles on the Gujarat coast[9]. Apart from engaging with the forest department staff and local fishermen, Sunderraj and Joshua also attempted to involve NGOs in the programme. This included Satish Trivedi of the Amateur Rangers Nature Club in Mithapur, who had escorted Jack Frazier many years before. One of the groups they encountered was the Kachba Unchar Kendra in Madhavpur, a small group of individuals who had shown great dedication in running the sea turtle hatchery at this site.

Another group was the Gir Nature Youth Club, with Amit Jethwa at its helm. Jethwa was an environmental activist who was fighting forest encroachments and poaching in Gir. He was instrumental in fighting many cases against illegal mining, but also in the page-three incident involving the shooting of a blackbuck by a Bollywood movie star. Jethwa persisted despite threats and assaults, but was tragically shot in 2010 by gunmen, allegedly paid by the mining lobby. While conducting environmental education programmes, Jethwa met Dinesh Goswami, a daily wage worker at Ambuja Cements Limited in Kodinar. In the late 1990s, Goswami started travelling the coast out of interest. He noted that there was a collection of eggs for sale in markets, both for consumption and for use as cattle feed (again, apparently for medicinal purposes).

Jethwa encouraged Goswami to start pursuing sea turtle conservation in his area, near Veraval along the Junagadh coast. Goswami and his friends started patrolling the coast on a daily basis. Their programme took a more definitive turn after a particular incident. Having learnt of the take of turtle eggs at a nearby village, Goswami visited the village masquerading as a cattle owner wanting to purchase eggs. On finding the culprit, they took him to the local range office, which marked

Goswami's introduction to the forest department. However, when the man was arrested and taken to court, the hapless judge did not know what a sea turtle was, let alone what to do about the offence. Jethwa, who was present, prompted a forest official to give the judge a copy of the Wildlife (Protection) Act, 1972 to determine the punishment for offences involving a Schedule I animal. At this juncture, Goswami and others realized that the punishment was fairly severe (it can include imprisonment), and persuaded the judge to order the accused to carry out two years of service in walking along the coast and spreading awareness about sea turtle conservation. And so Tappu Bhai became one of the first volunteers of Goswami's group.

Like other groups along the coast, Goswami and his partner, Jignesh Gohil, worked with little knowledge to begin with. When they found a nest, they would mark it and record the date of nesting, sometimes on a 'washed-up rubber slipper' and bury this near the nest. Not knowing when the nest would hatch, they would monitor it till they saw hatchling tracks. Unable to locate the eggs, they started erasing tracks on the advice of some WWF personnel they had met. Goswami also persuaded members of the community, who frequently combed the beach for all manner of things that washed ashore, to look out for turtle tracks and inform him.

Goswami formally established his organization, initially called Prakruti Parivar, in 2004. He had also become involved in the conservation of whale sharks, inspired by the film made on these animals by Mike Pandey. Despite his position as a daily wage employee, Goswami believed that his constant presence in the area gave him an advantage over other environmental activists who visited briefly and left. In fact, he believed that the fact of his being a poorly paid employee conducting a pro-bono environmental campaign sent out a message. Indeed, the activities of Prakruti Nature Club (PNC) became well known throughout the region, and even in his own company, which began to support his efforts. PNC began to employ volunteers in several coastal villages. They also approached the forest department and offered to help them maintain the Madhavpur hatchery with their volunteers.

PNC has established a strong rapport with the fishing communities along the coast. Much of this likely has to do with how they approach issues. Goswami says that when they find an offender, they take him to the village chief rather than the forest department. He said, 'It wins over the trust of the villagers, including the key decision makers in the village. And because of this, people are not usually opposed to us working in their area because they know that we're not going to get them into trouble with the police or forest department.'

Over the last decade, PNC has rescued over 200 whale sharks, numerous snakes, conducted migratory bird censuses, and released over 20,000 sea turtle hatchlings[10]. Through engagement and perseverance, PNC has managed to get the support of the forest department at all levels, quite an achievement for a community-based NGO. They have managed to establish a relationship with many of the industries along the coast by trying to work with them, and have received support for their work. About his own company, he says that they could easily have fired him for taking time off, but they saw it as their contribution and have become quite proud of their involvement. Ambuja Cements was one of the supporters of the sea turtle symposium in Goa.

Friends of the turtles in Maharashtra

Maharashtra has a coastline of 720 km. Little is documented about the sea turtles of this coast, possibly because Satish Bhaskar did not conduct surveys there. However, there are many brief notes in the *Marine Fisheries Information Service*, including of a leatherback[11]. The first of these, in 1981, documents the capture of a hawksbill turtle in a drift net, near the Elephanta caves, south of Mumbai[12]. The turtle was brought alive to Trombay, and held in captivity 'in the ante-room' of a large house belonging to the fisherman, who 'regarded the capture of the turtle as auspicious, and labelled it as "Sea God", duly worshipping it by the offering of rice and vermillion'. Needless to say, the turtle declined this unfamiliar diet, was considered to be starving itself in protest, and was released a few days after capture.

At the CMFRI conference in 1984, Kafeel Shaikh of the forest department of Maharashtra provided a remarkable account of sea turtles in the state[13]. Unfortunately, this has remained somewhat obscure. As deputy conservator of forests of Thane, Shaikh carried out a survey of the Maharashtra coast, interviewing old fishermen and 'specimen collectors' of the fisheries department, and documented nesting sites of green and olive ridley turtles in various districts. Nesting sites were also recorded in Mumbai itself and its suburbs, quite proximate to locations that were already heavily used by the public, such as 'Chowpati and the point of road diversion to Malabar Hill and Babulnath Mandir'. Shaikh also documents the consumption of eggs, which he says were made into omelettes or cakes with coconut milk, jaggery and sugar. Only the yolk was used for human consumption, and the whites were fed to cattle.

In 1992, Bhau Katdare and colleagues started the Sahyadri Nisarga Mitra (SNM) in Chiplun in Ratnagiri district in Maharashtra. The organization had about fifty members and worked mainly on birds, and their activities centred around education and awareness programmes. However, interest waned over the next few years, and in 1996, the group decided to renew interest by conducting status surveys for the white bellied sea eagle on the Maharashtra coast. Walking the beaches each Sunday, they managed to survey the entire Maharashtra coast. And in 2002, quite by chance, they discovered sea turtle nesting in Velas in southern Maharashtra. Due to some smuggling activity, the police had issued a curfew at night, and the beach was deserted when one of their members encountered a track.

Around this time, a systematic survey was carried out in 2000-01 by Varad Giri, of the Bombay Natural History Society as part of the GoI-UNDP Sea Turtle Project[14]. Quite independently of this effort, Katdare and colleagues collected information about sea turtles and discovered that there were a fair number of ridleys nesting in the region. Around this time, Katdare also visited Chennai and met with members of the SSTCN as well as with me. In late 2002, SNM formally initiated their sea turtle programme. Given the distance of Velas from their headquarters, Katdare realized that they would need the support of the local community if this operation was to work. While the community was sceptical at first, they were gradually swayed by the commitment of this group that would return time and again to talk to them about sea turtles and their importance. Unlike other villages in Maharashtra, there was no killing of adults here, just collection of eggs. Eventually, supported by the village headman, one person was appointed to monitor the beach and paid an honorarium by SNM. A small hatchery was built and fifty nests were translocated during the first season.

More and more people in the village community were intrigued by this programme. And one day, when hatchlings began to emerge, Katdare recalls: 'The entire village assembled there to witness it, they had never seen hatchlings emerging before. Even some of the older people, probably in their sixties or seventies told us about how they were collecting eggs since they were children, but had never once seen hatchlings emerging out of the nests and making their way to the sea.'

Over the next few seasons, SNM conducted awareness programmes in villages on the Ratnagiri coast[15]. With a little media publicity, the support for sea turtle conservation started to grow. SNM then started expanding their programme along the Maharashtra coast. Typically, most nests were collected by humans, and in many areas, nesting turtles were killed after they finished laying eggs. As in Velas, there was resistance to conservation in many villages, and little awareness of the law. SNM started by co-opting the egg poachers, and assuring them a small monthly income if they protected the nests instead of selling the eggs.

In 2004-2005, SNM instituted a Kasav Mitra Puraskar (Turtle Friend Award). The first award was given to head of the village panchayat in Velas. The following year, an award for NGOs was also instituted. In 2007, SNM conducted their first turtle festival which included awareness programmes, screening of films and other activities to publicize Velas as a turtle nesting site. Buoyed by the success of the event, SNM helped form a Kasav Mitra Mandal (Friends of the Turtle) in the village, to take the lead in organizing the festival and in sea turtle conservation activities. The village benefits by providing accommodation to tourists who attend the festival and visit the nesting sites. The members of the Kasav Mitra Mandal set aside 10 per cent of their total income during the festival for marine turtle conservation in Velas. In one of his frequent updates in *IOTN*, Katdare wrote that 'nearly 250 visitors comprising of housewives, school and college students, doctors, engineers, journalists, photographers, social workers, nature researchers and teachers attended the festival in 2008'[16].Over 400 tourists visited the village during the festival in 2012.

Meanwhile, SNM increased awareness across the coast by printing posters, pamphlets and articles in local newspapers. They organized cricket tournaments (Kasav Chashak) at Velas and Kolthare and gave out a 'turtle trophy' to the winner. In 2008, SNM published a booklet on marine turtle protection and conservation in Marathi for distribution to their volunteers and others.

Over the years, SNM has surveyed nearly 180 villages on the Maharashtra coast and initiated sea turtle conservation programmes in thirty-five villages. They have also worked closely with the state forest department, and in 2012, initiated a collaborative sea turtle conservation programme in twenty villages, funded by the environment department's Green Ideas programme. They have received several awards for their work in Maharashtra, including the Vasundhara Mitra Award to Bhau Katdare at the Vasundhara International Film Festival in 2008.

Tourists and turtles in Goa

Goa on the west coast of India has many sandy beaches (roughly 160 km), which, apart from providing nesting habitats for olive ridley turtles, are a major attraction for tourists. Many well-known beaches such as Calangute, Miramar, Colva and Majorda are popular amongst Indian and foreign tourists, and are highly developed with hotels and resorts. Bhaskar, who eventually moved to Goa after retiring from active sea turtle research, provided some notes about sea turtles along the Goa coast[ii].

In the early 1990s, Captain Gerard Fernandes moved back to his native Morjim after taking early retirement from the armed forces. Fernandes was distressed at the changes in his village and along the coast due to rapid development and environmental degradation. Comprising both Hindus and Christians, the small fishing community there consisted of about a 150 families who were also involved in agriculture, toddy tapping and liquor distilling, particularly feni, the local brew usually made from coconut or cashew. While fishing remained the major occupation, many traditional fishermen who once owned boats worked as labour on mechanized trawlers due to the depletion of fish stocks. Due to the lack of fishing income, many families began to erect tourist shacks on the beach.

Concerned with the collection of eggs and sale of meat in markets, Fernandes decided to initiate sea turtle conservation in this area[17]. Helped by his wife, Merle, bricklayers Domio D'Silva and Prakash Saptoji, and former fishermen, Dominic and Gilbert Fernandes, a conservation initiative was launched involving the local communities in 1995. Initially, Fernandes and his wife spread awareness within the local community about the importance of sea turtles. Fernandes also met Claude Alvares, a noted environmental activist in Goa, and was advised to involve local communities in the conservation programme. Nirmal Kulkarni, now a well-known Goan conservationist, was then in school and assisted them with the programme as well.

Working with the forest department, Fernandes's group persuaded local youth that they could gain income from the turtle nesting beaches by protecting the eggs and using the sea turtles and their hatchlings to attract tourists, especially foreigners. Despite facing a number of challenges, including a lack of amenities on the beach, and having to pay licence fees to the Department of Tourism, several individuals became involved in this programme. Individuals patrolled the beaches at night and escorted tourists onto the beach to observe sea turtle nesting. When a nest was laid, forest officials were notified and a protective fence was erected around the nest, accompanied by a sign that indicated protection by the forest department.

Initially, financial incentives were provided for not poaching turtle eggs by Captain Fernandes, but as the movement gathered momentum, there was disapproval within the community for poaching. This achieved even greater success as senior members of the community stopped collecting eggs. The attention received by turtle tourism attracted other members of the community who started setting up temporary shacks on the beach. While these were initially accompanied by loud music and lighting, they rapidly realized that this was actually deterring the tourists who visited the beaches to see turtles. Local residents also provided accommodation to tourists in their houses, for which separate rooms or floors were added. Captain Fernandes was recognized for his efforts by the Government of Goa in 1998.

The turtle conservation programme caught the attention of the local media as well as the state forest department, including forest officers such as C.A. Reddy, Richard D'Souza and Paresh Porob (then a volunteer and now a range forest officer with the Goa Forest Department) who enthusiastically supported these efforts. In 1996-97, the forest department became involved in the programme by deploying two guards to patrol the beach during the nesting season and assist the village youth in deterring depredation of eggs. Later, the department started paying some of the youth on a regular basis. A Turtle Study Centre was set up at Pernem within the campus of the range forest office.

Hearing about poaching from a local journalist at Galgibaga, Kulkarni and his colleagues, Harvey D'Souza and Neil Alvares, then visited the area and found extensive collection of eggs. The forest department, at this time, also established a protection programme at Galgibaga. About fifteen to twenty nests were protected at each site. By the early 2000s, there had been some success, but cracks were beginning to show. As a consequence of the attention, the beach had attracted not just tourists but also local politicians and bureaucrats, who started developing local infrastructure such as street lights which would have negative impacts for nesting. As in many tourism-based hatcheries (in Sri Lanka, for example), hatchlings were retained in basins for the tourists to see, which is detrimental to their survival. Most importantly, competition between shack-owners and differing perspectives – conservation versus economic – had started creating a rift within the community. The programme largely declined in the 2000s due to lack of support and internal conflict within the community. The forest department has continued off and on to operate hatcheries at Morjim and Galgibaga, two of the principal nesting sites for olive ridleys in Goa. In 2007-08, the collaboration between local communities and the forest department at Morjim was revived through the efforts of the Centre for Environmental Education in Goa[18]. Local shack owners agreed to help the forest department in patrolling and protecting nests, and also in implementing turtle-friendly measures such as reduced lighting.

Over the last five years, the beach of Mandrem has received some attention, with Denzil Sequeira, a well-known fashion photographer leading the conservation initiative with support from the Goa Foundation. Similarly, Agonda beach in south Goa has good nesting and is protected to some extent by local communities and forest department staff, mainly through the efforts of Porob when he was posted in the region.

Small coast, small beginnings in Karnataka

The Karnataka coast has received scant attention over the years. However, even though it is relatively small (about 260 km), it has some of the prettiest beaches on the Indian mainland, with hills undulating down to the coast. Even Bhaskar had little knowledge of this coast. By 1984, at the CMFRI workshop, Bhaskar had heard from CMFRI that there were long stretches of sandy beach at Karwar, but he admitted having 'no first-hand' data from this state[19]. The assistant conservator of forests of Karwar attended the meeting, and though he had little information, he committed to returning to the state and developing a sea turtle conservation programme[20].

At the invitation of the chief wildlife warden of the state, Jack Frazier visited the Karnataka coast and the turtle hatcheries at Mangalore, Kundapur and Bhatkal in March 1987[21]. He was told that the Karnataka forest department had run nine hatcheries the previous year but three were in operation that year[22]. In Bhatkal, the range forest officer, Vasant Kulkarni, informed him that the hatchery programme had been initiated in 1984 (perhaps in response to the information that had filtered back from the CMFRI workshop). The forest department staff had popularized marine turtles with local fishermen, using the popular ruse that reduced fish catch was related to decline of sea turtles, and that helping one would help the other. The programme had evolved towards an in-situ conservation programme with protection in the wild using metal enclosures, and only some nests were transported to the hatchery. Kulkarni also told Frazier that the only way to ensure sustainability of the programme was to gain the full cooperation of local residents and communities, as forest department staff were continually being transferred. Frazier was impressed by the knowledge and attitude of Kulkarni. Nearly two decades later, an NGO was initiated not far from Bhatkal, in Honnavar, and one of the key members of this group was a local employee of the forest department.

The Canara Green Academy (CGA) was formed in 2005 with the mission of conserving the flora and fauna of the north Kanara district. Led by N.D. Bhat, Ravi Pandit and Sankara Hegde, the organization became involved in the conservation of mangroves, medicinal plants and sea turtles[23]. In 2006, CGA collected information on sea turtles along the north Karnataka coast, and established temporary hatcheries at a number of sites. They carried out both ex-situ and in-situ conservation, depending on the safety of the nests, with the help of local communities. They also instituted an honorarium for those who informed them of sea turtle nests as an incentive to promote conservation and the participation of local people. Word started to spread and they slowly began to receive information about poaching of eggs. With the help of the forest department, CGA would visit the site and collect the eggs from the poachers.

CGA was also involved in the rescue and release of juvenile green and ridley turtles in Honnavar in 2006. A green turtle had been caught accidentally in a fishing net about a year-and-a-half earlier, while the ridley had been obtained from the hatchery at Gangavali. The range officer, Anand Udar, had been instrumental in setting up the rescue centre. A family from Gangavalli village had looked after the animals. Pandit wrote[24]:

> Members of our group, the Tandel family and I watched along with dozens of curious locals as the turtles moved seawards. On reaching the tidal area, Gulabi Tandel approached the turtles, put her hand on their carapaces and said, 'Both of you take care.'

CGA hosted the third annual meeting of the Turtle Action Group meeting in Kumta in 2010, inviting a number of the local forest department staff who participated enthusiastically in the meeting.

Meanwhile, another group was getting somewhat involved with sea turtles on the Kundapur coast. Field Services and Learning (FSL) was established in 2000 to provide an opportunity for foreign youth to participate and volunteer in community service and conservation[25]. The programme offered these youngsters a chance for interact directly with local youth, women, farmers, fishermen and other sections of society. One of the options was to work with sea turtles on the coast. The international volunteers and FSL staff conducted awareness programmes at local schools, including art exhibitions, theatre performances, and so on. In addition, the volunteers built temporary information centres and hatcheries. FSL currently works in about seventeen villages along 60 km of the northern Udipi coast, and plans to extend their work southward to the Kerala coast.

The turtle people of God's own country

Kerala has a 590-km-long coast, with relatively narrow beaches, interspersed with rocks, seawalls, inlets and backwaters. Though sea turtles were consumed regularly on the southern Kerala coast, little information was available apart from a few sporadic reports. Bhaskar had noted much of the coast was walled and unsuitable for nesting from the 1980s onwards, and both turtles and eggs were taken for local consumption[26]. Olive ridley turtles were reported to nest frequently along the coast, but there were also more leatherback records here than from other parts of the coast. Green turtles were also reported on the coast south of Quilon (now Kollam)[27].

Jayakumar and Dileepkumar carried out a survey as part of the GoI-UNDP project in 2000-01, and documented some nesting sites along the coast, mostly near river mouths[28]. They reported that Christian communities throughout the coast consumed turtle meat, particularly those in southern Kerala. In fact, this is one of the few places in the world where leatherback meat is consumed[29]. There is no evidence of a turtle fishery though, and this was likely to have been opportunistic take of accidentally caught or nesting turtles. Meat consumption was also linked to migrant fishing communities from southern Tamil Nadu. Muslim communities in northern Kerala generally did not consume turtle meat.

Interest in sea turtle conservation in Kerala dates back to at least the early 1980s, when the Calicut (now Kozhikode) branch of the Kerala Natural History Society recommended to the state forest department that the 'coastal stretch between Ponnani and Tellicherry' (or Thalasserry on the north Kerala coast) be protected[30]. In the 1990s, Jayakumar briefly led an organization called Marine Turtle Conservation Action, which aimed to promote community conservation[31]. One of the groups they interacted with were soon to become well known for sea turtle conservation in the state. This group of young fishermen in north Kerala, led by Surendra Babu, formed the Theeram Prakriti Samrakshana Samiti (Coastal Ecosystem Protection Committee) in 1992 in Kolavipalam, Kerala[32]. The idea originated from an article in a newspaper, *The Hindu*, about olive ridley turtles. Realizing that the marine turtles which nested on their beach were the same species, they formed a group for the protection of these turtles.

Kolavipalam is a small fishing village near Payyoli, in Kozhikode district, north Kerala. Much of the 8 km beach was walled to protect against erosion and only a one-kilometre stretch near the river mouth remained fully accessible. Although fishing predominated, several fishermen had given it up or supplemented fishing with other sources of income. This included self-employment as trained electricians, auto-rickshaw drivers and casual labour.

During the nesting season, the group members monitored the beach each night for nesting turtles and relocated nests to a hatchery. Before the involvement of the forest department, funds for paying a watchman to guard the hatchery were generated by donations in cash and kind from within the group and the community. Initially, the group was ignorant about even the basic biology of sea turtles. For example, Theeram members kept watch over the eggs for about thirty days (assuming that it took the same time to hatch as a poultry egg) and then extracted an egg every week to check the stage of development as well as to ensure that the eggs were still alive. Over time, their knowledge improved and they became more familiar with sea turtle nesting biology. They even began to recognize some of their 'regulars', in particular a female with a damaged flipper that always nested at the same spot.

The group finally received support in 1997 from the divisional forest officer, Prakriti Srivasatava, who encouraged the local youth to keep watch over the beach through payment of daily wages to some of the members during the nesting season from October to March. The members pooled their wages and used it to fund Theeram's conservation activities. The members met in a small thatched hut that was constructed with financial aid from the forest department with exhibits of turtles, hatchlings, and their protection efforts displayed on the walls. Media publicity about Theeram's protection efforts helped spread awareness among the neighbouring coastal villages. Fishermen from other villages called Theeram members or brought turtle eggs to Kolavipalam for protection. Theeram members encouraged these fishermen to protect the nests in situ and advised them on protection measures.

Theeram also became involved in a campaign against sand mining in the nearby estuary, initiating legal action at considerable expense. Despite legal restrictions laid down by the Kerala High Court as well as by the state government, sand mining continued, although not as openly as before. As a result of Theeram's petition appealing for a ban on sand mining in the Kottapuzha estuary, the Kerala High Court issued orders that restricted extraction of sand from the estuary. A total ban on sand mining in inland rivers was also issued by the courts in 2001. The failure of implementation of these orders led to repeated legal court battles between the sand mining lobby and Theeram, and their members faced many threats and attacks from the mining lobby.

Due to the problems of coastal erosion, Theeram realized the importance of mangroves in protecting habitats. They used turtles as charismatic icons to engage students in environmental issues and to motivate them to participate in the mangrove reforestation. The afforestation programme of mangroves in about five acres of the estuary was initiated in 1998 in collaboration with the forest department, who supplied Theeram members with the initial batch of mangrove seedlings and initiated a ban on cutting the original mangrove trees.

Theeram's work received both general acclaim and local support. The community felt that the quality of water in the village and fish catch in the estuary had improved thanks to the mangrove afforestation programme. Theeram was awarded the P.V. Thampy award in November 2000 for environmental protection through community participation. A film *Aamakaar: The Turtle People* was made to document their history and efforts. On the other hand, there were also challenges such as the lack of financial and human resources. Fighting the sand mining lobby continued to be a losing battle despite some legal victories. The group chose to restrict membership to prevent political groups from becoming involved and taking over their agenda. However, this meant that a relatively small group had to shoulder responsibility, including patrolling the beaches at night during the nesting season. When Roshni Kutty from Kalpavriksh carried out her study in 2001, the community was still optimistic. Though they had been limited by various factors, they felt they had been able to use turtles to motivate the community and promote environmental awareness. One Theeram member said[33]: 'We began by saying that we are protecting the turtles. Now, it is the turtles that are protecting us'.

Recently other groups have also sprung up along the coast such as Neythal in Kasargod on the north Kerala coast[34]. Sudheer Kumar, one of the leaders of this group, is a school teacher. Neythal was formed in 2001 by a team of youth at Thaikadappuram village to address environmental problems, in particular coastal conservation. Neythal also initiated a hatchery programme in 2002-2003, after seeing the success of Theeram. Neythal has also been involved in other environmental campaigns and received the P.V. Thampy Memorial Environmental Award in 2004 for its work on sea turtles.

Green Habitat was formed in 2002 by a group of high school science teachers in Guruvayoor[35]. Working on environmental conservation in Chavakkad district, this group carried out projects on mangroves, birds and sea turtles. The turtle conservation programme was run primarily in collaboration with Seethi Sahib V.H.S. School in Edakkazhiyur whose National Green Corps members served as the main volunteers. Most of these members were from the local fishing community. In 2010, Green Habitat started the *Kadalamakale Samrashikkuka* (or 'Save the sea turtles') programme. The group works closely with schools, clubs and local government bodies. It aimed to establish a hatchery and a turtle exhibition centre and make the beaches of the district 'turtle friendly'.

Conservation and conflict in the coral atolls

As usual, our knowledge of sea turtles of the Lakshadweep comes from Satish Bhaskar's visits. Bhaskar visited many islands on this trip, and as I described earlier, had himself dropped off for several months on Suheli Island, where he lived alone for five months. He had no way of communicating with the outside world, barring the assurance that the coast guard would occasionally look out for distress flares[36].

Lakshadweep is part of the Chagos-Laccadive ridge (which also includes the Maldives). It consists of thirty-six atolls, of which ten are inhabited and fourteen are uninhabited islands, while the rest are reefs, submerged banks and sand bars. Bhaskar counted over 200 green turtle nests in Suheli Valiyakara, one of six uninhabited islands where he found nesting; he also found some nesting on inhabited islands such as Kadmat, Androth, Agatti and Minicoy[37]. At the time, green turtles were harpooned in the lagoon on a regular basis for their fat, used as caulking for the boats. Prior to 1978, hawksbill turtles were also killed for tortoiseshell, which was sold to traders on the mainland. The population of these islands being almost entirely Islamic, there was, however, no consumption of meat or eggs. Even in the early 2000s, Basudev Tripathy found carapaces of green turtles in many uninhabited and seasonally inhabited islands[38]. He also found that juvenile hawksbills were killed and stuffed to be sold to tourists in Mangalore, Calicut and Cochin. Since fishing in the Lakshadweep is mostly pole and line for tuna, there has been little incidental mortality. The biggest threat to sea turtles and their nesting beaches comes from beach armouring, human habitations, and lighting.

In 2001, Tripathy spent nine months surveying the islands as part of the GoI-UNDP project[39]. Tripathy found some degree of nesting on all islands, a pattern confirmed by his interviews of local residents and fishermen. He largely confirmed Bhaskar's findings that green turtle nesting was higher on the uninhabited islands, with an order of magnitude higher nesting on Suheli (more than 300 nests) than the other islands.

When I visited Lakshadweep in 2001 to spend a few days with Basudev Tripathy, what was most striking was the extraordinary number of green turtles in the Agatti lagoon. Kavaratti did not have nearly as many. Nor did Bangaram, a mere stone's throw from Agatti, and being a small island with only a high end resort on it, with little disturbance. I did not visit many other islands, but Tripathy did, estimating the relative abundance of green turtles in lagoons. It was true, Agatti had a very large number of foraging green turtles. Adult males and females, sub-adults, juveniles.

Rohan Arthur was then completing his Ph.D on coral reefs in the Lakshadweep and not thinking about sea turtles at all. A few years later, continuing his research on coral reefs of the Lakshadweep, he became interested in the seagrass communities in the lagoons, and the 'ugly things' that ate them. He started monitoring the abundance of the green turtles and stumbled upon an interesting problem. Since the laws protecting sea turtles had been implemented, fishers believed that the numbers of green turtles had increased. Certainly in Agatti, one could stand on the jetty and look into the lagoon and see a turtle swimming or surfacing to breathe every few minutes. The fishermen were, by this time, fairly agitated by these large numbers of turtles in the lagoon. They believed that their declining fish catches were due to the turtles. Apart from occasionally tearing their nets, these turtles, they believed, were decimating the sea grass patches in the lagoon, and thereby depriving fish of feeding, breeding and resting sites. The fishermen believed that the turtle populations had been earlier kept in check by the occasional killing for their fat. Arthur was interested in exploring whether this purported ecological chain had any truth to it[40].

Arthur's surveys began in 2005, and they recorded large numbers in Agatti, spatially in the same locations as the sea grass and the fishing. Simultaneously, they found relatively low numbers in Kavaratti and Kadmat, the other islands where they worked on a regular basis. Aparna Lal, a Master's student from WII working with B.C. Choudhury, collaborated with Arthur and his spouse, Teresa Alcoverro, a seagrass biologist, and carried out a study on the effect of the green turtles on the seagrass. She found that these turtles were indeed modifying the ecosystem in a significant way, and that the meadows in Agatti were severely impacted or depleted[41]. Subsequent work confirmed that these turtles were causing species shifts in seagrass meadows[42]. In addition, Sruthi Kumar, a research scholar with B.C. Choudhury, initiated a research project on habitat use by green and hawksbill turtles and spent several months tagging turtles, but the project was closed midway[43].

While Arthur and other sea turtle biologists were considering this situation, the turtles decided to confound us. They moved, or are believed to have moved. Over the next few years, the population in the lagoons of Agatti fell drastically. Fortunately, no major conservation NGO was stationed here or many elegies would have been written about the impending extinction of green turtles in the Lakshadweep. Arthur and his team, including Nachiket Kelkar, a Master's in Wildlife Conservation from Bengaluru, conducted lagoon surveys, which showed counts of green turtles declining from over 500 to less than fifty within a few years. But at the same time, green turtles were on the rise in the lagoons of Kavaratti and Kadmat (and perhaps other islands as well). Note that in the early 2000s, Tripathy had noted these high densities at Agatti, but Bhaskar, with his keen eye, had not. Perhaps the green turtles were not hanging about Agatti in the late 1970s. Arthur's surveys, showing decreasing numbers at one lagoon and concomitant increases in other lagoons, indicated that the turtles had wiped out one seagrass patch and moved on to another one. Perhaps this was a cyclical process. Currently, Arthur, Alcoverro and their team are continuing their research on the impact of sea turtles on seagrass meadows, and the resulting conflict with fishing communities.

Conservation of sea turtles in the islands has mostly centred around the Lakshadweep Marine Resource Conservation Centre (LMRCC)[44]. Started by Jafer Hisham, and Idrees Babu in 2008, LMRCC comprised of young islanders who came together from various islands to promote community-based marine conservation. LMRCC's vision is to achieve 'a sustainably progressing Lakshadweep where marine ecosystems are healthy and well managed'.

Turtles by the Bay

Research and resources in Tamil Nadu

Tamil Nadu has a coastline of 980 km (including the Union Territory of Pondicherry), largely east facing (900 km) with a small stretch of west facing coast (80 km). Sea turtles have been documented on the Tamil Nadu coast for a long time, due to the turtle fishery in the Gulf of Mannar. Some of the earliest accounts of sea turtles in India come from this region. In southern Tamil Nadu, all the sea turtles have several common names, indicating a fair degree of interaction with people[iii].

Tamil Nadu is probably the one state in India where all five species (loggerheads, ridleys, greens, hawksbills and leatherbacks) have been recorded consistently[45], mostly in the Gulf of Mannar[iv]. Even here, loggerhead records are pretty rare. Though the main turtle fishery in the Gulf of Mannar declined due to protection, the take of turtles continued in a clandestine manner. In the 1960s, green turtles constituted 90 per cent of the catch (though this may represent preference rather than proportions). However, by the 1990s, green turtles constituted 65–70 per cent of incidental capture and in the market, and by the 2000s, there were roughly equal numbers of green and ridley turtles, and a few loggerhead and hawksbill turtles[46]. A more recent assessment of exploitation in the Gulf of Mannar also reported that greens constituted 60 per cent of the take, while ridleys constituted 35 per cent and hawksbills the rest[47]. Over the last few decades, there also appears to have been a clear reduction in the larger size classes, namely sub-adults and adults, likely a result of the exploitation[48].

Much has already been written about the activities of various NGOs on the Chennai coast. Elsewhere on the coast, S. Bhupathy of SACON helped establish two students' groups along the Nagapattinam coast during their 2003-04 survey, one at the TBML College in Tranquebar, and the other at Poompuhar. National Service Scheme (NSS) volunteers in the college helped spread awareness about sea turtles in the region.

The forest department was also active along this coast from the 1980s, when several hatcheries were established. In the early 1980s, hatcheries were established at Point Calimere and nearby sites and about 250 nests were collected each year[49]. The hatcheries were maintained by the forest department but also involved Abdul Rahaman and his colleagues at the Department of Zoology, AVVM Sri Pushpam College in Poondi. Rahaman and his colleagues conducted some studies on the growth of hatchlings, and clutch size variation[50]. The hatcheries were closed due to lack of funds in 1987, but were revived in 2000 by A.D. Baruah, the wildlife warden of Point Calimere Wildlife Sanctuary[51]. With the help of the sanctuary biologist, they constructed a hatchery and collected fourteen nests and released about a 1,000 hatchlings. Nearly a decade later, another wildlife warden, Velusamy reported that the forest department had set up hatcheries and olive ridley protection camps at several sites along the coast between 2005 and 2009[52]. The camps, manned by two staff each, helped protect the eggs from depredation and collected nests for hatcheries. Quite notably, this officer also removed the coastal casuarina plantations in a few locations as they were hindering sea turtles.

The other major contribution along this coast came from the TREE Foundation, led by the tireless Supraja Dharini. A sculptor by profession, she became interested in sea turtle conservation in the early 2000s. Supraja was also inspired by the ecologist Jane Goodall, and her organization is a part of Goodall's Roots & Shoots international network. Goodall visited Chennai and was the centrepiece of one of TREE's events in 2009.

TREE Foundation started by creating youth groups in the fishing villages along the southern Chennai coast, called 'Kadal Aamai Paadukavalargal' (Sea turtle protectors), or KAP. TREE started its campaign in 2002 with a few villages south of Nilankarai (where SSTCN's 'territory' ended)[53]. At a workshop at the crocodile bank, youth from two villages were trained in patrolling and hatchery maintenance by SSTCN members. Initially disillusioned at seeing a large number of dead turtles, the members saw their first nesting turtle in January 2003. A small enclosure was erected around the nest to protect the eggs. Seventeen nest sites were protected that year, and later in the season, KAP members rescued hatchlings from a wild nest as they moved towards a street light. The youth were sufficiently inspired to continue their work with sea turtle conservation. By 2003, TREE was covering several villages along the coast and their environmental education work had also expanded[54]. That year, they held programmes in eleven schools, reaching nearly 5,000 children. They also conducted a painting competitions, a sand modelling competition on the beach for fishing community youth, and street plays using folk theatre forms by college students about sea turtles, the environment and fishing communities.

Concerned by the large number of dead turtles on the coast (over seventy in the 2006 season), TREE and their KAP members decided to spread awareness about sea turtles in the Kasimedu fishing harbour, Chennai's largest harbour for trawlers and other mechanized fishing boats[55]. KAP members interacted with the unions to collect information about turtle mortality in fishing nets. As part of the coastal cleanup day organized by the coast guard in September 2006, KAP members cleaned the sea around the harbour using several catamarans and fibre boats, removing over two tonnes of marine debris. This helped the KAP members establish a rapport with the unions. In 2012, TREE conducted another meeting jointly with the wildlife wing of the forest department, fisheries department, Chennai Trawl Mechanized Boat Fishermen Welfare Association, and the Indian Coast Guard[56]. Sea turtle stickers with rescue and release methodology were stuck on more than 2,000 trawl boats.

TREE has now expanded its activities to over 100 km along the Tamil Nadu coast, and 30 km of the Nellore coast in Andhra Pradesh. They have implemented awareness and training programmes for officials of the environment and forest and fisheries departments, Indian Coast Guard and marine police. TREE now conducts on an annual basis a one-day teachers' conference, a coastal cleanup, a youth eco-summit over two days, and a festival on sea turtles for school and college students, called Flipper Fest[57]. Most recently, they have established a turtle rescue and rehabilitation centre and released several rescued turtles. In 2010, a number of turtles were brought to the rehabilitation centre with damaged or missing flippers and other injuries; the turtles were named Karuna, Hope, Sagari, Abdhi, Olivia, Adhira and Sagarika (all olive ridleys), Greenie (a green turtle) and Sukruti (a hawksbill turtle)[58]. Four of these turtles, Hope, Greenie, Olivia and Sukruti were released in the presence of more than 2,000 students, fisherfolk and members of the general public.

Animal rights activists in Andhra

The Andhra coast extends from the Bahuda, on the southern Odisha border, not far from the nesting beaches at Rushikulya, to the Pulicat estuary at the northern Tamil Nadu border, spanning about 1,000 km. The Andhra coast possesses, at Srikurmam near Srikakulam, the only temple dedicated to the 'kurma avatara' (turtle incarnation) of Lord Vishnu, known as Kurmanaswamy. There is even a caste in this region named after turtles (kurma kulam).

Bhaskar carried out the first survey here in the winter of 1982, a year where the arribada failed in Gahirmatha, covering about 225 km in northern Andhra Pradesh and counting 444 nests of olive ridley turtles[59]. Interestingly, he located a hawksbill hatchling in a shore seine net. Juvenile hawksbills have been frequently reported (usually dead) on the east coast from Odisha, Andhra and Tamil Nadu, but not hatchlings, and Bhaskar thought that this might imply that nesting occurred. He found no evidence of arribadas on the northern Andhra or southern Odisha coasts. He noted that Andhra fishermen had made camps as far north as Rushikulya river but did not find evidence of the arribadas that were discovered there a decade later.

In the early 1980s, B.C. Choudhury was visiting the Godavari and Krishna deltas as part of his crocodile surveys and found evidence of nesting at Sacramento, near Kakinada. Sacramento is a sand shoal near the mouth of the river Godavari, and named after a ship that grounded on the shoal in 1868. Subsequently, a lighthouse was built there in the 1890s. Several surveys were carried out in the years afterwards, including by Tripathy as part of the GoI-UNDP project[v]. Many of these confirmed Sacramento as an important nesting beach.

Despite the general reverence for sea turtles along this coast, Bhaskar had noted that sea turtle eggs were collected and eaten along much of the coast; adult turtles captured at sea were shipped by rail and lorry to markets in West Bengal as they were from Odisha[60]. They were also apparently supplied to a Bangladeshi refugee camp in Raipur in Madhya Pradesh. The capture and incidental mortality of sea turtles continued from the 1980s through the 2000s[vi].

An interesting conservation controversy arose in the late 1990s at Visakhapatnam, a coastal town just south of the Odisha–Andhra Pradesh border on the east coast of India, where the headquarters of the eastern naval command of the Indian Navy were located. Led by Pradeep Nath, the Visakha Society for the Prevention of Cruelty to Animals (VSPCA), which was set up 'for the promotion of welfare of all animals which are an essential part of the environment and society' and was 'dedicated to prevent cruelty and alleviate suffering of all animals', became involved in sea turtle conservation[61]. In 1999, the eastern command of the Indian Navy proposed to create a museum on a beach within the town using a decommissioned submarine. Shortly after, VSPCA, protesting against the impact on sea turtle nesting, filed a public interest litigation against the navy claiming that the submarine museum would be a violation of the Coastal Regulation Zone (CRZ)[62]. They attempted to enlist the support of national agencies (Wildlife Institute of India (WII, Dehradun), non-government agencies (Wildlife Protection Society of India, New Delhi) and international groups to achieve their goal. While the navy and state government claimed that the area was CRZ-II (substantially developed areas), the VSPCA claimed that it should be CRZ-I, which includes 'areas that are ecologically sensitive and important such as National Parks, Sanctuaries, Reserved Forests, Mangroves, Estuaries, Corals, areas close to breeding grounds of fish and other marine life' (The Environment [Protection] Act, 1986). VSPCA received letters of support from the Mediterranean Association to Save the Sea Turtles (including a letter to the High Commissioner of the UK) and *Animal People*, an international magazine, amongst others[63]. They also claimed that the submarine museum violated international conservation agreements that India had endorsed.

In June 2000, the Ministry of Environment and Forests received a letter from the principal chief conservator of forests (PCCF) regarding this particular issue[64]. In this letter, the PCCF pointed out that the VSPCA did not have the necessary permits to work on sea turtles for the year 1999-2000 unlike previous years. With regard to the VSPCA's petition against the Indian Navy, he pointed out that the beach where the proposed museum was to be set up 'has been the recreational area for the urbanites of Visakhapatnam and there is so much traffic and disturbance throughout the year that it is not at all a suitable site for conservation activities'. He also said that the navy 'red-handedly' caught members of the VSPCA trying to transplant eggs to the museum site, to make it appear that it is a nesting beach. VSPCA countered that they had been set up so that they could be discredited.

I argued that these campaigns were motivated more by animal rights motives than by conservation sense[65]. If all beaches with low-density ridley nesting were to be classified as CRZ-I, most sandy beaches on the east and west coasts of India (which total over 6,000 km) would have to be designated as CRZ-I, which hardly seemed practical. The disputed beach in Visakhapatnam, about 5 km long, was not even remarkable within the state. I also argued that the sacrifice of this nesting space could have been compensated by using the museum for furthering education and awareness about sea turtles by involving the navy[66].

The VSPCA nevertheless persisted with efforts to have this beach protected in the name of sea turtle conservation. This seemed to reflect a dichotomy between the animal welfare groups, clearly concerned with every turtle nest on the beach, and biologists, who did not think it necessary or sustainable to expend effort on a low-density beach of relatively low priority in terms of the sea turtle nesting population along the coast. Nath wrote about the organization[67]:

> We love all animals and for us, to save and protect sea turtles is not based on the number of sea turtles coming to nest on our beaches. This is the motive which drives our commitment to support such conservation efforts for sea turtles even if it is just one sea turtle!

Other NGOs were also active along this coast[vii]. In 1996, Green Mercy started their Save Sea Turtles mission in Srikakulam district and in 1999, extended it to Visakhapatanam district[68]. The group started its work by working closely with the Department of Environment of Andhra University at Visakhapatnam (mainly with P.S. Rajasekhar) and a WWF Nature Club. Led by K.V. Ramana Murthy, this group protected 150 to 200 nests each year along about 80 km of the Srikakulam coast between 1996 and 2000. Green Mercy continues to be active, conducting turtle walks and spreading awareness about coastal and marine conservation. Another group, the Marine Turtle Preservation Group, was started in 1991, and was active along the northern Andhra coast. In 1999, they launched a community-based effort called Operation Angel at Sacramento sand shoal[69]. Along the southern Andhra coast, TREE Foundation extended its work from Tamil Nadu[70].

Community groups in Odisha

Like the Chennai coast, Odisha has a rich tradition of NGO involvement in sea turtle conservation. At Rushikulya, Bivash Pandav had employed a number of local field assistants. Many of these field assistants became more and more involved with sea turtles. After Pandav's project ended, they worked with Operation Kachhapa and, subsequently, with several researchers including Basudev Tripathy, Suresh Kumar and others. Several of them are still involved in field research projects with the Indian Institute of Science and Dakshin Foundation.

One of these field assistants, Rabindranath Sahu (Rabi), had worked with Bivash Pandav since the mid-1990s. In fact, Pandav's field camp was set up on Sahu's family land. Pandav encouraged Sahu and his other assistants to set up the Rushikulya Sea Turtle Protection Centre (RSTPC) for the protection of sea turtles at this rookery[71]. Rabi and his companions registered the NGO in 2003 and used the little building as the centre of their activities. From the beginning, volunteers included Rabi himself, Damburu Behera, Shankar Rao, Mohan Behera, Gouranga Behera and Ganapati Sahu. When Basudev Tripathy conducted field research for his Ph.D at Rushikulya, he employed several of the field assistants in his project, but he also helped Rabi raise funds for RSTPC. Funds from the Vasant J. Sheth Memorial Trust helped build a new building on the plot that could serve as an interpretation centre. Additional grants from the World Turtle Trust, Wildlife Trust of India and Rufford Foundation helped created posters and other material to raise awareness about sea turtles in the area.

RSTPC grew to a membership of over thirty members from several villages along the coast. They have been involved in beach cleaning, beach profiling, monitoring turtles and nest protection during the nesting and hatching seasons. They have also been involved in awareness programmes for the local communities and school children, and have participated actively in the Gopalpur beach festival, setting up stalls and sand sculptures of turtles. RSTPC members remain loosely involved in sea turtle monitoring during the nesting season. Some are employed by researchers, while others get temporary employment from the forest department or as guides.

Many of them continue to work on research projects as field assistants. Ganapati helped Divya Karnad with her Master's dissertation, and continues to work with sea turtle conservation. Suresh Kumar, who also conducted doctoral research on ridleys in Odisha, had a dedicated group of field assistants, including Damburu, Shankar, Surendra, Madhu and Kedar. His boatman, Sri Ramalu, worked for several years on turtle projects. Despite being hearing and speech impaired, he was an invaluable assistant and boatman, with an uncanny ability to get the boat through the river mouths and negotiate rough seas non-stop for several hours. The trust that these field researchers (Basu, Bivash and Suresh) placed in their boatmen and field assistants is remarkable, especially since none of them could swim.

One of the major problems in Rushikulya is light pollution from a nearby highway, aquaculture farms and a chemical factory. Each year, when hatchlings emerged after mass nesting (millions at a time), most would be misoriented and end up in the vegetation behind the beach. RSTPC members were joined by others, including the forest department, in collecting the hatchlings in buckets and then releasing them in the ocean[72], but thousands of hatchlings still died, and even those that were rescued were probably weakened, soon to be devoured. The idea of using a net along the back of the beach to prevent hatchlings from straying landward originated from the group (principally, Damburu Behera), and has been effective in years when misorientation was significant. This was briefly controversial as several NGOs and the forest department attempted to take credit for this conservation measure[73]. With funding from WWF and other agencies, this barrier has served to prevent hatchlings from straying away from the beach and getting killed.

Another group in Purunabandha, the Maa Ganga Devi Santi Maitri Juvak Sangha, was formed with a broader vision of community development[74]. While many members of this group worked with Operation Kachhapa on sea turtle conservation, they were also involved in social activities such as cleanliness campaigns, as well as other environmental programmes with larger organizations such as the Nehru Yuva Kendra Sangathan – a national youth organization.

Further north, the Podampeta Ecotourism and Olive Ridley Protection Club was founded in 2008 to conserve sea turtles[75]. Comprising local community members who had either been working previously with other NGOs or the forest department, this NGO attempted to work with the women's self-help groups in their village. Like the other NGOs, this group also worked mainly to increase awareness about sea turtle conservation and participated in nest protection at Rushikulya rookery. The group also addressed the issue of coastal erosion which had a very direct bearing on their community. The seaward village Kontiagada that is part of Podompeta lost a fair amount of land to erosion and almost fifty families had to be relocated.

One of the groups that has worked at the crossroads of livelihoods and conservation and also helped coordinate the activities of other groups at this site is the United Artists' Association (UAA)[76]. Formed in 1965 when a number of youth groups came together to work under a common umbrella, this group works with a network of grassroots organizations and builds their capacity to work in a range of areas including education, health and environment. Headed by Mangaraj Panda for the last decade, UAA has been a key member of the OMRCC collective, serving as its secretariat, helping to administer the programme and keep both fishing communities and local conservation groups engaged in the process.

In central Odisha, a different set of groups were active at the Devi river mouth rookery. Pandav had employed local youth from the local community during his monitoring programmes in the 1990s including Sovakar Behera (Tuku) and Bishnu. In addition, a young boy called Bichitrananda Biswal, or Bichi, would join them after school on the walks, helping with their work. In the years that passed, Bichi became more and more active with sea turtles, and eventually started his own group, Sea Turtle Action Programme (STAP). Bichi and STAP conduct surveys for nesting and mortality during the nesting season and assist the forest department in monitoring.

Meanwhile, Tuku had been helping Robert Sutcliffe, a retired dentist who had come to India intending to offer pro-bono services as a dentist, and become fascinated with sea turtles in Odisha after a casual visit a few years earlier. Sutcliffe spent several weeks during the season at Gundalba village near the Devi rookery and supported Tuku's work (and Bichi's as well, occasionally). Sutcliffe monitored the beach for nesting and mortality and had a vision of forming the Odisha Turtle Trust. He also started creating some infrastructure for turtle tourism at Devi but left due to a combination of logistical and health issues.

Meanwhile, Tuku formed an organization called Green Life Rural Association (GLRA) in 1993 which was, according to him, 'different from most [other] organizations in that it relies on funds generated at the grass root level'[77]. GLRA monitors the beaches between Devi and Chilika mouth to the south, and Paradip to the north, recording both nesting and mortality. They started a project called 'turtle friends' in 2002, mobilizing the support of Devi's traditional community and school children. GLRA also used song and dance to communicate messages about sea turtle biology and conservation.

At the northern end of the mass-nesting range, little direct action was possible as the nesting beaches at Gahirmatha lie inside the Bhitarkanika and are inaccessible without forest department permits, and lie a fair distance from the nearest settlements. In 1999, Bijoya Kumar Kabi started the Action for the Protection of Wild Animals (APOWA) with the goal of 'building a community of people to support ... the welfare of animals and the environment and improve relationships between man, animal and environment'.[78] APOWA started their work on sea turtles along the periphery and buffer zone of the Gahirmatha Marine Wildlife Sanctuary and has been involved in nest protection, awareness programmes and beach cleanup activities. They have appointed volunteers from local coastal villages as 'turtle guides' to protect nests and turtles along the coast.

Another group in this region, Alacrity, was founded in April 1995 to promote conservation and awareness of the marine ecosystem in Kendrapara, Odisha, through the participation of local community and self-help women's groups[79]. Led by Kalpana Mallik, this group explicitly aimed to 'to serve down-trodden sections of society and to facilitate women's empowerment and promote social justice'. They established eco-development groups around the periphery of Gahirmatha for the protection of turtle habitats and mangrove forests. By 2011, over thirty such groups had been established within the fishing communities around the park. They have worked both with Odiya and Bengali communities (the latter including migrants from West Bengal and Bangladesh) whose access to forest and fishery resources has been negatively impacted by the expansion of agriculture and encroachment of forest lands.

In addition to these groups, Project Swarajya worked with fishing communities, both trawlers and artisanal communities, throughout the state. Founded in 1988, the organization's larger mandate was social welfare activities, but they became involved in sea turtle conservation in the 1990s. Chitta Behera, the leader of the group, was actively involved in documenting mortality of sea turtles, and in arranging capacity building workshops for the use of TEDs[80]. He later became a champion of the indigenously developed trawl guard. His colleague, Ashis Senapati, became a member of OMRCC and a strong voice for the inclusion of all sectors of fisheries in dialogues about marine conservation.

Battling trade in West Bengal

West Bengal has a relatively small mainland coast, mainly in Medinipore district. However, there are several coastal islands in the Sundarbans in 24 Parganas district. This state is best known for being the market for all the turtles shipped from Odisha and Andhra Pradesh. Though ridley nesting occurs in some of the islands in the Sundarbans, it is the landing centres at Digha and Junput that have a more prominent place in history[81].

Nesting along this coast occurs in Medinipore district (including Digha, Shankarpur, Junput and so on), and in the islands of the Sundarbans, both in the 24 Parganas district of West Bengal and in

Bangladesh[82]. The largest mangrove forest in the world, the Sundarbans consists of over fifty deltaic islands, many of which have been reclaimed for settlement. Well known for its population of tigers, this may be one of the few (or only) places where tigers predate sea turtles.

By the 1980s, there was considerable interest in Kolkata in the conservation of ridleys. WWF-Kolkata had become involved through the efforts of Bonani Kakkar, and the young Indraneil Das, who had just produced his book, *Indian Turtles: A Field Guide*. Kakkar discovered a rural scroll painter, Ranjit Chitrakar, at a fair and began to involve him in the environmental education programmes of WWF[83]. The scrolls, up to seven metres long, were taken from village to village, with songs about epics and legends. Chitrakar worked with WWF for several years, creating scrolls about environment and conservation. He was taken to the turtle markets to create scrolls and songs about sea turtle conservation. WWF also produced a pamphlet in Bengali to distribute amongst prospective consumers of sea turtle meat and eggs.

In 2000-01, Nature, Environment and Wildlife Society (NEWS) in Kolkata conducted a survey as part of the GoI-UNDP project[84]. They found olive ridley nests in the sea-facing beaches of several of the deltaic islands. By this time, the trade in turtles had clearly declined due to the vigilance of forest officials, and incidentally caught turtles were generally thrown back into the sea. However, they did find that turtle meat was still being sold at several markets in the Sundarbans. Turtles were caught along the coastline and landed at sites away from the Digha and Shankarpur landing centres which were inspected by forest department officials. Incidental catch resulted in the stranding of several hundred turtles along the Sundarbans coast.

Little else has been done on sea turtles in West Bengal. On the other side of the border, conservation groups have been working on sea turtle conservation in Bangladesh, particularly on St. Martin's Island. In recent years, the Marinelife Alliance, led by Zahirul Islam, has been carrying out monitoring along with the wildlife department, and has also worked on reducing the take of sea turtle eggs, and increasing awareness of issues[85].

The conservation conundrum

Choosing the right model[86]

As described above, a vast number of small groups have been involved in sea turtle conservation in India. These groups have been motivated by an array of diverse factors, including conservation, animal welfare, ecotourism, environmental education and community involvement. Though the motivation may have been diverse, the stated objective in most cases has been conservation. From this perspective, many of these programmes may have simply failed to achieve their objectives. The SSTCN could not prevent development and lighting from increasing along the Chennai beach, nor could they reduce sea turtle mortality. In Kerala, Theeram has fought a losing battle against sand mining, and the beach has declined over the past few years. The VSPCA was unable to prevent the submarine museum from being established on the beach. However, if the objective is seen as raising awareness through the conservation of these turtle populations, then at least some of these groups have achieved far more than they envisioned. The students' group in Chennai instigated the initiation of several other such groups throughout the country. And Theeram's efforts became a flagship for community-based conservation and an inspiration for groups in Kerala. Other groups have had similar impacts in their respective states over the last decade. But what actions serve turtle conservation best?

The idea of forsaking the hatchery in favour of beach management programmes took seed in the Chennai group in the early 1990s. It is not clear how this idea originated, but discussions indicated the recognition of the need for long-term solutions. The hatchery was seen as a short-term measure, and it was generally agreed that a wild species could not (or should not) be entirely dependent on human support. Hence, it was recognized to some degree that the hatchery was an educational device. However, even in an educational programme, it was recognized that the right models of conservation should be conveyed, and that beach management was a more suitable long-term option than hatcheries.

The groups in Goa and Maharashtra are amongst the few that

have attempted to combine conservation and tourism. Ecotourism has been widely touted as being an alternative source of income for local communities dependent on natural resources. In many parts of the world, turtle conservation and ecotourism have been successfully combined, for example, in Costa Rica[87] and Brazil[88]. When monetary benefits are involved, there are other negotiations within the community that have to be resolved. While problems arose in the programme in Goa, the homestays for turtle tourism in southern Maharashtra seem to be working well.

Similarly the VSPCA attempted to promote sea turtle conservation using the legal regime of the CRZ. However, their framing of the issue implied that each individual sea turtle on the Indian coast must be protected. The model is clearly based on animal welfare approach rather than conservation. Whether this benefits the conservation of the species remains to be seen. The attention given to the issue does raise the profile of the species as well as coastal regulation issues for the public, but whether the effect is positive or not is debatable.

Since limited resources are available for conservation, conservation biologists generally consider it necessary to prioritize habitats, taxa and populations for conservation action. Some populations are considered to be globally significant i.e. their extirpation would have consequences for the survival of the species or they have significant ecological or evolutionary value. The IUCN Red List attempts to provide such a classification, though there has been considerable criticism of both varying standards as well as quality of data. The Marine Turtle Specialist Group (MTSG) has also attempted to come up with lists of important or threatened populations.

In India, olive ridleys in Odisha or leatherback turtles in the Nicobars might have regional or even global significance. However, other populations such as green turtles in Gujarat, Lakshadweep and Andaman islands might have limited global significance, but are the only green turtle populations found in the country. And what of small sea turtle nesting beaches, which are not critical to the survival of ridley populations in India, let alone the world? If the stakeholders on a particular beach believe that the beach or species is of value to them

– culturally or economically – they will work to protect it regardless of its global or even national priority (or lack thereof). Furthermore, the efforts of various small groups indicate that even if their conservation programmes are unsuccessful in actually saving the population or site, the wider impacts might benefit the conservation of the species. Thus, 'local' conservation even for otherwise 'unimportant' populations might have considerable 'global' impact.

The consequences of conservation

The conflict in the Lakshadweep gives rise to an interesting conundrum. Why did the conflict arise only in the late 1990s? Were the fishermen unaffected by the sea turtles or did the numbers actually go up at that time because of reduced killing? In the early 1990s, Peter Richardson and Sue Ranger, visiting Sri Lanka as part of a volunteer programme, helped set up the Turtle Conservation Project (TCP). TCP, led for many years by Thushan Kapurusinghe and Lalith Ekanayake, is currently Sri Lanka's longest-run and most successful project with community-based sea turtle protection programmes at many beaches on the west coast[89]. During that period, it is likely that the production of hatchlings from these beaches increased. Richardson and colleagues at TCP satellite tagged adult green turtles in 2006 and 2007 and three of these animals found their way to the west coast of India, two to Netrani Island on the Karnakata coast, and one to the Lakshadweep Islands[90]. On other occasions, divers saw tagged green turtles that were later identified as being from Sri Lanka. Could the conservation programmes in Sri Lanka have led to an increase in foraging turtles in the Lakshadweep? And if this were true, it raises the intriguing possibility that conservation in one location creates conflict (and therefore negative conservation consequences) in another.

Is small beautiful?

While it is clear that there is a great deal of local effort, there is a need for greater coordination of sea turtle conservation in India. Especially since many of the issues such as coastal development, beach armouring and fisheries cut across state boundaries. National level programmes such as the GoI-UNDP project and international instruments would seem to serve this end best. The UNDP project served to initiate standardization of methodology for data collection, and brought numerous government, non-government and community groups on to a common platform. However, the inputs required were large ($300,000) and unsustainable without further funding. Organizations and individuals were involved in the project on a professional basis. The continuation of these activities under a new project from CMS-UNEP also required large amounts of funds ($35,000). International instruments, also expensive to develop, while promoting interest at the government level, have had little impact at the local level.

On the contrary, the various small programmes have been remarkably effective and inexpensive. Most have survived on annual budgets of less than Rs 50,000 (about $1000), many on Rs 10,000–20,000 per year. These numerous small programmes have also shown remarkable resilience over the last ten or twenty years. However, the small groups have been motivated by varying factors, something that has often been frustrating for conservationists, especially when the actions of animal welfare organizations conflict with the more scientifically-based principles of conservation, and in other subtle ways. While the question of resource allocation is apposite, it may actually not be possible to initiate small programmes equivalent to a single large programme, i.e. these programmes arise when the internal motivation of a group reaches some critical threshold and has less to do with availability of funds or resources. As we have seen, most of the small groups started with a minimum of funds, resources and training and many individuals spent personal funds to support the programmes till other sources became available.

Eventually, it may be a combination of these that will serve the conservation of marine turtles and their habitats best. The creation of networks that can bring the standardized protocols of national programmes together with the internal motivation and cost effectiveness of small programmes, combined with the bottom-up approach of community-based programmes, may be the answer. National and international programmes may serve best to build capacity at the government level, foster cooperation between government departments, and support local groups in achieving their objectives.

In 2009, a group of us initiated the Turtle Action Group (TAG) as a consortium of NGOs working on sea turtle conservation and coastal issues in India. With a membership of over twenty organizations, TAG's broader goal is to collaborate on and coordinate sea turtle conservation across the coast. The group comes together at national meetings, which have been held across the country in Chennai, Bhubaneswar, Kumta, Mahabalipuram, Jamnagar and Visakhapatnam. TAG has also supported smaller workshops for training local community groups and forest department personnel. The groups recognize that sea turtle conservation needs to move beyond the concerns of the species alone, and serve as flagships for the protection of their habitats, and that collective action is needed for this.

Notes

i Bhaskar's surveys provide a detailed account of nesting on the Gujarat coast[91]. He showed that some of the best nesting beaches for olive ridleys were on Bhaidar Island where he encountered peak nesting of seventy ridleys on one night, and over 100 nests on 11 and 12 August 1981. He also recorded significant green turtle nesting on the Kutch and western Saurashtra coasts, enumerating over 800 nests on the latter coastline during his surveys in 1981. On the northern Saurashtra coast, he noted seagrass beds and coral reefs, particularly near Karumbhar Island, where he found a considerable number of feeding green turtles. Bhaskar speculated that these green turtles might nest in Sandspit Beach and Hawkes Bay in Pakistan.

ii In his paper at the 1984 CMFRI workshop[92], Bhaskar noted that there were previous records of ridley nesting, and encountered live and dead ridleys during his survey. He also heard that some ridleys near some neighbouring islands were resident and thought it merited investigation. A large green turtle was caught and sold; interestingly, according to the newspaper, it was caught several kilometres upriver from the mouth of the Mandovi River. Bhaskar chatted with a fishermen called Pedro who said that thirty to forty turtles had been caught near his village that year and he admitted to relishing turtle meat and eggs like most fisherfolk in Goa.

iii The ridley is known as yeth aamai (turtle that comes up), panchal aamai (after a village near Kanyakumari), and vakatta aamai (poor turtle, probably for the small amount of meat). The green is known as perunthalai aamai (large headed turtle), then aamai (honey turtle), panguni aamai (after the Tamil calender month in March-April) and per aamai (large turtle). The hawksbill is called yeli or yelithalai aamai (rat or rat-headed turtle), alungu aamai (pangolin turtle) and kili mooku aamai (parakeet beaked turtle). As for the leatherback, it is compared to dolphins (oongal aamai) as well as to pigs (panni aamai). It is also known as the ezhuvari aamai (seven line turtle), ooduvetti aamai (after the powerful flippers that can injure a person) and as thoni aamai (bullock cart wheel turtle).

iv There have been several records of sea turtles in the Gulf of Mannar in *Marine Fisheries Information Service*, both of strandings of ridleys, hawksbills and leatherbacks and landings of ridleys and green turtles. The only record of a hawksbill nest was noted by CMFRI in Manapad on the Tirunelveli coast in 1980[93]. The eggs were transported to the Tuticorin Research Centre and hatched there. There does appear to have been a fair amount of interest in sea turtles at the Mandapam research station of CMFRI through the 1970s and 1980s. Agastheepillai and Thiagarajan provide a fine account of the biology of green turtles in the Gulf of Mannar, based on turtles caught during 1971 to 1976[94]. Perumal Kannal, a Ph.D student of M. Rajagopal, carried out a study of green turtles and olive ridleys in the Gulf of Mannar, looking at catch, trade and accumulation of heavy metals in their tissues, finding significant levels of heavy metals in both species[95].

v P.S. Raja Sekhar of Andhra University carried out surveys between 1983 and 1987[96], as part of his doctoral research, and K.V.R. Priyadarshini, as part of the Worldwide Fund for Nature-India's Conservation Corps Volunteer Fellowship Programme[97]. Priyadarshini also documented a high intensity of nesting at Sacramento Island, and during the project, the Andhra Forest Department initiated conservation of sea turtles at the site. Basudev Tripathy carried out a complete survey of the coast in 2000-01 as part of the GoI-UNDP project[98]. He estimated that about 10,000 nests were laid annually on this coast, second only to Odisha. Most interestingly, the density of nesting near four major river mouths was significantly higher than the average, with these areas receiving fifty to 100 nests per km per season, while other areas averaged around 10 nests per km per season[99]. Fishermen in Kalingapatnam in northern Andhra told him that a 10–15 km stretch of sandy beach between Vamsadhara and Nagavali river mouths had large numbers of turtles nest during the Sivaratri mela, a festival in February/March. Similarly, he was told that ridleys nested in large numbers in the Sacramento shoal beach during Antarvedi Theertham, a village festival celebrated in January/February.

vi Raja Sekhar documented the capture of hundreds of turtles during courtship and mating in the 1980s on the northern Andhra coast, and also enumerated over 500 dead turtles on the coast[100]. His surveys in the mid-1990s found that nearly 2,000 turtles were caught at sea, and about 300 on land for sale in markets. And nearly 1,000 dead turtles were washed ashore each year on the northern coast alone. Shortly afterwards, Tripathy documented considerable mortality from incidental catch in fisheries, with about 800 dead turtles counted during the survey[101]. Adult turtles were occasionally killed for medicinal use; the calipee and liver of turtles were used to treat pregnancy-related and rheumatic problems.

vii As part of his project, Tripathy worked with several local NGOs, including Green Mercy in Visakhapatnam, Sravanti Association in Rajahmundry, Coastal Community Development Programme (CCDP) in Machilipatnam, and Society for National Integration through Rural Development (SNIRD) in Ongole. The NGOs helped carry out surveys, but also carried out education and awareness programmes[102]. Other NGOs that were also involved included the Association for Rural Development and Action Research in Vizianagaram, Mother India International in East Godavari District, Sruthi in Bheemavaram and Tribal Community Development Society (TCDS) in Visakhapatnam.

ISLANDS OF HOPE

The sound of a leatherback

ON A WARM NOVEMBER afternoon, the skies clouded over as was the pattern at this time of the year on Great Nicobar Island. A late afternoon shower was in the offing. The sun would set at about 5 p.m. anyway, since the island followed Indian Standard Time, even though it was closer to Indonesia. On this occasion, the shower turned out to be a storm that raged for four days, confining us to camp. The thatched roof held up, but the mud floors absorbed water till our feet sank into the soil. Less than fifty metres away, the sea battered against the shore as if intent on submerging the entire island.

I had been waiting nearly fifteen years to see a leatherback turtle. From the moment I started working on sea turtles, it had been my dream (and that of every fellow turtler I know) to see this amazing animal. Sea turtles have now become part of polite drawing-room conversation, though it is species such as the olive ridley and green turtles that are more familiar to most.

The aboriginal tribals of Great Nicobar include the Nicobarese, who occupy most islands in the Nicobars, and have differing levels of integration with mainstream society. They used to traverse vast expanses of open ocean between islands in their small canoes, though now most rely on the ship like everyone else. The other tribal group, the Shompen, number a few hundred and are found in small groups throughout Great Nicobar. The rest are the 'settlers', largely ex-army servicemen who were moved here in the 1960s and 1970s and given land to farm. Since then, there has been a small influx of farmers and fishermen from the mainland which has increased after the tsunami. Great Nicobar has two roads, one originally running 51 km south from its only port, Campbell Bay, to Indira Point. Following landslides, the road stopped a few kilometres short of the southernmost point. The only other road runs west, originally up to the coast 43 km away, but landslides and entropy have destroyed much of it.

Our journey to the Nicobars was arduous and included a long wait for permits at Port Blair and a four-day journey aboard the Shipping Corporation of India's *MV Sentinel*. Once at Campbell Bay, the gateway to Great Nicobar Island, we discovered that the bus no longer ran to Galathea, the nesting beach, since a landslide washed away the road four kilometres from the beach. Though we were assured that there would be tractors, cycles and other means of covering this gap, we found one of Campbell Bay's three auto-rickshaws to ferry us over. Galathea was then a small two-kilometre beach in a south-facing cove and we were lodged in a camp of the forest department. We did not have to wait long for our first evidence of leatherbacks. We began patrolling the beach and soon came across one of the huge tractor-like treads that we would get very used to in the days to come.

Having worked much of my life on ridley nesting beaches with a low intensity of nesting, I was used to looking intently at the beach in front of my feet for tracks. The first thing to do on finding one track was to look for another one, which would indicate that the turtle had returned to the sea. Time, then, to start searching for the nest and eggs. Occasionally, a track would be fresh and we would creep up slowly on an unsuspecting ridley and watch her dig her nest or lay her eggs.

On a leatherback beach, however, the entire ethos of turtle walking is different. Apart from tracks, one's eyes had to be peeled for beaching leatherbacks, their mountainous forms heaving out of the sea. We were with our Karen field assistants, Agu and Glen; the Karen are originally a community from Myanmar, now settled in the Andamans. Meera, my partner and colleague, and Agu were far better than I was at spotting the beasts as they left the breakers, a good half hour or more before reaching the high-tide line and the nesting site. For those with less than 20/20 vision, there were noisy uncouth grunts to listen for, or the slap of the front flippers against their side as they flung sand to clear nesting space. And one always had to be careful not to break an ankle stepping into a huge body pit or abandoned nest hole.

Leatherbacks that come ashore late at night are often still on the beach in the morning. In the harsh light of day, they sometimes give the impression of not being the brightest of animals, especially one that came ashore at noon, and attempted to make a nest in the middle of a tidal pool! But they have an undeniable gentleness and elephantine beauty about them, with their hanging folds of skin and ponderous outlook on life.

Until the previous year, the Galathea river had flowed around a huge sand dune for more than a kilometre before entering the sea. This high dune was a major leatherback nesting area. During a violent storm in 2001, the river cut right through the dune and washed away the entire nesting beach. Turtles now only nested north of the river and during high tide, there was often less than a metre (if that) between the high water mark and the vegetation. Leatherbacks were forced under the pandanus and scaevola, to the very edge of the nesting beach. Worse, they often dug at the high water mark, only to find water at the bottom of the nesting pit. Many turtles abandoned these nests, but those who could not retain their eggs any longer, nested anyway. Most of these eggs were eventually destroyed.

We were in the Nicobars to collect tissue samples from leatherbacks and other sea turtles for a genetics project. For all the years I waited, it did not take too much time for me to get my fill of them! During the peak season, ten or more leatherbacks came ashore each night on the two-kilometre beach. And each one would spend hours on the beach; leatherbacks take an hour or more just to haul themselves ashore and then spend another hour clearing away the surface sand and digging a nest, often only to decide that the site is unsuitable. That means starting afresh somewhere else. The egg laying itself is quick and lasts just about ten minutes, but we had to quickly tag turtles and collect samples during that window. This often meant running back and forth between turtles to ensure that we did not miss our opportunity with either one. The next morning, the beach looked as if it had been bombed, and the line of tracks looked like a field that had been ploughed.

Other molecular studies suggested that Indo–Pacific leatherbacks may be ancestral to global leatherback populations, which our work confirmed. The origin of many sea turtle species in the Indian Ocean may be due to more conducive climatic conditions in the region in the recent geological past. We believed that the large leatherback nesting populations in Great Nicobar would provide some insight into the evolutionary history of this species.

Turtle Islands

The Andaman and Nicobar Islands are divided into the northern Andaman group, and the southern Nicobar group. The Andamans, a part of the Arakan Yoma mountain chain of Myanmar, comprise three main islands, South, Middle and North, and are surrounded by over 300 small islands, islets and outcrops (together known as Great Andaman). The islands were referred to in the Chola dynasty as Timmaitivu (impure islands), but earlier Arab geographers referred to Andaman-al-kabir (Great Andaman). A more recent theory is that it derives from the Malay word (Handuman) for Hanuman of Indian mythology. Some of the iconic islands include the still volcanic Barren Island, Narcondam to the north, famous for its endemic hornbill, South Sentinel famous for its robber crabs, and North Sentinel, peopled by the Sentinelese, still a no-contact tribe.

While there were numerous aboriginal tribes in the Andamans, many of them have now been extirpated, and only the Jarawas, who live in the reserve set aside for them in the South and Middle Andamans, remain in any numbers, still only about 400. The Great Andamanese, a collection of related tribes, who were estimated to be up to 5,000-strong in the nineteenth century, now number only about fifty. They were originally distinguished as ten separate tribes spread across the Great

Andaman and with very different dialects. Of the surviving tribes, the Onge-Jarawa are described as a separate linguistic group, and the current Great Andamanese is a patois of more than two dialects[1]. The last native speaker of one of the dialects, 'Bo', died recently; the late Mrs Boa was the last speaker of the dialect from 1956 when she, along with nineteen remaining Andamanese of various remnant tribes, were relocated to Strait Island.

These tribes were thought to be of African origin due to their Negrito features, and were believed to have occupied the islands 50,000–60,000 years ago, as part of the first wave of migrations out of Africa. However, more recent genetic evidence suggest that the tribes originated from Paleolithic groups living in south or south-east Asia, and likely migrated to the islands from north-east India about 20,000 years ago[2].

South of this group is Little Andaman Island, originally peopled by the Onge, and between the Andamans and Nicobar is Car Nicobar Island. The Nicobars include the central group which consists of the picturesque Nancowry group of islands and the southern most Nicobar group, comprising Great and Little Nicobar and a few smaller islands. The Nicobars were known as far back as the third and fourth century CE, and are mentioned in Sri Lankan Buddhist pali manuscripts as Naggadipa (nagga = naked and dipa = island). The current name probably owes to the Chola dynasty, who conquered the islands and referred to it as Nakkavaram (naked land in Tamil)[3]. The Shompen, a mongoloid tribe genetically related to Sumatrans and linguistically part of the Austroasiatic people, live on Great Nicobar Island, and now number only about hundred. The coast of the Nicobar islands are occupied by Nicobarese, who are of south-east Asian origin, and have probably occupied the islands in multiple waves over the last several hundred years or more.

Owing to its greater proximity to the south-east Asian region, the biogeographic affinity of the Nicobars is largely to that region (the distance between the southern tip of Great Nicobar and the Sumatran mainland is about 100 km)[4]. The Nicobars host a high diversity of flora and fauna, many of which are endemic. Species level endemism is very high, but many genera and families are shared with south-east Asia. Although only a small percentage of forests have been lost till date (about 14 per cent by various estimates), habitat conversion poses potential dangers to the area. The biogeographic importance of these islands were debated by British naturalists. In one of the earliest accounts, Blyth wrote a paper titled 'Notes on the Fauna of the Nicobar Islands' in 1846 and documented the fauna of the islands, noting the paucity of mammals and amphibians[5]. In his appendix to Frederic Mouat's book[6], he wrote that little was known of the islands at the time, but commented on the similarity of the Nicobar to south-east Asia in this respect.

Most of the information came from the collections of Captain T.H. Hodge, captain of the guard-ship *Sesostris*, in 1859-60, and from Dr G. von Leibig, and Lieutenant-Colonel Tytler, governor of Port Blair, who was also an amateur naturalist[7]. Later, in 1915, Nelson Annandale remarked on Thomas Blanford's comment that the islands were 'of small importance zoologically'[8]. A collection of lizards from Narcondam which was presented to the Indian Museum led him to 'doubt whether this summary dismissal by so catholic and liberal an authority [did] not perhaps unduly minimize' the zoological importance of the region. Annandale emphasized the relationship between the Andamans and the Burmese region, while the Nicobars were faunistically similar to the south-east Asian island groups, particularly Sumatra, a point emphasized by Malcolm Smith in his work on reptiles in the *Fauna of British India* series. Smith wrote in 1931 that the Andaman contained 'an impoverished Burmese fauna; that of the Nicobar Island approximate to the Sumatran type'[9]. Today, this biogeographic distinction is well established and the two island groups form parts of different global biodiversity hotspots, the Andamans being part of the Indo-Burma hotspot, while the Nicobars form part of the Sundaland hotspot.

Turtling in the islands

The earliest references to sea turtles in the Andaman and Nicobar

Islands are in the accounts of British administrators, naturalists, and sociologists from the 1800s. In one of the earliest references, Mackey reported that green turtles were common in Car Nicobar Island[10]. According to him, the 'natives' would turn green turtles on their 'carapans' and leave them there to be collected the next day.

In 1857, following the Indian mutiny (or the first war of Independence, as it is now known), Frederic John Mouat, a British surgeon posted with the Indian Medical Service, was asked to investigate the possibility of the Andaman Islands as a penal colony. Mouat wrote of this trip to islands in a book titled *Adventures and Researches among the Andaman Islanders* on board the steam frigate the *Semiramis*, including in their company several amateur Goan musicians[11]. Mouat provides a review of knowledge about the Andamans at the time and several entertaining stories about their visit. Mouat, of course, believed that 'the civilization both of the East and the West [had] passed them by without in any way affecting the condition of the inhabitants'. He considered their state 'lower even than that of the natives of those remote islands of the sea that lie far beyond the bounds of civilization'. He did not believe that they had law and took their lack of knowledge of a supreme being or 'even a rude system of religious faith' as evidence that they were worse than savages.

As Mouat says, little reliable knowledge was available about the Andamans before the late 1700s. Two men, Captain Blair, an early hydrographer of the East India Company and Colonel Robert Hyde Colebrook, the Surveyor General of India, surveyed the islands carefully, completing their work in 1789. Blair's report provided, for the first time, a much more favourable account of the inhabitants. This in fact led to the establishment of the first settlement in the South Andamans, first called Port Cornwallis, and later changed to Port Blair. The settlement was moved as Port Cornwallis to the North Andamans for a few years, and abandoned in 1796 mainly due to disease. Port Blair was re-established at its current location in the mid-1800s.

Blair returned to survey the islands in 1789 and reassessed his opinion of the islanders, whose hostility was generally ascribed to their having been slaves at one time. Different theories abounded, including that they were slaves from a Portuguese ship sailing from the Mozambique, but Mouat dismisses this saying that the Andaman Islands were known to have been inhabited before the tenth century, and that they were likely responding to cruel incursions by East Asian traders.

Mouat considered Colebrooke's account, published in 1795, to be even superior to Blair's report, but it was not widely accessible even at the time. His original watercolour sketches of the islands as well as their inhabitants are deemed to be of great quality (I found one such watercolour titled 'View in the interior part of Port Cornwallis, Andaman Islands' for sale in Christie's for £750. Another one was available for £2000). Colonel Colebrook had described the abundance of turtles and the appreciation for their meat near Diamond Islands on the Tenessarim coast of Burma. Mouat wrote:

> The manuscript account of Colebrooke's survey is diversified by many light and amusing details. In the course of his voyage, the vessel visited the Diamond Island, lying near the Tenessarim coast, which have always been remarkable for the number of turtles frequenting their shores, offering valuable prizes to the adventurous mariner ... In the short period of three days, Colonel Colebrooke's crew secured no fewer than one hundred and two of those valuable natives of the Eastern seas.

According to him, a single turtle had enough meat to feed the ship's company of a hundred and eighty-five men for a day. With reluctance, they left behind another fifty or sixty turtles that they could easily have captured, but did not have space for. Mouat also wrote of their learning of the customs of the natives from a Brahmin sepoy, an incarcerated mutineer who had escaped and spent time with the local tribes. In his account, it is noted that turtle oil was often applied during festive occasions, including after childbirth where it was mixed with earth and used as an unguent. Mouat was aware that turtles were consumed, which they captured using nets, and wrote of turtle heads being used to adorn their huts.

Beyond this, Mouat does not provide very much information. That was left to Edward Horace Man, who arrived in Port Blair in 1869, and spent the rest of his life working on the islands, occupying various positions including that of chief commissioner. Man was an amateur anthropologist and an insatiable collector of objects and artifacts, which he sent to western institutions and experts. He was described at the time by an Oxford professor as '... the sort of man who might send four or five entire Nicobar villages with all the inhabitants'[12].

Man also made copious notes and observations, which he published in 1883 as *The Aboriginal Inhabitants of the Andaman Islands*, including detailed accounts of the hunting customs and rituals of sea turtles by these tribes[13]. Writing about the Andamanese, Man mentioned hunting of both green and hawksbill turtles, with the average size of the former ranging from eighty to hundred pounds, indicating that they must have caught a large number of juveniles. Turtle hunts generally took place during high tide, typically after sunset. With the aid of bioluminescence in the water, sparked by the movements of the boat, they were able to discover and harpoon the turtles.

Man says that they were aware that sea turtles nested on sandy beaches, but for reasons he did not fully understand, they did not capture these turtles while nesting. He believes that they may have considered it unsportsmanlike or that they just 'thoroughly enjoy[ed] their own methods of procedure'. Turtles were speared when they surfaced to breathe, often detected by just the sound.

Their love of green turtle meat is evident from the following lines:

> They do not preserve the carapace of either description of turtles, but having removed the flesh, place the shell over the fire, that all the remaining fat may be melted down, when – with an appreciation worthy of a city alderman – they ladle it into their mouths with *Cyrena* shells, which thus serve as spoons. So great a delicacy do they consider this that the shell is finally broken up and divided, that no particle may be lost!

The fat was highly valued as an unguent. Recent work suggests that it may have had a role in the Andamanese ethnoanemological belief of playing hide and seek with spirits based on the aroma of foods eaten and foods of the spirit world[14]. Turtle eggs were sometimes eaten raw and the blood was also consumed, after being boiled in the shell till it was thick. Generally, turtles were not consumed during mourning.

Alcock, in his travels aboard the *Investigator* also wrote of the turtles of Diamond Island, where the ship docked in late 1889[15]. He says, 'It is a great breeding resort of turtles and the Burmese who farm the eggs objected that our boat-party would frighten all the turtles away'.

In 1899, Portman wrote that 'the sea round the Islands swarm with fish and turtles'[16]. He had seen large turtle shells around Andamanese camps. The Andamanese travelled to other islands to hunt turtles as well. For example, the Great Andamanese from South Andaman Island went to Rutland Island and Tarmugli to hunt turtles, and Rutland Islanders travelled to Cinque and Passage Islands to catch turtles and large fish. Portman also found hundreds of skulls of turtles on Temple Island near Diglipur in North Andamans. The importance of sea turtles in Andamanese culture is evident from the fact that the surviving Andamanese still sing turtle hunting songs.

Portman also noted that the Onge from Little Andaman Island also visited South Brother Island for 'turtling'. On 27 October 1880, he wrote that 'there were traces of a number of people having recently been on the island and from the amount of tortoise-shell about I should imagine their reason for coming to procure turtle …' He wrote that it was their custom to visit islands as far as Cinque Island during calm weather. Portman also wrote that dogs were introduced to the islands in the 1860s, and the South Andamanese started using them for pig hunting, a trick apparently learned from some escaped Burmese convicts. Today, egg depredation by feral and domestic dogs is a major threat to sea turtle populations on the islands.

In 1903, Cecil Boden Kloss, an English zoologist, and later director of the Raffles Museum in Singapore, published a book titled *In the Andaman and Nicobars* and as was the custom at the time, added a subtitle of considerable length – 'The narrative of a cruise in the schooner "Terrapin" with notices of the islands, their fauna, ethnology, etc'[17]. The *Terrapin* was captained and owned by the American naturalist, William Louis Abbot, and set sail from Singapore in 1900 to explore the islands. Writing about the Nicobars, Kloss says that turtles were common in the islands, and skulls could often be seen in the houses of the Nicobarese, where 'they were used for the expulsion of demons'[i]. He wrote that they were captured while floating on the surface, by a harpoon with a 'skewer-shaped iron head, which, when fast in the shell, detaches from the shaft and remains connected by a short piece of cord only'.

Later, in 1932, Charles Suydam Cutting of American Museum wrote of the spearing and hunting of sea turtles and dugongs by the coastal tribes in the Andamans[18]. In particular, he was impressed by the ease with which they pierced the hard shells of turtles with their spears. Cutting wrote that the turtle was cooked in its own shell and that the entrails were the choicest tidbits. Bonington described turtle nesting beaches and turtle eggs in net bags hanging inside Onge huts along the West Bay beach in Little Andaman Island[19]. During the 1930s, the British administration had an enclosure for holding turtles, to be slaughtered later for food. Located at the edge of the sea at South Point, now in Port Blair town, the tide could come into the enclosure (a partial remnant of which can still be seen there) to keep the turtles alive as long as required.

Few authors wrote about sea turtles in the Andaman and Nicobar Islands in the decades following Independence. In 1966, the Zoological Survey of India embarked on what was called the Great Nicobar Expedition. Biswas and Sanyal reported on the reptile collections from this expedition[20]. They were accompanied by Humayun Abdulali of the Bombay Natural History Society. Even at this time, Nicobar was little known though the reptiles of the Andamans had been better documented thanks to the work of Blyth, Stoliczka, Annandale and Smith. Biswas and Sanyal recorded a hawksbill turtle head at Campbell Bay and encountered green turtles at Galathea Bay, but surprisingly made no mention of leatherback turtles.

In the 1970s, the multifaceted T.A. Davis, who had written

previously about olive ridley turtles at Gahirmatha, visited South Sentinel Island. He accompanied Rudolf Altevogt, one of the world's leading authorities on crabs from the University of Munster, who wanted to study the robber crabs. Titled 'Giant Turtles and Robber Crabs of the South Sentinel', their article detailed their two expeditions in 1973 and 1974[21]. The island had received few visitors (other than Alcock who had visited in the early 1900s) because of its isolation, and the suspicion that the hostile Sentinelese visited frequently from North Sentinel Island. Davis and Altevogt recorded a fair bit of nesting by green turtles on this island.

Travels in the islands

Satish Bhaskar, who I introduced in the very first chapter, is a pioneer of sea turtle biology and conservation in India. From the late 1970s onwards, Bhaskar conducted the first surveys for sea turtles in most parts of India including the Lakshadweep Islands, Gulf of Mannar, Gujarat, Kerala, Goa, Andhra Pradesh and Odisha[22]. In 1982, he famously revisited the Lakshadweep Islands, spending several weeks alone on an uninhabited island[23]. In 1984-85, he spent some months in West Papua, then called Irian Jaya, surveying Jamursba Medi and Wermon beaches, and was the first outsider to visit some of the villages on that coast[24].

But it is for his work in the Andaman and Nicobar Islands that he will be best remembered. His surveys and sojourns on many uninhabited islands in these islands provided the first (and in some cases, only) information on sea turtle nesting on these beaches. Satish's long association with the Andamans is both legendary and inspiring. His compiled report on the 1979 survey, which covered an eight month period from September 1978 and May 1979, was astonishing, to say the least. The editor's note preceding *Hamadryad* 4.3 said the following[25]:

> Satish Bhaskar has been with the (Madras) Snake Park for over three years, involved mainly in the study of sea turtles on the Indian

sub-continent. He is a self-made marine biologist, in his element when being circled by sharks or stung by jellyfish …

So, since we can't think about anything but the Andamans, and because we feel Bhaskar's report is a valuable contribution to science, it seems appropriate to devote the entire September '79 issue to his survey.

Satish's accounts begin with the remarkable words[26]: 'In the interest of brevity, it will be better to list the areas and islands not surveyed.' After a brief listing that included the North Andamans, the west coast of the Andamans that was part of the Jarawa reserve, and the southern Nicobar group, he says: 'Practically all the remaining sandy coasts in the Andaman and Nicobars were visited.'

Satish arrived in the Andamans by ship in September 1978, and set about surveying the islands in the South Andamans. He was hosted initially by Captain Dennis Beale, who started tourism in the islands, but soon found lodging in the YMCA. On one of his first trips, he had a boatman drop him off at Tarmugli Island. Despite the boatman's warnings, Satish attempted to cross the mangrove swamp at one end of the island after walking around the island and surveying its beaches. He got hopelessly lost as night fell, but eventually managed to find his way to the beach. Finding his way back to the point where his luggage was stowed, he ended up having to cross a couple of creeks. In one of his first reports back to the croc bank, he wrote[27]:

> Some bad news. The Camera's had it. While it's not yet 'done for' (the shutter mechanism is fine) the spring connected to the cocking mechanism appears to have got rusted because of a brief dousing it received while I was swimming a creek which came in my way.

From September 1978 to May 1979, he covered South Andamans, Little Andamans, and most islands in the Andamans group (except North), central Nicobar and Great Nicobar. Bhaskar first surveyed South Andamans between 7 October and 4 November 1978. He found that killing of sea turtles had been rampant before the implementation of the Wildlife Act in 1972. Sea turtles were actively hunted by fishing communities in Wandoor and other small towns. Wandoor[ii] was the principal 'turtle depot' and Bhaskar found thirty-four green turtle skulls there even a year after the ban. Local estimates of the catch ranged from five to twenty during fishing days before the ban. Most turtles were caught by harpooning by the Bengali fishing community[28].

During that survey, he covered much of the coast of the main island of South Andamans (apart from the Jarawa coast), most of Rutland, twelve small islands off Wandoor (the Labyrinth group) and the Twins. On the Twins, he found twenty-three sets of fresh hawksbill tracks, and a green turtle carcass. The islands, then known as 'kachua tikeri' (turtle island) were remote, but fishermen made the journey of several hours from Wandoor to collect turtles and eggs. Bhaskar also found evidence of nesting on Rutland and several of the smaller islands[29].

In a series of letters from this trip, Satish narrated a variety of fascinating encounters[30], including with a shell diver called Tuna from Havelock. Tuna and his crewmates had been caught in a cyclone and drifted after the boat engine failed. They reached Burma after twenty-three days without food, all unconscious by then. Satish questioned him about sharks:

> He's encountered sharks on numerous occasions while diving and has seen a Bengali dismembered to death off Herbertabad. No.4, but when I questioned him about sharks, the first incident he related was about a co-diver who grabbed his neck and hung on in terror when approached by a 'monster' – which was a dugong! This was in the channel where I saw the croc bubbles.

Bhaskar surveyed Little Andaman Island between 29 December 1978 and 5 January 1979[31]. He covered much of the coast looking for tracks and excavations and also interviewed the settlers and Onges. He found a large number of leatherback excavations and small numbers of the other species. Bhaskar finally observed his first leatherback nesting on 31 December at West Bay beach, a beach that is today our primary monitoring site. He wrote[32]:

> I've just returned from Little Andaman today after 8 days there and it was unbelievable … And two hours before New Year I saw my first nesting leatherback! She had finished egg laying and was camouflaging the nest area when I saw her. I tried barring her path to the sea by standing in front of her but she kept coming on and I had to step aside.

In a chapter in a CMFRI collection on the 'Mariculture Potential of Andaman and Nicobar Islands' that he co-authored with Whitaker[33], he proudly stated: 'This is the first occasion in 50 years that leatherback nesting has been observed and recorded on Indian soil'.

What is notable is the excitement this discovery of leatherback nesting generated. Given the disappearance of scattered leatherback nesting from the mainland, and the lack of knowledge of the islands, this was very exciting at the time. Bhaskar noted then that the main nesting beaches were located on South Bay and West Bay, which is largely the case today as well, with monitoring programmes at both beaches. On the interaction with the Onge tribal communities, Bhaskar believed that traditional hunting may have no impact on the population 'because of the antiquity of this interaction, and the small scale' on which it occurred[34].

Another legendary Satish story occurred on this trip. In general, Satish travelled light. Apart from a few clothes and bedding, he was usually well stocked on biscuits. Often he would run pretty low on food, but the biscuits did come in useful during his survey of the Little Andamans. He wrote[35]:

> Most of the leatherback nesting takes place on 8 km long west bay which is about 1½ days walk from the nearest human presence at the lighthouse construction site which in itself is quite remote. I had travelled 3½ days without seeing a human footprint. Coming back, I ran into an Onge tribal camp, occupants missing, and green turtle meat roasted and just begging to be eaten, which I surreptitiously did (never having eaten turtle before and being quite famished – had

eaten only biscuits and vitamins for the past four days). I left two biscuit packs for the Onges to salve my conscience, actually mainly to reduce chances of stopping a spear if found out!

Apart from various aspects of sea turtle nesting biology, Satish remained a keen observer of the local culture, and often wrote about their customs. On one occasion, he wrote[36]:

A hunting party of Onges had set up camp by the seashore. The Onges were not present. A green turtle had been freshly roasted over a fire and carved up. Many portions had been carefully wrapped among the leaves of a *Crinum asiaticum* plant which grew nearby, and placed over the still-warm embers of the fire. A few choice parts – portions of what appeared to be liver and flippers – had been roasted (the liver incompletely so) and placed on an elevated grating made of sticks below which another fire had been made. The turtles' carapace was found nearby. Its outer aspect was charred black. A monitor lizard prowled about the site and entered the hollow space below the empty carapace.

Satish also wrote about human exploitation of turtle meat and eggs. It is not clear how he felt about it at the time; on the one hand, he was quite happy to partake of a few eggs and meat when required and therefore did not consider it fundamentally wrong; on the other, he did feel that wildlife laws had to be enforced or turtle populations would decline beyond redemption. At Betapur, he also met a man named Rosappa, who told him proudly that few sea turtles nested there because he had removed every single sea turtle nest he had encountered on the beach in the past ten years[37]. Prior to the legal ban in 1977, turtle eggs were reportedly on sale at Port Blair and Diglipur. About exploitation of turtles for meat, he said[38]:

The Green turtle and the Hawksbill are the species usually eaten. People of most communities will eat turtle. They include Bengalis, 'Ranchis', Tamilians, Andhraites and Karens as well as the descendants of settlers and the tribals – Nicobarese, Onges, and Great Andamanese. The Leatherback is apparently killed only for its oil – whether for medicinal purposes or for fuel was not ascertained. Turtles found nesting do not often escape slaughter. Two labourers found and killed a nesting Hawksbill on 4 or 5 April, 1979. On my arrival at Pygmalion Point on 6 April the Hawksbill was hidden under the sand, where I found it accidentally while digging for eggs.

Satish also visited central Nicobar and Great Nicobar on this trip[39]. He found the Nicobars absolutely fascinating. On one occasion, he was camping at Trinkat Island and was woken at 5 a.m. by a small crocodile gaping at him through the mosquito net. He wondered if there were instances of crocodiles attacking sleeping persons and noted that if the crocodiles that woke him got any bigger, he might head home. In central Nicobar, Satish injured his knee, and nearly passed out '10 km from anywhere'. The knee became sore and stiff, and unable to bend it, Satish spent several frustrated days unable to dive or jump on boats going to interesting islands.

Satish was rightly concerned about human habitation and development, which constitute the principal threats to sea turtles in the islands today[40]. He felt that the leatherback beaches near the mouths of the Dagmar and Alexandria Rivers had become vulnerable because of the construction of the East–West road from Campbell Bay to Kophen Heat. Satish would have been happy to know that the road fell into disuse after a series of landslides and when we visited, we had to go there by boat. The Nicobari hamlet was still there, as was Mathe budda, a 'ranchi' who had arrived with the General Reserve Engineer Force (GREF) and stayed. Unfortunately none of them would survive the 2004 tsunami.

He also wrote of the 'Galathea river, about 40 km by road south of Campbell Bay, there exists a one kilometre long sandy beach which straddles both sides of the river mouth. Substantial nesting by three species – Leatherbacks, Greens and Ridleys (or Hawksbills) occurs there.'[41] The beach would become the site of study and tragedy many years later. When I visited Great Nicobar, local people still remembered him fondly. We worked for several months at Galathea, though no green turtles nested there any longer.

Despite this monumental survey, Satish felt there was still much to be done. Many of these lacunae, he would fill in the coming years. In 1981, he returned to visit Great Nicobar and Little Andaman Islands again with his Karen team (including Saw Uncle Pa-Aung, Saw Ladi and Saw Nelson) with WWF providing the princely sum of Rs 6,750[42]. Bhaskar visited these islands between January and March of that year. Given the logistics involved, he actually only spent nine days on Great Nicobar and six on Little Andamans, but this was sufficient to confirm these as significant leatherback nesting beaches. On Great Nicobar, he found that many leatherback nests and most ridley nests had been marked by stakes by the Nicobarese, indicating that these had been predated by animals or collected by the Nicobarese themselves. On Little Andamans, he found the same number of excavations on West Bay (eighty) and South Bay (ten) as he had on his previous visit. He also recorded an immature leatherback of an intermediate size in Hut Bay on 1 March 1981, a rare record for any part of the world even today. That turtle was killed and eaten by settlers (another rare occurrence).

In 1984, funded again by WWF, he returned to the islands and finally covered the North Andamans and other Andaman islands[43]. During this trip, he visited Landfall, the northernmost of the Andaman Islands, but did not find much evidence of nesting (thought some nesting has been recorded there on subsequent surveys). Alcock, in 1902, had recorded nesting in Little Coco Island, part of Burmese territory and only 38 km away. Bhaskar visited a number of islands on this trip including Interview where he recorded the capture and consumption of hawksbill turtles by Telugu and Karen fishermen. In one instance, the Telugu fishermen presented the hawksbill to the Karen in exchange for getting their boat towed to Mayabunder. The Karen apparently also ate boiled hatchlings when they could get them.

Bhaskar also visited South Reef Island for the first time during this visit; this was already known as 'turtle island' amongst the locals. Bhaskar documented hawksbill nesting during his survey and established the island as a significant hawksbill and green turtle nesting beach; he found over 350 hawksbill nests on ten islands with nearly half on South Reef and North Brother[44]. Apart from revisiting a number of islands in the southern Andamans group, Bhaskar also returned to Little Andaman Island where he documented sixty-four nests on West Bay; on this trip, he also discovered leatherback nesting at Jackson Creek, north along the coast from West Bay[45].

A few years passed before he could return to the islands, but he remained occupied with his surveys in Papua, and mentored the SSTCN in the late 1980s. In 1991-92, he returned for a survey of Great Nicobar Island[46] with Manjula Tiwari, a young enthusiastic field researcher with the Croc Bank. That year, he started his monitoring on South Reef Island, which he would continue for several years[47]. South Reef is a small boat-shaped island, about 500 m long and 100 m wide, located a kilometre and a half from Interview Island. The island was mostly forested and had a few clearings with camps established by divers and fishermen. Only small boats could land, and even that only during relatively calm weather. Bhaskar would be dropped off with supplies and he'd spend several weeks by himself on the islands. On occasion, he swam the entire kilometre back to Interview Island for freshwater. Between 1993 and 1995, Bhaskar continued his monitoring of South Reef Island sand and surveyed other islands periodically[iii]. Between May and November of 1995, Bhaskar spent most of his time on South Reef Island. That was to be his last trip to the islands.

Satish's time in the Andamans is unparalleled. His published and unpublished reports have formed the basis for current sea turtle conservation initiatives and it is thanks to his data that interventions were made possible to protect beaches in the Andaman Islands, which were otherwise slated for tourism development. He wrote prolifically, detailing all his trips in his reports and articles for *Hamadryad*. He also wrote widely and engagingly in other journals and magazines. In 2012, we carried a profile of his work[48] in the *Indian Ocean Turtle Newsletter*, which included a tribute by Rom Whitaker[49] along with a list of his publications and surveys[50].

Turtle surveys afresh

Following these surveys, there had been little work and Harry Andrews was keen to start a monitoring programme for leatherback turtles. Andrews started surveying the islands with Manish Chandi and a team of Karen field assistants from the mid-1990s. Each year from 1996, they made a round trip of all the Andaman Islands, except Barren and Narcondam. During these surveys, they travelled by boat to all the nesting beaches mentioned by Bhaskar and attempted a few others as well; as usual signs of green and hawksbill nesting were common across nearly all beaches.

The opportunity to expand the scope of the project arrived when funding came for surveys from the GoI-UNDP sea turtle project. Armed with these funds, Andrews carried out surveys and initiated a monitoring programme[51]. The team set up field camps at Jahaji beach in Rutland Island, and Cuthbert Bay Beach in July 2000, and in Galathea in November 2000. Between July 2000 and April 2001, the team carried out surveys of all the beaches and islands in the Andaman group and established that sporadic leatherback nesting was occurring at several beaches such as Cuthbert Bay and Jahaji beach, but the West Bay beach in Little Andaman continued to be the most important site in the Andamans for leatherback turtles. In the Nicobar, the important beaches remained the ones on Great and Little Nicobar identified by Bhaskar and Tiwari, with hundreds of nests each season.

Two field assistants, Saw Glen and Saw Agu were sent to the Galathea beach along with a researcher, Shreyas Krishnan. Krishnan spent seven months with Glen and Agu on Great Nicobar as part of the first monitoring season at Galathea. Known to doze off while tagging turtles on the beach, it is a wonder though that he did not end up as some crocodile's dinner! The funds also allowed them to purchase Passive Internal Transponders (PIT) tags – which are injected under the skin and read using scanner-like bar codes. In the first year, they tagged about 146 animals at this beach alone. In addition to PIT tagging leatherback turtles, olive ridley, green and hawksbill turtles were notched on the carapace with hacksaws at Galathea, as well as at Cuthbert Bay and Jahaji beach when field camps were maintained there. In 2001-02, I spent a nesting season at Galathea as part of my genetics project; we tagged about 150 leatherbacks that year. We also used Inconel tags to tag the smaller turtles. Monitoring continued for the next couple of years, but no tagging was carried out as we did not have the required permits.

In December 2004, tragedy struck in the form of a tsunami. That year, Andrews and I had raised additional funds for the monitoring programme and were excited by the prospect of restarting tagging. We hired a researcher, Ambika Tripathy from Odisha, and sent him to Galathea with Agu, by now a completely competent field hand. Agu was the only survivor.

With the surveys of Great and Little Nicobar, we were able to estimate that about a thousand leatherback turtles may be using these islands to nest[52]. With the decline of Pacific populations, these Indian Ocean populations gained importance. While these populations are reasonably large and did not face any major direct threat, the beaches were destroyed by the tsunami in 2004. In the days before the tsunami, Galathea, though inaccessible to tourists from the mainland, did attract a large number of locals who flocked to the beach on weekends to catch a glimpse of the turtles. Popular myth had it, of course, that the turtles came all the way from Australia, which might have been part of the attraction. Though this was indeed possible, there was no evidence for or against this hypothesis when we worked there.

After the tsunami

The Indian Ocean tsunami, which struck on 26 December 2004 after a massive earthquake off the coast of Sumatra, resulted in the death of more than 2,00,000 people in fourteen countries, and is considered one of the biggest natural disasters ever. The earthquake itself was the third largest ever recorded and the longest in duration. The southeast coast of India was severely affected, with waves as high as one to two metres striking the coast, and resulting in substantial loss of lives, habitations, boats and ports. Over 2,000 sq km of land was affected by inundation, with the effects ranging from 300 m to 3 km inland. With the epicentre of the earthquake just over a 100 km from the southern tip of Great Nicobar, the island was hit by waves over ten metres in height. The coasts of most islands in the Nicobar group were completely devastated and over 6,000 people died across the islands. There was large scale inundation of coastal areas, and it is estimated that 50 to 100 per cent of mangroves were destroyed in the Nicobar Islands. Some islands such as Trinkat and Pulomilo were divided into two or three parts by the subsidence and inundation. The islands are believed to have moved 100 feet and tilted, while Indira Point (the erstwhile Pygmalion Point and the southernmost tip) subsided by 4.25 metres. The lighthouse was destroyed.

After the 2004 tsunami, Harry Andrews, Allen Vaughn and Manish Chandi carried out surveys of sea turtle nesting beaches in the Andaman and Nicobar Islands[53]. The subsidence of many islands in the Nicobar group, and upheaval in the Andaman group had caused dramatic changes in the coastal topography. Some beaches had become inaccessible due to the upheaval of the reefs, but nesting continued on other beaches, but seemingly at a lower intensity. When they surveyed beaches in 2005, the major nesting beaches on Little Andaman Island had all been washed away and there was no evidence of nesting, but the following year, in 2006, the beaches appeared to have formed again and there was nesting by several species, including leatherbacks. During one of their surveys, they also met some Onge who said they had gone turtle hunting near Dugong Creek[54]. They returned with several lean and undernourished green turtles, and realized that the sea grass beds had been damaged and the turtles did not have enough to eat. The Onge women then said that, despite their craving, they should not eat turtles until the turtles had had a chance to feed themselves.

The beaches in Great and Little Nicobar had been completely destroyed by the tsunami, the waves reaching more than 1.5 km inland[55]. The coast was covered with the debris of fallen trees and other flotsam. A few beaches had begun to form again at this time, and surveys a few years later indicated that sea turtles were beginning to nest again at these sites. Chandi visited Galathea, Indira point and the north-western coast of Great Nicobar and the west coast of Little Nicobar in 2008 and 2009 with construction labour and Nicobarese, and found beach formation as well as sea turtle nesting.

The site was visited by Naveen Namboothri and Saw Agu in 2011, who was returning to the beach for the first time since the tsunami. Nesting beaches had formed again on the south side of the river, and it appeared that a large number of turtles were nesting there again. However, the river mouth was exceedingly large and difficult to cross, rendering the beach inaccessible and the logistics of a field camp prohibitively difficult. Manish Chandi and Adhith Swaminathan returned to the beach the following year, and were greeted by a mini tsunami, courtesy of another earthquake in Indonesia.

Leatherbacks in LA

The West Bay and South Bay beaches in Little Andaman Island were discovered by Satish Bhaskar during his surveys in the 1970s and 1980s. One of the first to carry out a research project there was Arjun Sivasundar, a Master's student at the Pondicherry University. Sivasundar made his first trip to the islands in the company of Bhaskar himself, travelling by the challenging bunk class on ship from Chennai to Port Blair, a journey of over three days. Working with the nascent ANET in 1995-96, Sivasundar studied the nest placement of leatherback turtles on West Bay and South Bay beaches on Little Andamans and Jahaji beach on Rutland Island[56].

The South Bay and West Bay beaches were confirmed as important nesting sites by Andrews and the ANET team during surveys. Following the tsunami, the team visited these sites again and found that the beaches had been washed away but started forming again the following year. Manish Chandi and I surveyed the South Bay and West Bay beaches in 2007 and established that nesting was occurring at these sites.

In January 2008, we started the first monitoring camp in South Bay with Devi Subramaniam, a junior researcher. Naveen Namboothri, a postdoctoral researcher at the Indian Institute of Science helped establish the camp along with our Karen boatmen, Uncle Pamwein and Uncle Berny and a group of Karen field assistants, including Agu. Monitoring continued at this site for a couple of years more, but the beach was difficult to access (the camp was separated from the beach by a creek that ran quite deep at high tide) and the number of nesting turtles was quite low. Adhith Swaminathan, who had been monitoring olive ridleys in Odisha, joined the project in 2011, and initiated a camp at West Bay. Once the logistics of landing was worked out (involving a precarious landing process at the rocks nearby), this camp turned out to be idyllic.

Over the last few years, about 100 leatherback nests have been recorded here each season as well as smaller numbers of other turtles. Naveen Namboothri and I visited the camp each year to fit satellite transmitters on the leatherback turtles. Of the ten turtles we tagged between 2011 and 2015, most travelled eastward, either along the coast of Sumatra or towards the Cocos (Keeling) Islands in the Indian Ocean. Two of these turtles even reached the north-western coast of Australia. So the myth was true. A few went westward and reached the central Indian Ocean and finally, in 2014, both tagged turtles went westward past Seychelles, reaching the Madagascar and Mozambique coasts. The IISc-Dakshin-ANET collaboration aims to monitor this nesting beach for several years to come.

Island people

Rom and Zahida 'Zai' Whitaker first planned a trip to the Andamans in 1975 but Rom could not go as he was not an Indian citizen. That year, Zai went on a trip with Annamalai, one of the Irulas from the Madras Snake Park, and conducted a brief survey of herpetofauna. More important was a serendipitous introduction to the Karen. Zai had made contact with Fred Burn who was just retired from the Andamans Bush Police and had taken over as manager of the Wimco match factory, part of the Western India Match Corporation, the Swedish match company. Burn and his wife Jean looked after Zai and introduced them to the Karen through Gerry Vaughn, the head radio operator for the Andamans Bush Police. Gerry Vaughn, also an Anglo-Indian from Calcutta, was married to Ruth, a Karen nurse at the government hospital in Mayabandar (and Uncle Pa-Aung's sister-in-law). Their son Allen, with whom Rom and Zai would strike a long friendship and partnership, would eventually became crocodile bank manager for a period.

The following year, Rom got his citizenship and travelled to the islands. He had been hired by Robert Bustard to collect crocodile eggs. Following up on their lead, he went straight to Mayabunder to meet Allen and the Karen. At the time, they were mostly into hunting, fishing and shell collection, but would go on to become the backbone of ecological research in the islands. Almost right away, they got news of a king cobra nest and were rewarded with the sight of king cobra babies hatching. Over the next few years, Rom worked on getting saltwater crocodile eggs back to the Crocodile Bank. He was finally successful, but the flight from Port Blair to Calcutta was via Rangoon, and he had to get a Burmese visa. Placing the box of eggs on the seat next to him, it was quite a challenge, but the eggs did arrive and thirty-three out of thirty-five hatched successfully.

Rom and Zai were deeply concerned about the timber operations in the Andamans. Recalling the period, Rom said:

> Everywhere that you went they were log rafting – everywhere. You couldn't go to any part of the islands without running into rafting. Hardwoods and softwoods lashed together with chains and being hauled by tug-boats, going here, going there. Chatham – the whole Port Blair harbor was just nothing but rafts of logs and people running across them and pushing them. Just vast quantities of logs with guys jumping from one log to the other, chaining them up, and elephants working, the Chatham saw mill going full steam. They were cutting mangroves to power the sawmills, to power SS Cholunga, the boat that used to run up to North Andaman.

In the 1980s, after the formation of the Island Development Authority and the interest in the islands shown by Rajiv Gandhi, deforestation was halted and the focus shifted to fisheries and tourism. The reduction and eventually cessation of the timber industry was in great part due to the efforts of prominent conservationists on the mainland, who recognized the tremendous biodiversity value and heritage of the islands. As Zai Whitaker wrote in one of her editorials in *Hamadryad*[57]:

> A kindly critic of *Hamadryad* wrote me a short and nasty post card some months ago. 'You and your Andamans!' he said. 'Can't you think about anything else?'
>
> No. All of us at the snake park have the Andaman and Nicobar fever, which is transmitted through contact with these fascinating islands. ... It is a bird and reptile paradise. When our field staff stagger back to Madras (with malaria or jaundice or just general decay) from the Islands, there are endless nocturnal story-telling sessions – of king cobra and crocodile nests, unrecorded sea turtle nesting beaches, *Laticauda* nesting colonies, and much more.

Rom and Zai would get increasingly involved in various issues. Dennis Beale, their host and mentor in the islands, connected them to various others in the islands, and found land for a research base. Their first boat was a trawler seized from Customs, brought to the islands by a naval vessel (courtesy their friend Vice Admiral Awati) and renamed *MV Aka Bea*. They took the first underwater photographs of the reefs using a Nikonos camera and black-and-white film. They hosted their first student visitors, a group from the Harvard Museum of Comparative Zoology.

Rom himself wrote several articles that were eventually compiled into a booklet. Various others wrote and spoke on behalf of the islands, and the Central government under Prime Minister Rajiv Gandhi was sensitive to environmental issues. Most critically, Rom and Zai would raise funds for their idea of setting up a base for ecological research and conservation in the islands, an idea that was to come to fruition under the leadership of their Curator at the Croc Bank, Harry Andrews.

Engineering surveys in the islands

In the mid-1970s, another engineering student became tired of the drudgery of courses, and started volunteering at the snake park. Harry Andrews became a regular at the park and on turtle walks, and started exploring the areas around Madras with Satish Bhaskar and others. Since he was one of the few who had a motorbike, he was able to get around easily, an invaluable asset. By the 1980s, Harry had become a central figure in the crocodile bank and eventually became the curator. As an engineer, Harry had skills that were particularly useful, and crafted thermistors that would be used for measuring nest temperatures. By the early 1990s, however, the Andamans bug had bitten Harry as well. The Andaman and Nicobar Islands Environmental Trust (later changed to Team) had been registered in 1989 with Zafar Futehally as the third trustee, but activities started in earnest in 1993. Rom did the begging rounds in New Delhi resulting in grants from the Netherlands, Norway and Sweden. Five acres of agricultural land near the sea was bought and four researchers' quarters were built at this time, and a series of surveys were initiated including sea turtle and crocodile surveys.

Harry went on a five-month long survey with Satish and Allen and was introduced to the other Karen. Harry attributes the tradition of working with the Karen to Satish's strong bond with them. Over the next few years, Harry conducted a series of crocodile surveys, often wading in chest deep water across creeks at night, using a torch to look for eye shine. On one occasion, surveying the west coast of the Andamans with Satish and Uncle Pa-Aung, they came across hundreds of mating turtles, the most Harry had seen, but they had to beat a hasty retreat because a group of Jarawas started shooting arrows at them. Harry was also present with Satish on Smith Island on a night when there was a mini arribada with about 200 ridleys coming ashore.

Between 1996 and 1999, Harry conducted yearly sea turtle and crocodile surveys along both coasts of the Andamans. He and Manish Chandi mapped and logged a variety of information and hoped to initiate multi-disciplinary surveys in future years. During this time, ANET's infrastructure also grew; the ANET office building was constructed and the kitchen refurbished with a grant from the Netherlands embassy.

In 1999, Harry did a reconnaissance survey of Galathea beach, which Satish had identified as one of the important beaches for leatherback monitoring. Along with Manish Chandi, Agu and Allen Vaughn, they visited the east and west coasts of Great Nicobar. Agu, according to Harry, fell immediately in love with Galathea and was delighted to set up camp and work there. Harry set up the long-term monitoring programme in Galathea with Agu, Glen and Shreyas. This programme would carry on for the next five years till the tsunami. He also set up camps at Rutland and Cuthbert Bay which would be operational for a few years.

Harry spent more than a decade working on the islands, and collected a number of stories from his friends and acquaintances. His friend Percy Myers told him about the British holding tanks for turtles, where his mother would buy turtles off one of the sentries. And he also heard stories about Galathea from his friend Susheel's father, perhaps the earliest information that one has on this beach:

> He used to be there in the late '50s or something. And he used to talk about thousands of turtles coming to Galathea. So, every season in December, they would come to Great Nic. And halfway they would go by truck and tractor and stuff like that, and then trek, with supplies and water carriers. And they used to stay on Galathea, during Christmas time. There were some nights when the entire beach would just be black with turtles. And he's the one who told me – you get these tunnels [in] waves ... You know the waves come up like that and surfers go through ... and he used to say you can see *huge* turtles.

Harry Andrews played an absolutely critical role in the growth and development of ANET. He planned and supported the sea turtle surveys by Satish Bhaskar, Manjula Tiwari and Arjun Sivasundar, and numerous other researchers who worked at ANET. Harry was the link between Satish's seminal surveys and the long-term monitoring programmes that would come later. The National Biodiversity Strategy plan for the islands was prepared by ANET under his leadership, as also the State of the Environment report and a compilation with Vasumathi Sankaran of the Indian Institute of Public Administration.

Manjula goes to Galathea

Growing up at the Sri Aurobindo International Centre of Education in Pondicherry in the 1980s, Manjula Tiwari spent many enjoyable days diving around the pillars of Pondicherry's piers and helping maintain the marine aquarium for the marine life captured on the dives. After completing her Bachelor's degree, Manjula wanted hands-on research experience and joined the Croc Bank. There she assisted with the crocodile research project on temperature sex determination and built marine tanks and caught fish for the tanks. Within a short while though, Manjula was busy planning a trip to the Nicobar Islands. It had been some time since Satish Bhaskar had last visited the islands, and many islands in the Nicobars had not been surveyed. Manjula prepared by discussing the trip and the islands and turtles with Satish. Still, this was a young woman in her early twenties heading out to these remote and challenging islands on her own. But once she was there, she says that she never felt uncomfortable and always had help from local assistants.

Manjula surveyed the southern Nicobars with a local field assistant and a local Nicobarese boat and captain[58]. When she didn't have paper, she inscribed her notes on scaveola leaves. She walked the beaches, crossed crocodile-infested creeks and lived in the tribal huts en route. The creeks were often broad and deep, and prime salty habitat. The assistant ensured they crossed at low tide during the day. And in addition, he told her, 'Madam, throw your life energy over to the other side before you cross the creek. Nothing can harm you then.'

There were of course some notable incidents. One night at Galathea, as Manjula looked for nesting leatherbacks, she turned on her flashlight as she approached the creek to ensure that she did not stumble into any salties. Suddenly she caught sight of two pairs of glinting eyes, huge crocodiles judging by the gap between the eyes. Suddenly, the eyes rose several feet off the ground and started walking towards her. Thoughts of monster crocs passed rapidly through her mind, till the creatures took shape – feral cows. A decade later, we encountered the beach cows at night by the creek and were equally startled by their presence.

Manjula carried out a second survey of Great Nicobar Island with Satish. Her surveys provided the basic information and impetus for the initiation of the monitoring programme there in 2000. Manjula went on to join the graduate program at the Archie Carr Center for Sea Turtle Research, University of Florida, Gainesville, and worked on sea turtles in South America and Europe for her M.Sc before settling into her Ph.D on green turtles at Tortuguero, Costa Rica. Manjula now works for National Oceanic and Atmospheric Administration (NOAA) in San Diego, California, and her projects are scattered around the world in Africa, the Middle East, and West Papua, including at Jamursba Medi, which her one-time mentor made famous. She is also the long serving secretary of the International Sea Turtle Society.

Manish of the Islands

In the early 1990s, Manish Chandi was finishing his Master's at the Madras Christian College, my alma mater, and the birthplace of the Students Sea Turtle Conservation Network. Manish had started a campus green movement along with other students of the time, called 'Scrub Society'. The society was formed to defend the natural scrub habitat from being removed to have, of all things, a water conservation fair. Bored with his Master's, Manish wandered to the Madras Crocodile Bank and ended up, like many others, in the 'islands'. He spent the first three or four years (1995-1998) almost entirely at ANET as his project (as a WWF conservation corp volunteer) was to reforest the entire property, create brakes and check erosion and initiate the rainwater harvesting system, all of course with Harry and the rest of the team. He spent the next several years wandering around both the

Andaman and Nicobar Islands, but mainly the latter, interested in the natural history and the culture of the native communities, especially their resource use and engagement with nature.

Manish's travels have made him an authority on Nicobari communities, and he is currently completing his Ph.D on the human ecology of these communities. Manish was involved in ANET from the outset, and was one of the key figures in its establishment and growth. He has conducted numerous projects on the socio-economics of local communities, and been involved in developing plans for conservation and sustainable livelihoods. He also accompanied the team on sea turtle surveys and was instrumental in establishing the monitoring programme in Little Andaman in 2007.

These pioneers paved the way for a vibrant new generation of researchers at ANET. In the last five years, ANET has blossomed under the enterprising leadership of Tasneem Khan. Today, there are a host of researchers at ANET each season from a range of institutions including Dakshin Foundation, Indian Institute of Science, National Centre for Biological Sciences, Nature Conservation Foundation, Wildlife Institute of India and other institutions. The centre also hosts courses such as a marine biology course for students of the Master's programme in Wildlife Conservation at NCBS with which ANET has an MoU. ANET is now abuzz with activity during the 'dry season' from October to April.

The Karen

The Karen are a community from south-eastern Myanmar who were brought to the Andamans by the British as forest labourers in the 1920s, and settled in a few villages near Mayabunder in the middle Andamans. Some of the older 'Uncles' as they are known still remember the Japanese occupation during World War II. In general, the Karen are spectacular hunters and fishermen which makes them perfect partners for the naturalists and ecologists who turned up half a century later. The long association with the Karen started when Rom and Zai Whitaker met Allen Vaughan and Saw (Uncle) Pa-Aung in

Mayabunder. While Allen started working with them and later with Croc Bank, Uncle Pa-Aung accompanied Satish on some of his first surveys. However, in the 1990s, once ANET was started, he became their indispensable boatman for all their surveys. In the late 1990s, another Karen boy became involved with ANET; Saw John soon became an indispensable asset as an organizer. John helped organize all the logistics for surveys, set up camps, occasionally accompanied researchers on surveys and has, for over fifteen years, run the ANET base in Wandoor as its manager.

In 2007, Manish and I surveyed the coast of Little Andaman, travelling by boat from Wandoor with the aging Uncle Pa-Aung and Agu. This was Uncle Pa-Aung's last field trip, and my abiding memory of him is sitting impassively at the rudder when a squall hit us on our return journey. As rain lashed the boat, and rough seas tossed the *dunghi* around, Uncle Pa-Aung still looked as if he were on a picnic on a lazy Sunday afternoon. When we started the field camps in the Little Andamans, it was two other senior citizens from the Karen community who stepped in to run the operations. Uncle Bernie, in his seventies already, was the boatman, and Uncle Pamwein, probably older, ran the camp for a year. Uncle Bernie continued as the boatman for the Little Andamans project for several years.

The other central figure in the sea turtle projects was Saw Agu. Agu's field talents were spectacular even by Karen standards. After the first survey of Galathea, Agu remained devoted to the place and worked there for several field seasons, from 1999 to 2004, for long stretches without researchers to guide him. In his first year at Galathea, Agu went through at least ten bouts of malaria, one every few weeks. In 2000, while Shwether and Manish surveyed the west coast for locations to set up another camp, they too caught malaria, and returned with Agu to Wandoor in December of that year. Tragically, Shwether died of malaria while Manish escaped narrowly. Agu was determined not to let the camp close down and promptly returned the following year with Glen. Agu had complete command of all the field techniques, and was able to maintain detailed data records.

When the tsunami struck on the morning of 26 December, Agu

was on the beach with Ambika Tripathy and some visiting researchers from Pune. Not heeding Agu's warnings, they remained on the beach when the water receded and were caught by the incoming wave while they tried to escape to higher ground.

Agu's tale of survival is legendary. He had been washed out into the bay in Galathea which had become a swamp flooded with water and fallen timber. Holding on to a log of wood, he survived for more than two weeks without food and water. When a monitor lizard started nibbling on him, he swam back to the mainland with broken collarbones realizing he had no other option. He was rescued by a helicopter sortie seventeen days after the tsunami; the story is recorded by Manish Chandi in an issue of *Current Conservation*[59].

Agu returned to this beach with Naveen Namboothri in 2011. After waiting for the fishing boat to pick them from the beach, now south of an extremely wide river mouth, they decided to raft across the mouth on some Styrofoam and drums lashed together. Agu was not amused at having to risk his life at this beach again, but remains as committed as ever to his work at ANET. More recently, a number of others have become dedicated members of the ANET sea turtle team. Agu's brother, Thesorow and others such as Kennik, Columbus, Isaac, Sabien, Uncle Berny and Uncle Mumo, have worked at the sea turtle camp in the Little Andamans. There have also been field assistants from the settler community, especially the Ranchi, but the Karen remain indispensible.

Island theory

Islands have played a disproportionately important role in the history of community ecology and biogeography. They have fascinated biologists from Charles Darwin's time and even earlier. The theory of island biogeography, developed in the early 1960s by Robert MacArthur and E.O. Wilson[60], generated a whole field of study on animal and plant communities on islands including aspects of diversity, community structure and population dynamics. The theory was later extended to fragmented and isolated habitats on land and gained importance in the context of fragmentation of forests all around the world. In this context, it also became an integral part of the field of conservation biology and was invoked to create rules for the design of natural reserves or refuges.

The theory essentially calculates species richness on islands as an equilibrium between the number of species colonizing the island and the number of species going extinct on the island. Islands further away from a mainland or species source have lower rates of colonization than those close to a mainland or species source, while larger islands have lower rates of extinction than small islands. Hence, large islands that are closer to the mainland have higher species richness at equilibrium than small, far islands.

Despite initial opposition, the theory became the backbone of conservation biology within a decade, with a series of papers advocating a set of rules for the design of refuges[61]. One of the most widely debated of these was the SLOSS ('Single Large or Several Small' reserves) dilemma[62,63], i.e. if a given area was to be set aside as a natural reserve or sanctuary, would it be better to have several small patches of forest or a single large patch? Island biogeography suggests that smaller islands would have higher rates of extinction and therefore would not support many of the species that large patches would. This implies that a single large patch would have more species than several small patches combined.

A related theory, metapopulation biology, became prominent in the 1990s. The term metapopulation was coined by Richard Levins as early as the 1970s to describe a population of populations[64]. Classical metapopulations are composed of a series of patches where a particular species could potentially survive, with some of the patches being occupied and others being empty at any given point in time[65]. Over time, the landscape changes with the extinction of populations in occupied patches and the recolonization of empty patches. The classical Levins metapopulation is a large network of similar small patches, with population processes within a patch occurring at a much faster rate than metapopulation processes, i.e., population processes across patches. Hence, births and deaths within a population in a patch occur at a faster rate than extinctions and recolonizations. This also implies that all the populations including the largest one have a significant risk of extinction. The simplest models of metapopulation dynamics, such as the classical model, only look at the occupancy of sites by a species (presence and absence) and the population sizes are ignored.

Metapopulation theory clearly demonstrates the importance of a larger number of sites, where extinction on some sites (some unoccupied sites are always present) is a natural part of the process of metapopulation persistence i.e. empty sites are necessary for the survival of the metapopulation. But how is this relevant to sea turtles? It may be. Sea turtle nesting populations may actually act as metapopulations and minor nesting beaches may act as sources for recolonization, as in the case of the ridley arribadas.

The perfect view

We spent several remarkable months on Great Nicobar Island, There was no shortage of excitement during our time there. My partner, Meera, spent her time observing tree shrews as part of a behaviour study on these fascinating animals. Tree shrews are a unique Order of small mammals found in south and Southeast Asia. They had been previously classified as primates and were at one time considered to be the 'missing link' in the evolution of primates. The most recent genetic studies suggest that they might actually be most closely related to rabbits and hares. They resemble squirrels in size, general appearance and some behavior traits, but can be distinguished by their long snouts and the absence of long facial whiskers. The Nicobar tree shrew, *Tupaia nicobarica*, is endemic to the islands and very little is known about the species.

We visited Indira Point, the southernmost point in India. There is a lighthouse here, and a helipad for visiting dignitaries, mostly past presidents and prime ministers, whose visits are commemorated by huge headstones. Local legend has it that most of them die or are run out of office shortly after they visit! We pitched our tents in the shade of some casuarina trees, and I set off to explore the small beach. Satish

Bhaskar had noted that hawksbills nested at Pygmalion Point, as it was then called; in one of his reports, he wrote that a hawksbill had nested on 4-5 April 1979 at Pygmalion Point[66]. I was on the beach trying to get a sense of the place as he must have seen it more than twenty years earlier. And as if time had stood still, there was a hawksbill turtle nesting under a scaveola bush. I considered that a special connection to Satish's time there.

She was an extremely small specimen, with a carapace of about 64 cm and laid only fifty eggs. At Indira Point, as with many of the other nesting beaches of hawksbill and green turtles, the offshore approach is littered with rocks and coral, and the turtles have to crawl over some very rough terrain before returning to sea. Ridleys and leatherbacks prefer open approaches and would never nest at beaches such as these.

Just a few days later, we were on a boat and sailing past Indira Point, the southernmost tip of India, with truly nothing south of us till the Antarctic. We stopped for lunch in a little bay called *saphed balu* or white sands, with blue water and corals, where leatherbacks, green turtles and hawksbills nest occasionally. Finally we reached Kophen Heat, on the west coast, one of the major leatherback nesting beaches of Great Nicobar. Here, we stayed near a Nicobarese settlement and were looked after by Mathe Budda[67], relic of the Ghat Road Engineering Force – someone who stayed behind because life was peaceful and the view perfect.

Despite its pristine beauty and untapped bounty of ecological information, few ecologists have spent much time in the Nicobars. The logistics of getting and staying here are not easy. Add to that the high incidence of cerebral malaria and poor prospects of medical attention, and it is not surprising that only a few intrepid explorers have made this their home. Those pioneers – Satish Bhaskar, Ravi Sankaran, Manish Chandi – travelled all over the Nicobars, paving the way for future generations of ecologists to seek answers in these remote outposts. Many more will arrive because of the lure of untapped knowledge and exotic biological mysteries. Some will leave because their projects are completed or their patience worn thin. And some will stay longer because life is peaceful and the view, indeed, perfect.

Notes

i Turtle skulls are sometimes kept as trophies and may have been seen alongside carved effigies or other artistic/spiritualized forms of animistic representation, which are called 'kareyou/Kareava' and also 'Hentakoi' which are constructed images of spirits, demons, ancestors etc used in ritual events. Turtle skulls are not part of this material culture, and may have been mistaken as such (Manish Chandi, pers. comm.)

ii New Wandoor beach was a turtle nesting site, and Satish Bhaskar, the local Bengalis, and the Onge who used to visit the beach to catch turtles in the 1940–50s are the last to have known and witnessed nesting here. The sand from this beach stretch was mined during the early 1990s for construction at Port Blair – including the forest department house at Haddo, which spelled the end of this beach as a nesting ground (Manish Chandi, pers. comm.)

iii In 1993, Bhaskar surveyed thirty-seven islands on the Andaman coast and Coffeedera, Karamatang and Cuthbert Bay in the middle Andamans[68]. Most of these were uninhabited and had mostly hawksbill and green turtle nesting. Later that year, he visited South Reef, Interview and Latouche islands. In early 1994, Bhaskar visited Little Nicobar, Great Nicobar and Cuthbert Bay and returned to South Reef between June and August[69]. Between September and December, he surveyed Snark, Kwangtung, Latouche and South Reef again. In 1995, he covered South Sentinel, North and South Brother, Sisters and Twin islands in the South Andamans[70].

REFERENCES

Breaking the surf

1 Sanjeeva Raj, PJ (1958) Egg laying habits of sea turtles described in the Tamil Sangam Literature. *Journal of the Bombay Natural History Society* 55: 361–363.

2 Hamilton, A (1727) *A New Account of the East Indies: Being the Observations of Captain Alexander Hamilton*. John Mosman, Edinburgh, UK. See: Hejmadi, P (2000) Earliest record of Gahirmatha turtles. *Marine Turtle Newsletter* 88: 11–12.

3 Stirling, A and J Peggs (1846) *Orissa: Its Geography, Statistics, History, Religion and Antiquities*. John Snow, London, UK.

4 Bennett, JW (1843) *Ceylon and its Capabilities*. Trumpet Publishers, Rajagriya, Sri Lanka. See also: De Silva, A (2006) Marine turtles of Sri Lanka: A historical account. In: *Marine Turtles of the Indian subcontinent* (Eds. K Shanker and BC Choudhury), pp. 324–353. Universities Press, Hyderabad, India.

5 Tennent, JE (1861) *Sketches of the Natural History of Ceylon with Narratives and Anecdotes*. Longman, Green, Longman, and Roberts, London UK.

6 Mouat, FJ (1863) *Adventures and researches among the Andaman Islanders*. Hurst and Blackett, London, UK.

7 Gunther, ACLG (1863) *The Reptiles of British India*. Robert Hardwicke, Piccadilly, UK.

8 Alcock, A (1902) *A Naturalist in Indian Seas: Or, Four Years with the Royal Indian Marine Survey Ship 'Investigator'*. Dutton, London, UK.

9 Frazier, JG (2011) Early accounts of marine turtles from India and neighbouring waters. *Indian Ocean Turtle Newsletter* 14: 1–4.; Henderson, JR (1913) *Guide to the Marine Aquarium*. Government Press, Madras, India (cited in Frazier 2011).

10 Mawson, N (1921) Breeding habits of the Green Turtle, *Chelonia mydas*. *Journal of the Bombay Natural History Society* 27(4): 956–957.

11 Greaves, JB (1933) Nesting of *Caretta olivacea*. *Journal of the Bombay Natural History* 37(1&2): 494–495.

12 Chacko, PI (1942) A note on the nesting habit of the olive loggerhead turtle, *Lepidochelys olivacea* (Eschscholtz) at Krusadai. *Current Science* 12: 60–61.
13 See n. 1.
14 Biswas, S and DP Sanyal (1977) Notes on the Reptilia collection from Great Nicobar Island during the Great Nicobar Expedition in 1966. *Records of the Zoological Survey of India* 72: 107–124.
15 Blyth, E (1846) Notices and descriptions of various new or little known species of birds. *Journal of the Asiatic Society of Bengal* 15(169): 1–54.
16 For a bibliography of his work, see De Silva, A (2006) An annotated bibliography of publications on marine turtles of Sri Lanka. *Indian Ocean Turtle Newsletter* 3: 12–26.
17 In particular, see Deraniyagala, PEP (1939) *The Tetrapod Reptiles of Ceylon. Vol. 1 Testudinates and Crocodilians.* The Director, Colombo Museum; London, Dulau and Co., Ltd. 412 pp.
18 Deraniyagala, PEP (1936) The nesting habit of leathery turtle, *Dermochelys coriacea*. *Spolia Zeylanica* 19(3): 331–336.
19 Cameron, TH (1923) Notes on turtles. *Journal Bombay Natural History Society* 29: 299–300.
20 Jones, S (1959) A leathery turtle *Dermochelys coriacea* Linnaeus coming ashore for laying eggs during the day. *Journal of the Bombay Natural History Society* 56(1): 137–138.
21 Shanker, K (2013) Leatherback turtles on the mainland coast of India. *Indian Ocean Turtle Newsletter* 15: 15–17.
22 Dutt, S (1976) The leatherback turtle. *Seafood Export Journal* 8(8): 36.
23 Anonymous (1982) Leatherback turtle washed ashore on Madras coast. *Hamadryad* 7(2): 3. (Credited to Ms. Reliable Ferret).
24 Bustard, HR (1976) World's largest sea turtle rookery. *Tiger Paper* 3(3): 25.
25 Summarized in MC Dash and CS Kar (1990) *The Turtle Paradise, Gahirmatha*. Interprint, New Delhi.
26 Whitaker, R (1979) Captive rearing of marine turtles. *Journal of the Bombay Natural History Society* 76(1): 163–166.
27 Rajagopalan, M (1984) Study on the growth of olive ridley *Lepidochelys olivacea* in captivity. *CMFRI Bulletin* 35: 49–54.
28 Dimond, MT and P Mohanty–Hejmadi (1983) Incubation temperature and sex determination in a sea turtle. *American Zoologist* 23: 1017.
29 Sahoo, G, BK Mohapatra, RK Sahoo and P Mohanty–Hejmadi (1996) Ultrastructure and characteristics of eggshells of the olive ridley turtle (*Lepidochelys olivacea*) from Gahirmatha, India. *Acta Anatomica* 156 (4): 261–267.
30 Sahoo, G, RK Sahoo and P Mohanty–Hejmadi (1998) Calcium metabolism in olive ridley turtle eggs during embryonic development. *Comparative Biochemistry and Physiology – Part A: Molecular & Integrative Physiology* 121 (1): 91–97.
31 Subramoniam, T (1981) Protandric hermaphroditism in a mole crab, *Emerita asiatica* (Decapoda: Anomura). *The Biological Bulletin* 160: 161–174.
32 Pandav, B, BC Choudhury and CS Kar (1994) Discovery of a new sea turtle

rookery in Orissa. *Marine Turtle Newsletter* 67: 15–16.
33 Pandav, B (2000) *Conservation & Management of Olive Ridley Sea Turtles on the Orissa Coast*. Ph.D thesis. Utkal University, India.
34 Pandav, B, BC Choudhury and CS Kar (1997) Mortality of olive ridley turtles *Lepidochelys olivacea* due to incidental capture in fishing nets along the Orissa coast, India. *Oryx* 31(1): 32–36.
35 Shanker, K and B Mohanty (1999) Guest editorial: Operation Kachhapa: in search of a solution for the olive ridleys of Orissa. *Marine Turtle Newsletter* 86: 1–3.
36 Shanker, K et al. (2004) Phylogeography of olive ridley turtles (*Lepidochelys olivacea*) on the east coast of India: implications for conservation theory. *Molecular Ecology* 13(7): 1899–1909.
37 Shanker, K and BC Choudhury (Eds.) (2006) *Marine Turtles of the Indian Subcontinent*. Universities Press, Hyderabad, India.
38 Shanker, K and HV Andrews (2006) *Towards an integrated and collaborative sea turtle conservation programme in India: a UNEP/CMS–IOSEA project report*. Centre for Herpetology/ Madras Crocodile Bank Trust, Mamallapuram.
39 Lal, A et al. (2010) Implications of conserving an ecosystem modifier: Increasing green turtle (*Chelonia mydas*) densities substantially alters seagrass meadows. *Biological Conservation* 143(11): 2730–2738.
40 Arthur, R, N Kelkar, T Alcoverro and MD Madhusudan (2013) Complex ecological pathways underlie perceptions of conflict between green turtles and fishers in the Lakshadweep Islands. *Biological Conservation* 167: 25–34.
41 Sridhar, A, S Mallick and K Shanker (2011) *The Nature of Conflict: Understanding Knowledge and Perception of and Attitudes towards Sea Turtle Conservation in Orissa*. Technical report prepared by ATREE, Bangalore, p. 53.
42 Anonymous (1977) The Gulf of Mannar Islands. *Hamadryad* 2(2): 12.
43 Bhaskar, S (1977) The Mannar Islands. *Hamadryad* 2(2): 12.
44 Bhaskar S (1978) Marine turtles in India's Lakshadweep Islands. *Marine Turtle Newsletter* 8: 5.
45 Bhaskar, S (1979) Notes from Lakshadweep. *Hamadryad* 4(1): 7–9.
46 Anonymous (1982) Sea turtle survey on Suheli Island. *Hamadryad* 7(3): 22.
47 Tripathy, B, K Shanker and BC Choudhury (2006) The status of sea turtles and their habitats in the Lakshadweep Archipelago, India. *Journal of the Bombay Natural History Society* 103(1): 33.
48 Bhaskar, S (1978) Notes from the Gulf of Kutch. *Hamadryad* 3(3): 9–10.
49 Bhaskar, S (1982) Turtle tracking in Gujarat. *Hamadryad* 7(1): 13–14.
50 Bhaskar, S (1979) Letters from the Andamans. *Hamadryad* 4(2): 3–6.
51 Bhaskar, S and M Tiwari (1992) *Andaman and Nicobar Sea Turtle Project. Phase–I. Great Nicobar Island*. Unpublished report for the Centre for Herpetology Madras Crocodile Bank Trust, Tamil Nadu, India.
52 Bhaskar, S (1996) Re–nesting intervals of the hawksbill turtle (*Eretmochelys imbricata*) on south Reef Island, Andaman Islands, India. *Hamadryad* 21: 19–22.
53 Whitaker, R (2006) Turtle Trekker: Satish Bhaskar. In: *Marine Turtles of the*

Indian subcontinent (Eds. K Shanker and BC Choudhury), pp. 17–21.Universities Press, Hyderabad, India.

54 Shanker, K (2010) Special profile: Satish Bhaskar. *Indian Ocean Turtle Newsletter* 12: 23.

55 Bhaskar, S (1985) Satish Bhaskar writes from Irian Jaya. *Hamadryad* 10(3): 19–20.

56 Bhaskar, S (1985) *Management and Research of Marine Turtle Nesting sites on the North Vogelkop coast of Irian Jaya Progress Report*. 5 Nov. 1984 – 30 April 1985. WWF/IUCN Project 1528. 14 pp.

57 Bhaskar, S (1985) Letters from the Dark Unknown Department. *Hamadryad* 10(1&2): 3–9.

58 See n. 55.

59 Ibid.

60 Betz, W and M Welch (1992) Once thriving colony of leatherback sea turtles declining at Irian Jaya, Indonesia. *Marine Turtle Newsletter* 56: 8–9.

61 Bhaskar, S (1981) Preliminary report on the status and distribution of sea turtles in Indian waters. *Indian Forester* 107(11): 707–711.

62 Bjorndal, KA (Ed.) (1982) *Biology and conservation of sea turtles*. Smithsonian Institute Press, Washington DC, USA.

63 Kar, CS and S Bhaskar (1982) Status of sea turtles in the eastern Indian Ocean. In: *Biology and Conservation of Sea Turtles* (Ed. KA Bjorndal), pp. 365–372. Smithsonian Institution Press, Washington DC, USA.

64 Bhaskar, S (1984) The distribution and status of sea turtles in India. In: *Proceedings of the Workshop Sea Turtle Conservation* (Ed. EG Silas), pp. 21–35. Central Marine Fisheries Research Institute Special Publication No. 18, Kochi, India.

65 This section is based on information from a number of sources. For further reading on sea turtle biology, the three volumes of 'Biology of Sea Turtles' (CRC Press) are recommended for advanced readers. Several popular books are also available, including Archie Carr's classics *The Windward Road* and *So Excellent a Fishe*, Carl Safina's *Voyage of the Leatherback*, and recent coffee table books by Blair Witherington and Jim Spotila.

66 Casey, JP, MC James and AS Williard (2014) Behavioral and metabolic contributions to thermoregulation in freely swimming leatherback turtles at high latitudes. *Journal of Experimental Biology* 217: 2331–2337.

67 Paladino, FV, MP O'Connor and JR Spotila (1990) Metabolism of leatherback turtles, gigantothermy, and thermoregulation of dinosaurs. *Nature* 344: 858–860.

68 Spotila, JR et al. (2000) Pacific leatherback turtles face extinction. *Nature* 405: 529–530.

69 Chan, EH and HC Liew (1996) Decline of the leatherback population in Terengganu, Malaysia, 1956–1995. *Chelonian Conservation and Biology* 2(2): 196–203.

70 Hughes, GR (1996) Nesting of the Leatherback Turtle (*Dermochelys coriacea*) in Tongaland, KwaZulu–Natal, South Africa, 1963–1995. *Chelonian conservation and biology* 2(2): 153–158.

71 Nel, R (2006) Turtle monitoring in South Africa: 42 years' worth of data. In:

Book of Abstracts. Twenty–sixth Annual Symposium on Sea Turtle Biology and Conservation (Compilers: M Frick., A Panagopoulou, AF Rees and K Williams) pp. 309–310. International Sea Turtle Society, Athens, Greece.

72 Andrews, HV, S Krishnan and P Biswas (2006) Distribution and status of marine turtles in the Andaman and Nicobar Islands. In: *Marine Turtles of the Indian Subcontinent* (Eds. K Shanker and BC Choudhury), pp. 33–57. Universities Press, Hyderabad, India.

73 Bjorndal, KA and JB Jackson (2003) Roles of sea turtles in marine ecosystems: reconstructing the past. In: *The Biology of Sea Turtles – Volume II*. (Eds. PL Lutz, J Musick and J Wyneken), pp. 259–273. CRC Press, Boca Raton, Florida, USA.

74 See n. 40.

75 See n. 52.

76 Carr, A (2013) *The Windward Road*. Random House, New York, USA.

77 Márquez, R et al. (1996) Good News! Rising numbers of kemps ridleys nest at Rancho Nuevo, Tamaulipas. Mexico. *Marine Turtle Newsletter* 73: 2–5.

78 Carr, A (1982) Notes on the behavioral ecology of sea turtles. In: *Biology and Conservation of Sea Turtles* (Ed. KA Bjorndal), pp. 19–26. Smithsonian Institution Press, Washington DC, USA.

79 Avens, L and ML Snover (2013) Age and age estimation in sea turtles. In: *The Biology of Sea Turtles – Volume III* (Eds. J Wyneken, KJ Lohmann and JA Musick), pp. 97–133. CRC Press, Florida, USA.

80 Frazier J (1985) Misidentification of marine turtles: *Caretta caretta* and *Lepidochelys olivacea* in the East Pacific. *Journal of Herpetology* 19: 111.

81 Bowen, BW and SA Karl (1996) Population genetics, phylogeography, and molecular evolution. In: *The Biology of Sea Turtles* (Eds. PL Lutz and JA Musick), pp. 29–50. CRC Press, Florida, USA.

82 Pillai, SK (2003) Instance of meat of leatherback turtle *Dermochelys coriacea* used as food. *Fishing Chimes* 23 (3): 46–47.

83 Pillai SK, KK Suresh and P Kannan (2003) Leatherback turtle released into the sea at Vizhinjam in Kerala, India. *Kachhapa* 9: 5–6.

84 Pillai, SK et al. (2003) Community participation in the release of a leatherback turtle in south Kerala. *Kachhapa* 9: 23.

85 Balachandran, S, P Sathiyaselvam and P Dhakshinamoorthy (2009) Rescue of a leatherback turtle (*Dermochelys coriacea*) at Manakudi beach, Kanniyakumari District, Tamil Nadu and the need for an awareness campaign. *Indian Ocean Turtle Newsletter* 10: 19–20.

86 Anil, MK et al. (2009) A note on the leatherback turtle *Dermochelys coriacea* (Vandelli 1761) rescued at Vizhinjam, Kerala. *Marine Fisheries Information Service T & E Series* 200: 23.

87 Shanker, K and BC Choudhury (2006) Marine Turtles of the Indian Subcontinent: A Brief History. In: *Marine Turtles of the Indian Subcontinent* (Eds. K Shanker and BC Choudhury), pp. 3–16.Universities Press, Hyderabad, India.

88 Choudhury, BC et al. (2008) *Determining the offshore distribution and migration*

pattern of Olive Ridley Sea Turtle (Lepidochelys olivacea) *along the east coast of India (Annual Report)*. Wildlife Institute of India, Dehradun, India.

89 Thomassen, J, J Linnell and K Skogen (2011) *Wildlife–Human Interactions: From Conflict to Coexistence in Sustainable Landscapes. Final report from a joint Indo–Norwegian project 2007–2011*. NINA Report 83 pp.

90 Arthur, R and K Shanker (2010) Olive and Green: Shades of Conflict between Turtles & Fishers in India. *Current Conservation* 4(4): 28–35.

91 Boopendranath, MR et al. (2006) Design and Development of the TED for Indian Fisheries. In: *Marine Turtles of the Indian Subcontinent* (Eds. K Shanker and BC Choudhury), pp. 244–261.Universities Press, Hyderabad, India.

92 Ramarao, SVS (1995) *Tour report on operation of turtle excluder device from FSI vessel 16–25 August 1995*. CIFT, Kochi, India.

93 Gopi, GV, B Pandav and BC Choudhury (2007) Estimated annual incidental captures of *Lepidochelys olivacea* (Eschscholtz, 1829) in trawl nets along the Orissa coast, India. *Hamadryad* 31 (2): 212–215.

94 WWF (2011) *Turtles and TEDs: Outcome of trials conducted off Dhamra, Orissa*. WWF–India, New Delhi, India.

95 Rajagopalan, M (1989) *Ecophysiological studies on the marine turtle* Lepidochelys olivacea. Ph.D thesis. University of Madras, India.

96 Venkatesan, S (2004) *Ecological and physiological studies on the green sea turtle* Chelonia mydas. Ph.D thesis. University of Madras, India.

97 Kannan, P (2004) *Studies on the biology and incidental catch of sea turtles in selected centres along the Indian coast*. Ph.D thesis. University of Madras, India.

98 Raja Sekhar, PS (1987) *Captive Propagation and Management of an Endangered Sea Turtle*, Lepidochelys olivacea (Eschschltz) *of the Northern Andhra Coast*. Ph.D thesis. Andhra University, India.

99 Kar, CS (1988) *Ecological studies on the Pacific Ridley Sea Turtle*, Lepidochelys olivacea (Eschscholtz, 1829) *in the Orissa Coast*. Ph.D thesis. Sambalpur University, India.

100 Ram, CK (2000) *Behavioural ecology of the olive ridley sea turtle* Lepidochelys olivacea (Eschscholtz, 1827) *during the breeding period*. M.Sc dissertation. Salim Ali School of Ecology, Pondicherry University, India.

101 Tripathy, B (2005) *A study on the ecology and distribution of olive ridley sea turtle* Lepidochelys olivacea *at the Rushikulya rookery of Orissa coast, India*. Ph.D thesis. Andhra University, India.

102 Sivasundar, A (1996) *Studies on the Nesting of Leatherback Sea Turtles* (Dermochelys coriacea) *in the Andaman Islands*. M.Sc dissertation. Salim Ali School of Ecology, Pondicherry University, India.

103 Sivasundar, A and KV Devi Prasad (1996) Placement and predation of nests of leatherback sea turtles in the Andaman Islands, India. *Hamadryad* 21: 36–42.

104 Banugoppan, K & P Davidar (1999) Status of sea turtles along the Pondicherry coast, India. *Hamadryad* 24: 43.

105 Subramanean, J (2004) *A study on the nesting and adult mortality of olive ridley sea turtle* (Lepidochelys olivacea) *along the Mammalapuram–Pondicherry coast,*

Tamil Nadu. M.Sc dissertation. Pondicherry University, India.

106 Karnad, D (2008) *The effect of lighting and temperature on the eggs and hatchlings of olive ridley turtles at Rushikulya, India*. M.Sc dissertation. Manipal University, India.

107 Karnad, D, K Isvaran, CS Kar and K Shanker (2009) Lighting the way: reducing the impact of light on misorientation of olive ridley turtle hatchlings at Rushikulya, India. *Biological Conservation* 142: 2083–2088.

108 Muralidharan, M (2009) *Nest site selection and effects of anthropogenic changes to the Rushikulya nesting beach, Orissa on olive ridley sea turtles*. M.Sc dissertation. Saurashtra University, India.

109 Fatima, E (2010) *Estimating Population Structure of Ridley Population Nesting On Indian Mainland*. M.Sc dissertation. Guru Gobind Singh Indraprastha University, India.

110 Mallick, S (2010) *A comparison of fishing practices, attitudes to and perception of sea turtle conservation along the southern Orissa coast*. M.Sc dissertation. Forest Research Institute, India.

111 Lal, A (2007) *An investigation of the relationship between Green Turtle* (Chelonia mydas) *herbivory and seagrass in the Agatti Lagoon, Lakshadweep*. M.Sc dissertation. Saurashtra University, India.

112 See n. 39.

Sea turtles: from fishe to flagship

1 Frazier, J (2003) Prehistoric and ancient historic interactions between humans and marine turtles. In: *The Biology of Sea Turtles – Volume II* (Eds. PL Lutz, J Musick and J Wyneken), pp. 1–38. CRC Press, Boca Raton, Florida, USA.

2 Burstein, SM (1989) *Agatharchides of Cnidus: On the Erythraean Sea*. The Hakluyt Society, London, UK.

3 Mathew, G (1975) The dating and significance of the Periplus of the Erythrean Sea. In: *East Africa and the Orient: Cultural Syntheses in Pre–Colonial Times*. (Eds. HN Chittick and RI Rotberg), pp.147–163. Africana, New York, USA.

4 Casson, L (1989) *The Periplus Maris Erythraei*. Princeton University Press, Princeton, USA.

5 Jones, HL (1929) *The Geography of Strabo*. Harvard University Press, Massachusetts, USA.

6 Philemon Holland, translator (1601): *C. Plinius Secundus The Historie of the World. Book VI* (pp. 114–151).

7 Bowrey, T (Unknown) A *geographical account of countries round the Bay of Bengal, 1669–1679*. Ed. RC Temple, reprinted in 1905 by the Hakluyt Society, Cambridge, UK.

8 Hamilton, A (1727) *A New Account of the East–Indies. Being the Observations and Remarks of Capt. Alexander Hamilton from the Year 1688–1723*. John Mosman, One of His Majesty's Printers. Edinburgh, Scotland.

9 Hejmadi, P (2000) Earliest record of Gahirmatha turtles. *Marine Turtle Newsletter* 88: 11–12.

10 Mouat, FJ (1863) *Adventures and researches among the Andaman Islanders*. Hurst and Blackett, London, UK.

11 Man, EH (1883) The Aboriginal Inhabitants of the Andaman Islands. Reprinted in 2001 by Mittal Publications, New Delhi, India.

12 Portman, MV (1899) A *history of our relations with the Andamanese* (Vol. 1). Office of the Superintendent of Government Print, India.

13 Kloss, CB (1902) *Andaman and Nicobars*. Reprint (1971) Vivek, New Delhi, India.

14 De Silva, A (2006) Marine turtles of Sri Lanka: A historical account. In: *Marine Turtles of the Indian subcontinent* (Eds. K Shanker and BC Choudhury), pp. 324–353. Universities Press, Hyderabad, India.

15 Bennett, JW (1843) *Ceylon and its Capabilities*. Trumpet Publishers, Rajagriya, Sri Lanka.

16 Tennent, JE (1861) *Sketches of the Natural History of Ceylon with Narratives and Anecdotes*. Longman, Green, Longman, and Roberts, London UK.

17 See n. 1.

18 Acharji, MN (1950) Edible chelonians and their products. *Journal of the Bombay Natural History Society* 49(3): 529–532.

19 Kuriyan, GK (1950) Turtle fishing in the sea around Krusadai Island. *Journal of the Bombay Natural History Society* 49(3): 509–512.

20 Krishnamoorthi, B (1957) Fishery resources of the Rameswaram Island. *Indian Journal of Fisheries* 4(2): 229–253.

21 Jones, S and AB Fernando (1973) Present status of the turtle fishery in the Gulf of Mannar and Palk Bay. In: *Proceedings of the Symposium on Living Resources of the Seas Around India*, pp. 712–715. Central Marine Fisheries Research Institute, Cochin, India.

22 Agastheesapillai, A (1996) Turtle export from south east coast of India during 1945–64 period. *Marine Fisheries Information Service T & E Series* 145: 16.

23 See n. 16.

24 Deraniyagala, PEP (1939) *The Tetrapod Reptiles of Ceylon. Vol.1.* Colombo Museum, Colombo, Sri Lanka.

25 Silas, EG and AB Fernando (1984) Turtle poisoning. *CMFRI Bulletin* 35: 62–75.

26 Bhaskar, S (1978) Turtle meat poisoning. *Hamadryad* 3(1): 6.

27 Bhupathy, S and S Saravanan (2006) Marine turtles of Tamil Nadu. In: *Marine Turtles of the Indian subcontinent* (Eds. K Shanker and BC Choudhury), pp. 58–67. Universities Press, Hyderabad, India.

28 Whitaker, Z (1981) The Riddled Ridley – Sea Turtles Eaten to Extinction. *The India Magazine* (June): 13–25.

29 Shantharam, B (1975) The marine edible turtles. *Seafood Export Journal* 7(10): 31–33.

30 Murthy, TSN (1981) Turtles: Their natural history, economic importance and conservation. *Zoologiana* 4: 57–65.

31 CMFRI (1984) *Proceedings of the Workshop on Sea Turtle Conservation* (Ed. EG

Silas). Central Marine Fisheries Research Institute Special Publication No. 18, Kochi, India.

32 Rajagopalan, M (1984) Value of sea turtles to India. In: *Proceedings of the Workshop on Sea Turtle Conservation* (Ed. EG Silas), pp. 49–58. Central Marine Fisheries Research Institute Special Publication No. 18, Kochi, India.

33 CMFRI (1973) *Proceedings of the Symposium on Living Resources of the Seas around India*. CMFRI, Cochin, India.

34 See n. 21.

35 Silas, EG (Eds.) (1985) *Proceedings of the Symposium on Endangered Marine Animals and Marine Parks*. Marine Biological Association of India, Cochin, India.

36 Groombridge, B (1985) India's sea turtles in world perspective. In: *Proceedings of the Symposium on Endangered Marine Animals and Marine Parks* (Ed. EG Silas), pp. 205–213. Marine Biological Association of India, Cochin, India.

37 Whitaker, R (1985) Rational use of estuarine and marine reptiles. In: *Proceedings of the Symposium on Endangered Marine Animals and Marine Parks* (Ed. EG Silas), pp. 298–303. Marine Biological Association of India, Cochin, India.

38 Davis, TA and R Bedi (1978) The sea turtle rookery of Orissa. *Environmental Awareness* 1(2): 63–66.

39 Mrosovsky, N (1982) Editorial. *Marine Turtle Newsletter* 23: 1–2.

40 Mrosovsky, N (1983) Editorial. *Marine Turtle Newsletter* 25: 1.

41 Vijaya, J (1982) Turtle slaughter in India. *Marine Turtle Newsletter* 23:2; Moll, EO, S Bhaskar and J Vijaya (1983) Update on the olive ridley on the east coast of India. *Marine Turtle Newsletter* 25: 2–4.

42 Bobb, D (1982) Massacre at Digha. *India Today* 31: 64–65.

43 Shanker, K (2003) Thirty years of sea turtle conservation on the Madras coast: A review. *Kachhapa* 8: 16–19.

44 Pandav, B, BC Choudhury and CS Kar (1997) Mortality of olive ridley turtles *Lepidochelys olivacea* due to incidental capture in fishing nets along the Orissa coast, India. *Oryx* 31(1): 32–36.

45 Pandav, B and BC Choudhury (1999) An update on the mortality of the olive ridley sea turtles in Orissa. *Marine Turtle Newsletter* 83: 10–12.

46 Shanker, K and B Mohanty (1999) Guest editorial: Operation Kachhapa: In search of a solution for the olive ridleys of Orissa. *Marine Turtle Newsletter* 86: 1–3.

47 IOTN (2005) Profile of NGOs working on sea turtle conservation and fisheries in Orissa. *Indian Ocean Turtle Newsletter* 1: 1–25.

48 Silas, EG, M Rajagopalan, AB Fernando, and SS Dan (1983) Marine turtle conservation and management: A survey of the situation in Orissa 1981/82 and 1982/83. *Marine Fisheries Information Service T & E Series* 50: 13–23.

49 National Research Council (1990) *Decline of the Sea Turtles: Causes and Prevention*. National Academy Press, Washington DC, USA.

50 Shanker, K, B Pandav and BC Choudhury (2004) An assessment of the olive ridley turtle (*Lepidochelys olivacea*) nesting population in Orissa on the east coast

of India. *Biological Conservation* 115: 149–160.

51 Gilman, E et al. (2010) Mitigating sea turtle by-catch in coastal passive net fisheries. *Fish and Fisheries* 11: 57–88.

52 Lewison, RL, SA Freeman & LB Crowder (2004) Quantifying the effects of fisheries on threatened species: the impact of pelagic longlines on loggerhead and leatherback sea turtles. *Ecology letters* 7: 221–231.

53 Epperly, SP (2003) Fisheries–related mortality and Turtle Excluder Devices. In: *The Biology of Sea Turtles – Volume II* (Eds. PL Lutz, J Musick and J Wyneken), pp. 339–353. CRC Press, Boca Raton, Florida, USA.

54 Jenkins, LD (2012) Reducing sea turtle bycatch in trawl nets: A history of NMFS' turtle excluder device research. *Marine Fisheries Review* 74(2): 26–44.

55 Jenkins, LD (2010) The evolution of a trading zone: a case study of the turtle excluder device. *Studies in History and Philosophy of Science* 41: 75–85.

56 Eckert, KL (1995) Anthropogenic Threats to Sea Turtles. In: *Biology and Conservation of Sea Turtles* (Ed. KA Bjorndal), pp. 611–613. Smithsonian Institute Press, Washington DC, USA.

57 Feagin, R.A., N. Mukherjee, K Shanker et al. (2010) Shelter from the storm? The use and misuse of 'bioshields' in managing for natural disasters on the coast. *Conservation Letters* 3: 1–11.

58 Witherington, BE and RE Martin (2003) *Understanding, assessing, and resolving light–pollution problems on sea turtle nesting beaches*. Florida Marine Research Institute Technical Report TR–2. 73 pp.

59 Van Houtan, KS, SK Hargrove and GH Balazs (2010) Land Use, Macroalgae, and a Tumor–Forming Disease in Marine Turtles. *PLoS ONE* 5(9): e12900.

60 Mortimer, JA (1999) Reducing threats to eggs and hatchlings: hatcheries. In: *Research and management techniques for the conservation of sea turtles* (Eds. KL Eckert, KA Bjorndal, FA Abreu–Grobois and M Donelly), pp. 175–178. IUCN/SSC Marine Turtle Specialist Group Publication 4.

61 Bjorndal, KA (Ed.) (1982) *Biology and conservation of sea turtles*. Smithsonian Institute Press, Washington DC, USA.

62 Mrosovsky, N (1976) Editorial. *Marine Turtle Newsletter* 1:1.

63 Shanker, K (1999) Editorial. Its turtle time in Orissa again. *Kachhapa* 1:1.

64 Shanker, K (2005) Editorial. *Indian Ocean Turtle Newsletter* 1:1.

65 Coyne, M (1996) Seaturtle.org. Retrieved September 10, 2014, from http://seaturtle.org.

66 Carr, A (1967) *So Excellent a Fishe: A Natural History of Sea Turtles*. The Natural History Press, New York, USA.

67 The Bermuda Assembly, 1620.

68 Carr, A (2013) *The Windward Road*. Random House, New York, USA.

69 Mrosovsky, N (1997) A general strategy for conservation through use of sea turtles. *Journal of Sustainable Use* 1: 42–46.

70 Mrosovsky, N (2000) *Sustainable use of hawksbill turtles: contemporary issues in conservation* (No. 1). Key Centre for Tropical Wildlife Management.

71 Campbell, LM (2002) Science and sustainable use: views of marine turtle conservation experts. *Ecological Applications* 12: 1229–1246.
72 Jackson, JB (1997) Reefs since Columbus. *Coral reefs* 16(1): S23–S32.
73 Jackson, JB et al. (2001) Historical overfishing and the recent collapse of coastal ecosystems. *Science* 293: 629–637.
74 Bjorndal, KA & JB Jackson (2003) Roles of sea turtles in marine ecosystems: reconstructing the past. In: *The Biology of Sea Turtles – Volume II* (Eds. PL Lutz, JA Musick and J Wynekken), pp. 259–273. CRC Press, USA.
75 Murthy, TSN and AGK Menon (1976) The turtle resources of India. *Seafood Export Journal* 81: 1–12.
76 Alagarswami, K (Ed.) (1983) *Mariculture potential of Andaman and Nicobar Islands: An indicative survey*. CMFRI Bulletin 34 CMFRI, Cochin, India.
77 Suseelan, C (Ed.) (1989) *Marine living resources of the union territory of Lakshadweep: An Indicative Survey with Suggestions for Development*. CMFRI Bulletin 43, CMFRI, Cochin, India.
78 Bhaskar, S and R Whitaker (1983) Sea turtle resources in the Andamans – Mariculture potential in Andaman and Nicobar Islands: An indicative survey. *Bulletin of Central Marine Fisheries Research Institute* 34: 94–97.
79 Lal Mohan, RS (1983) Marine turtle resources. *Bulletin of Central Marine Fisheries Research Institute* 34: 102–103.
80 Silas, EG, M Rajagopalan and AB Fernando (1983) Sea turtles of India – Need for a crash programme on conservation and effective management of the resource. *Marine Fisheries Information Service T & E Series* 50: 1–12.
81 Eckert, KL, KA Bjorndal, FA Abreu–Grobois and M Donnelly (Eds) (1999) *Research and management techniques for the conservation of sea turtles*. IUCN/SSC Marine Turtle Specialist Group Publication.
82 Wallace, BP et al. (2011) Global Conservation Priorities for Marine Turtles. PLOS One DOI: 10.1371/journal.pone.0024510.

From the ports of Odisha to the pans of Kolkata

1 Chadha, S & CS Kar (1999) *Bhitarkanika: Myth and Reality*. Natraj Publishers, Dehradun, India.
2 Hamilton, A (1727) A *New Account of the East Indies: Being the Observations of Captain Alexander Hamilton*. John Mosman, Edinburgh, UK.
3 Stirling, A and J Peggs (1846) *Orissa: Its Geography, Statistics, History, Religion and Antiquities*. John Snow, London, UK.
4 Hejmadi, P (2000) Earliest record of Gahirmatha turtles. *Marine Turtle Newsletter* 88: 11–12.
5 Dash, MC & CS Kar (1990) *The Turtle Paradise, Gahirmatha*. Interprint, New Delhi.
6 Information derived from various records of the Zamindar and revenue department by CS Kar.

7 Bustard, HR (1976) World's largest sea turtle rookery. *Tiger Paper* 3(3): 25.
8 Biswas, S (1982) A report on the olive ridley, *Lepidochelys olivacea* (Eschscholtz) (Testudines: Chelonidae) of the Bay of Bengal. *Records of the Zoological Survey of India* 79: 275–302.
9 Two issues of the Marine Fisheries Information Service (No. 50 in 1983, and No. 64 in 1985) were devoted exclusively to sea turtles.
10 Silas, EG, M Rajagopalan, AB Fernando & SS Dan (1983) Marine turtle conservation & management: A survey of the situation in Orissa 1981/82 & 1982/83. *Marine Fisheries Information Service T & E Series* 50: 13–23.
11 Ibid.
12 Ibid.
13 Moll, EO (1983) Turtle Survey Update. *Hamadryad* 8(2): 5–17.
14 Moll, EO, S Bhaskar & J Vijaya (1984) Update on the olive ridley on the east coast of India. *Marine Turtle Newsletter* 25: 2–4.
15 Appendix 1 of Silas, EG, M Rajagopalan & SS Dan (1983) Marine turtle conservation and management: A survey of the situation in West Bengal 1981/82 & 1982/83. *Marine Fisheries Information Service T & E Series* 50: 24–33.
16 Appendix 3, Ibid.
17 Appendix 2 of Silas, EG, M Rajagopalan, AB Fernando, and SS Dan (1983) Marine turtle conservation and management: A survey of the situation in Orissa 1981/82 and 1982/83. *Marine Fisheries Information Service T & E Series* 50: 13–23.
18 Appendix 1 of Silas, EG, M Rajagopalan, SS Dan & AB Fernando (1985) On the large and mini arribada of the olive ridley *Lepidochelys olivacea* at Gahirmatha, Orissa during the 1985 season. *Marine Fisheries Information Service T & E Series* 64: 1–16.
19 Appendix 2, Ibid.
20 Appendix 3, Ibid.
21 Bobb, D (1982) Massacre at Digha. *India Today* 31: 64–65.
22 Mrosovsky, N (1982) Editorial. *Marine Turtle Newsletter* 23:1–2.
23 Mrosovsky, N (1983) Olive ridleys in India. *Marine Turtle Newsletter* 24: 17.
24 Mrosovsky, N (1983) Editorial. *Marine Turtle Newsletter* 25:1.
25 Anonymous (1984) Mrs. Gandhi writes about turtles. *Hamadryad* 9(3): 21.
26 For a detailed interview with Anne Wright, see Sanctuary Asia, Vol XXXII No. 4, August 2012 (http://www.sanctuaryasia.com/people/interviews/8945–meet–anne–wright.html).
27 Bustard, HR (1976) World's largest sea turtle rookery. *Tiger paper* 3(3): 25.
28 Bustard, HR (1972) *Australian sea turtles*. Collins, London, UK.
29 Vijaya, J (1982) Turtle Slaughter in India. *Marine Turtle Newsletter* 23: 2.
30 Vijaya, J (1982) Rediscovery of the forest cane turtle (*Heosemys silvatica*) in Kerala. *Hamadryad* 7(3): 2–3.
31 Posthumously published as Vijaya, J (1989) Kadars – People of the forest. *Hamadryad* 14(1): 10.

32 Vijaya, J (1982) Rediscovery of the forest cane turtle (*Heosemys silvatica*) in Kerala. *Hamadryad* 7(3): 2–3.
33 Vijaya, J (1983) The Travancore tortoise, *Geochelone travancorica*. *Hamadryad* 8(3): 11–13.
34 Vijaya, J (1984) Cane turtle (*Heosemys silvatica*) study project in Kerala. *Hamadryad* 9(1): 4
35 Vijaya, J (1983) World's rarest turtle (we think) lays eggs in captivity. *Hamadryad* 8(1): 13.
36 Lenin, J (2006) Vijaya, India's first woman herpetologist. *Indian Ocean Turtle Newsletter* 4: 29–32.
37 Jagannathan, P (2006) My sister Viji. *Indian Ocean Turtle Newsletter* 4: 32–34.
38 See Indian Ocean Turtle Newsletter 21 for tributes to CS Kar by Sudhakar Kar, BC Choudhury and Basudev Tripathy.
39 Ganguly, D (1980) An appeal to save Pacific ridley turtles from mass–killing in West Bengal and Orissa. In: *Proceedings of the Workshop on Wild Life Ecology, Dehra Dun, Jan. 1978* (Vol. 35, p. 35). Zoological Survey of India.
40 Davis, TA, R Bedi and GM Oza (1978) Sea Turtle faces extinction in India. *Environmental Conservation* 5: 211–212.
41 Davis, TA and R Bedi (1978) The sea turtle rookery of Orissa. *Environmental Awareness* 1(2): 63–66.
42 Anonymous (1978) Mass slaughter of sea turtles. *Hamadryad* 3(3): 8; Anonymous (1977) Notes on turtle conservation in India. *Marine Turtle Newsletter* 5: 3.
43 Frazier, JG (1980) Sea Turtle faces extinction in India: Crying 'wolf' or saving sea turtles? *Environmental Conservation* 7: 239–240.
44 Kar, CS (1980) The Gahirmatha turtle rookery along the coast of Orissa, India. *Marine Turtle Newsletter* 15: 2–3.
45 Bustard, HR (1980) Should sea turtles be exploited? *Marine Turtle Newsletter* 15: 3–5.
46 See n. 23.
47 Bhaskar, S (1991) Turtles: India's heritage from the sea. *Swagat* (October): 49–52.
48 See n. 5.
49 Hughes, GR (1979) Conservation, utilization, antelopes and turtles. *Marine Turtle Newsletter* 13: 13–14.
50 Mrosovsky, N (2001) Guest editorial: The future of ridley arribadas in Orissa: From triple waste to triple win? *Kachhapa* 5: 1–3.
51 Godfrey, MH, LM Campbell, K Shanker & C Tambiah (2003) Report from the 'Research on Use' Session at the 23rd Symposium on Sea Turtle Biology and Conservation, Kuala Lumpur, Malaysia. *Marine Turtle Newsletter* 101: 33–34.
52 Pilcher, NJ (2006) *Proceedings of the Twenty Third Annual Symposium on Sea Turtle Biology and Conservation (2003): Living With Turtles*. NOAA Technical Memorandum, Kuala Lumpur, Malaysia.
53 Shanker, K (2001) Guest editorial: The swampland of sea turtle conservation: In search of a philosophy. *Marine Turtle Newsletter* 95: 1–4.

54 Callicott, JB (1980) Animal Liberation: A Triangular Affair. *Environmental Ethics* 2: 311–338.
55 Hargrove, EC (1989) An overview of conservation and human values: Are conservation goals merely cultural attitudes? In: *Conservation for the Twenty First Century* (Eds. D Western & M Pearl), pp. 227-231. Oxford University Press, UK.
56 Norton, BG (1984) Environmental Ethics and Weak Anthropocentrism. *Environmental Ethics* 6: 131–147.
57 Norton, BG (1989) The cultural approach to conservation biology. In: *Conservation for the Twenty First Century* (Eds. D Western & M Pearl), pp. 241-246. Oxford University Press, UK.
58 Silas, EG, M Rajagopalan & SS Dan (1983) Marine turtle conservation and management: A survey of the situation in West Bengal 1981/82 & 1982/83. *Marine Fisheries Information Service T & E Series* 50: 24–33.
59 Silas, EG, M Rajagopalan, SS Dan & AB Fernando (1985) On the large and mini arribada of the olive ridley *Lepidochelys olivacea* at Gahirmatha, Orissa during the 1985 season. *Marine Fisheries Information Service T & E Series* 64: 1–16.
60 Raut, SK. & NC Nandi (1986) Net–bound death of marine turtle *Lepidochelys olivacea* off west Bengal coast during 1984–85. *Journal of Bombay Natural History Society* 83(1): 223–224.
61 See n. 10.
62 Moll, EO, S Bhaskar & J Vijaya (1984) Update on the olive ridley on the east coast of India. *Marine Turtle Newsletter* 25: 2–4.
63 Kar, CS & MC Dash (1984) Conservation and status of sea turtles in Orissa. In: *Proceedings of the Symposium on Endangered Marine Animals and Marine Parks* (Ed. EG Silas), pp. 93–107. Marine Biological Association of India, Cochin, India.

Flagging ships of conservation

1 Kar, CS & MC Dash (1984) Mass nesting beaches of the olive ridley *Lepidochelys olivacea* (Eschscholtz, 1829) in Orissa and the behaviour during an arribada. In: *Proceedings of the Workshop on Sea turtle Conservation* (Ed. E.G. Silas), pp. 21–35. Central Marine Fisheries Research Institute Special Publication 18, Cochin, India.
2 Kar, CS and MC Dash (1984) Conservation and status of sea turtles in Orissa. In: *Proceedings of the Symposium on Endangered Marine Animals and Marine Parks* (Ed. EG Silas), pp. 93–107. Marine Biological Association of India Cochin, India
3 Anonymous (1984) Arribada – the arrival. *Hamadryad* 9(2): 12.
4 James, PSBR, M Rajagopalan, SS Dan, AB Fernando & V Selvaraj (1989) On the mortality of marine mammals and turtles at Gahirmatha, Orissa from 1983 to 1987. *Journal of the Marine Biological Association of India* 31: 28–35.
5 Mohanty, B (2000) Operation Kachhapa: First work report for 1999–2000

turtle season. Reporting period: 1st November 1999 to 24th November 1999. *Kachhapa* 2: 2–3.

6 Wright, B & B Mohanty (2006) Operation Kachhapa: An NGO initiative for sea turtle conservation in Orissa. In: *Marine turtles of the Indian Subcontinent* (Eds. Shanker, K & BC Choudhury), pp. 290–302. Universities Press, Hyderabad, India.

7 Kar, S (1997) *Saving the turtle from a soup*. Indian Express Newspapers, 12 November 1997, Bombay, India (reprinted in Marine Turtle Newsletter 79).

8 See n. 6.

9 Mohanty, B & B Wright (2001) The wandering minstrels of Orissa – Singing to save sea turtles. *Kachhapa* 5: 21–22.

10 *The Asian Age*, 3 February 1999.

11 *The New Indian Express*, Bhubaneswar, 3 February 1999.

12 *The New Indian Express*, Bhubaneswar, 11 March 1999.

13 Rajaram Satapathy, *The Times of India*, 4 February 1999.

14 *The Asian Age*, 3 May 1999.

15 *The Asian Age*, 30 January 1999.

16 Wright, B, B Mohanty & S Matheson (2001) An update on sea turtle conservation activities in Orissa. *Kachhapa* 4: 10–13.

17 The Central Empowered Committee was constituted by the Supreme Court of India by order dated 9.5.2002 in Writ Petitions (Civil) Nos. 202/95 & 171/96.

18 Central Empowered Committee, Interim Orders, 7th March 2003, Application No 46, Alok Krishna Agarwal versus Union of India, State of Orissa and others.

19 Centre for Environmental Law, Worldwide Fund for Nature, India vs State of Orissa and Ors. [O.J.C. No. 3128 of 1994 dated 14.05.1998].

20 Wright, B & B Mohanty (2002) Olive ridley mortality in gill nets in Orissa. *Kachhapa* 6: 20.

21 Proceedings of High Powered Committee Meeting on Conservation of Olive Ridley Sea Turtles held on 10 October 2003 in Rajiv Bhavan Conference Hall, Bhubaneshwar, Orissa.

22 Central Empowered Committee, 2004, 'Visit of Central Empowered Committee to Orissa from February 10–14, 2004'.

23 Bache, SJ & JG Frazier (2006) International Instruments and Marine Turtle Conservation. In: *Marine turtles of the Indian Subcontinent* (Eds. K Shanker & BC Choudhury), pp. 324–356. Universities Press, Hyderabad, India.

24 Department of Commerce, in 'Ruling Seen Barring Most Shrimp Imports to US', *Reuters* (3 May 1996) available in LEXIS, News Library, Wires file.

25 For a review of TEDs in Orissa, see Behera, C (2006) Beyond TEDs: The TED controversy from the perspective of Orissa's trawling industry. In: *Marine turtles of the Indian Subcontinent* (Eds. K Shanker & BC Choudhury), pp. 238–243. Universities Press, Hyderabad, India.

26 Anonymous (1996) *Workshop–cum–demonstration on TED at Paradeep during 11–14 November 1996: A report*. Department of Fisheries, Government of Orissa and Project Swarajya, Orissa.

27 For a review of CIFT's involvement in developing TEDs, see Boopendranath et al. (2006) Design and Development of the TED for Indian Fisheries. In: *Marine turtles of the Indian Subcontinent* (Eds. K Shanker & BC Choudhury), pp. 244–262. Universities Press, Hyderabad, India.

28 Tucker, A, J Robins, and D McPhee (1997) Adopting Turtle Exclusion Devices in Australia and the United States: What are the Differences in Technology Transfer, Promotion, and Acceptance? *Coastal Management* 25: 405–421.

29 Anonymous (2002) *Workshop–cum–demonstration on TED at Paradeep during 9–12 February, 2002: A report*. Department of Fisheries, Government of Orissa and Project Swarajya, Orissa.

30 See n. 28.

31 Jenkins, LD (2008) The end game is diffusion: Adoption of turtle excluder devices and the diffusion process. In: *Proceedings of the Twenty–Fifth Annual Symposium on Sea Turtle Biology and Conservation* (Compilers: KH Rodhe, K Gayheart, and K Shanker) NOAA Technical Memorandum NMFS–SEFSC–582: 14.

32 Jenkins, LD (2010) Profile and influence of the successful fisher–inventor of marine conservation technology. *Conservation & Society* 8: 44–54.

33 See n. 25.

34 Anonymous (2006) Turning Turtle (http://www.greenpeace.org/international/en/news/features/turning–turtle/).

35 Greenpeace (2006) *I witness: Turtle witness camp*. Greenpeace, India.

36 Wright, B and B Mohanty (2005) Operation Kachhapa – A NGO initiative to protect olive ridleys in Orissa. Testudo 6(2).

37 Anonymous (2006) Fisherman shot dead Gahirmatha sea coast. (http://www.orissadiary.com/CurrentNews.asp?id=3534).

38 Letter from Orissa Traditional Fishworkers' Union (OTFWU) to P.V. Jayakrishnan, Chairman, Central Empowered Committee (CEC) 19 February 2004 titled 'Prayer of the traditional fishermen of Orissa for hearing on turtle issues'. The letter responds to the Interim Orders of the CEC dated 7 March 2003, and objects to the complete ban on fishing in several zones off the Orissa coast.

39 Aleya, K (2005) Perspectives of the traditional fishworkers on sea turtle conservation in Orissa. *Indian Ocean Turtle Newsletter* 1: 7–8.

40 Shanker, K, B Tripathy & B Pandav (2005) Biological studies on sea turtles on the coast of Orissa. *Indian Ocean Turtle Newsletter* 1: 10–11.

41 Project Swarajya (2005) Views of the trawler owners and their association on sea turtle conservation in Orissa. *Indian Ocean Turtle Newsletter* 1: 8–9.

42 Kar, M (2014) Tension after fisherman death. *The Telegraph*, January 14, 2014.

43 Kutty, R (2006) Community–based Conservation of Sea Turtle Nesting Sites in India: Some Case Studies. In: *Marine Turtles of the Indian subcontinent* (Eds. K Shanker and BC Choudhury), pp. 271–290. Universities Press, Hyderabad, India.

44 See n. 39.

45 See www.omrcc.org.

46 Shanker, K and BC Choudhury (Eds.) (2006) *Marine Turtles of the Indian Subcontinent.* Universities Press, Hyderabad, India.
47 Shanker, K & R Kutty (2005) Sailing the flagship fantastic: myth and reality of sea turtle conservation in India. *Maritime Studies* 3(2) and 4(1): 213–240.
48 See articles on the use of sea turtles as flagships in a special issue of the journal Maritime Studies (MAST), edited by Jack Frazier.
49 Ram, K. (2000) *Behavioral ecology of the olive ridley sea turtle* Lepidochelys olivacea *(Eschscholtz, 1827) during the breeding period.* M.Sc dissertation. Pondicherry University, India.
50 Pandav, B. (2000) *Conservation & management of olive ridley sea turtles on the Orissa coast*. Ph.D thesis. Utkal University, India.
51 Presented in the Valedictory Session of the Workshop–cum–Demonstration on TEDs held at Employees Recreation Centre, Paradip on 12th Feb. 2002 under the joint auspices of Directorate of Fisheries, Orissa and Project Swarajya.
52 Chadha, S & CS Kar (1999) *Bhitarkanika: Myth and reality*. Natraj Publishers, Dehradun, India.
53 Thomassen, J and R Sukumar (2011) *Wildlife Human Interactions: From Conflict to Coexistence in Sustainable Landscapes*. Norwegian Institute for Nature Research and Centre for Ecological Sciences, Indian Institute of Science, Bangalore, India.
54 Sridhar, A, S Mallick and K Shanker (2011) *The Nature of Conflict: Understanding Knowledge and Perception of and Attitudes towards Sea Turtle Conservation in Orissa*. Technical report prepared by ATREE, Bangalore, p 53.
55 Arthur, R, N Kelkar, T Alcoverro and MD Madhusudan (2013) Complex ecological pathways underlie perceptions of conflict between green turtles and fishers in the Lakshadweep Islands. *Biological Conservation* 167: 25–34.
56 See www.seaturtlesofindia.org.
57 Pandav, B. & BC Choudhury (1999) An update on the mortality of the olive ridley sea turtles in Orissa. *Marine Turtle Newsletter* 83: 10–12.
58 See n. 6.
59 Gopi, GV, B Pandav and BC Choudhury (2007) Estimated annual incidental captures of *Lepidochelys olivacea* (Eschscholtz, 1829) in trawl nets along the Orissa coast, India. *Hamadryad* 31 (2): 212–215.
60 Gopi, GV, B Pandav and BC Choudhury (2006) Incidental capture and mortality of olive ridley turtles (*Lepidochelys olivacea*) in commercial trawl fisheries in coastal waters of Orissa, India. *Chelonian Conservation and Biology* 5: 276–80.
61 Memo No 1895/ FARD, issued by the Fisheries and Animal Resource Development Department, Government of Orissa dated 4th February 2005.
62 The 'public interest letter' petition dated 19th October 1998 was drawn up by lawyer Raj Panjwani (who represented WWF in the earlier case). The subject line of this letter was 'the prevention of mass slaughter of olive ridley sea turtles'. The citation for the PIL itself was O.C.J. No 14889 of 1998.
63 See n. 18.
64 Memo No 402, dated 6th December 1998 from the Director of Fisheries to the

Assistant Directors of Fisheries of Kujang, Balassore, Puri and Ganjam.
65 Sridhar, A, B Tripathy and K Shanker (2005) A Review of Legislation and Conservation Measures for Sea Turtles in Orissa, India. *Indian Ocean Turtle Newsletter* 1: 1–6.
66 Bhavani Sankar, O & M Ananth Raju (2003) Implementation of the Turtle Excluder Device in Andhra Pradesh. *Kachhapa* (8): 3–6.
67 Gopi, GV, B Pandav and BC Choudhury (2007) Estimated annual incidental captures of *Lepidochelys olivacea* (Eschscholtz, 1829) in trawl nets along the Orissa coast, India. *Hamadryad* 31 (2): 212–215.
68 WWF (2011) *Turtles and TEDs: Outcome of trials conducted off Dhamra, Orissa.* WWF–India, New Delhi, India.
69 Interim orders of the CEC dated February 2004.

Hype and Hypocrisy in Las Tortugas

1 Mrosovsky, N (2002) Editorial: Hype. *Marine Turtle Newsletter* 96: 1–4.
2 Davis, TA, R Bedi and GM Oza (1978) Sea Turtle faces extinction in India. *Environmental Conservation* 5: 211–212.
3 Frazier, JG (1980) Sea turtle faces extinction in India: Crying 'wolf' or saving sea turtles? *Environmental Conservation* 7: 239–240.
4 Oza, GM (1993) Last chance to save olive ridley turtles in India. *Environmental Conservation* 20: 367.
5 Das, I (1986) Marine turtle conservation: the tribal connection. *Marine Turtle Newsletter* 36: 2–3.
6 Andrews, H (1993) Olive ridleys threatened in India: letters needed. *Marine Turtle Newsletter* 61: 5–6.
7 Eckert, KL and SA Eckert (1994) Editorial. *Marine Turtle Newsletter* 64: 1–3.
8 Mrosovsky, N (1993) World's Largest Aggregation of Sea Turtles to be Jettisoned. *Marine Turtle Newsletter* 63: 32–33.
9 Bjorndal, KA (1993) Future of the Gahrimatha Arribadas a Matter of International Concern. *Marine Turtle Newsletter* 63:S3.
10 Anonymous (1993) Nesting ground of turtles threatened. Excerpted from *Indian Express*, 10 September and reprinted in *Marine Turtle Newsletter* 63 (Supplement): 3–4.
11 Anonymous (1993) Concern rises over threat to Indian turtles. *Emirates News*, 28 November 1993. Reprinted in *Marine Turtle Newsletter* 64: 1–2.
12 Anonymous (1994) Urgent Notice. *Marine Turtle Newsletter* 64: 2.
13 Mohanty–Hejmadi, P (1994) Latest word on the Talachua Jetty, Orissa, India. *Marine Turtle Newsletter* 67: 1.
14 Mohanty–Hejmadi, P (1995) Legal Briefs: Good news from Bhitarkanika. *Marine Turtle Newsletter* 69: 25.
15 Eckert, SA and KL Eckert (1996) Editors' Note. *Marine Turtle Newsletter* 73: 2.
16 Marquez–M, R et al. (1996) Good News! Rising Numbers of Kemp's Ridleys

Nest at Rancho Nuevo, Tamaulipas, Mexico. *Marine Turtle Newsletter* 73: 2–5.
17 Marquez–M, R, C Peñaflores & J Vasconcelos (1996) Olive Ridley Turtles (*Lepidochelys olivacea*) Show Signs of Recovery at La Escobilla, Oaxaca. *Marine Turtle Newsletter* 73: 5–7.
18 Das, BB (1997) Struggle to protect the Bhitarkanika ecosystem is ongoing. *Marine Turtle Newsletter* 76: 18–20.
19 Das, BB (1998) Present Status of Gahirmatha beach in Bhitarakanika Sanctuary, Orissa. *Marine Turtle Newsletter* 79: 1–2.
20 Centre for Environmental Law, Worldwide Fund for Nature, India vs State of Orissa and Ors. [O.J.C. No. 3128 of 1994 dated 14.05.1998].
21 Godfrey, MH and C Schauble (2008) Special Theme Section: The Dhamra Port Debate – Perspectives and Lessons. *Indian Ocean Turtle Newsletter* 8:1.
22 Anonymous (2005) A Century of Trust Threatens Centuries of Turtles – DON'T SAY TATA TO TURTLES. Greenpeace.org, 26 April 2005 (http://www.greenpeace.org/india/en/news/don–t–say–tata–to–turtles/).
23 IUCN (2006) *Scoping Mission to the Dhamra Port Project*, p.11. Bangkok.
24 IUCN–DPCL (2007) Partnership Promoting Sustainable Environmental Alliances.
25 See Hykle, D (2008) Editorial: India's Dhamra Port controversy heats up again. (See http://www.ioseaturtles.org/feature_detail.php?id=246 with appended emails).
26 Email from Nicolas Pilcher, Co–Chair of IUCN SSC Marine Turtle Specialist Group, posted on CTURTLE, 27 March 2008.
27 Email from Romulus Whitaker, Basudev Tripathy and Bivash Pandav, MTSG members; posted on CTURTLE 27 March.
28 Email from Romulus Whitaker, MTSG member; posted on CTURTLE, 6 April 2008.
29 Reprinted in *Indian Ocean Turtle Newsletter* 8: 3–4.
30 Harry Andrews, BC Choudhury, Kartik Shanker, Wesley Sunderraj, Basudev Tripathy and Rom Whitaker.
31 Bombay Natural History Society, Foundation for Ecological Security, Gujarat Institute of Desert Ecology, Indian National Trust for Art & Cultural Heritage, Salim Ali Centre for Ornithology & Natural History and Wildlife Protection Society of India.
32 Reprinted in *Indian Ocean Turtle Newsletter* 8: 3–4.
33 Lenin, J and R Whitaker (2008) Membership Excluder Devices. *Indian Ocean Turtle Newsletter* 8: 10–15.
34 Pilcher, NJ (2008) The MTSG and its involvement in the Dhamra Port, Orissa, India. *Indian Ocean Turtle Newsletter* 8: 24–25.
35 Hawk, E (2009) The Continuing shame of Orissa. *Marine Turtle Newsletter* 124: 1–3.
36 Shanker, K et al. (2009) A Little Learning… The Price of Ignoring Politics and History. *Marine Turtle Newsletter* 124: 3–5.
37 Letter to Tata Steel, DPCL and Larson and Toubro from Kartik Shanker,

Member, MTSG; Dhrubajyoti Basu, Gharial Conservation Alliance; Subir Chowfin, Gharial Conservation Alliance; Manish Chandi, Andaman and Nicobar Environmental Team; Nikhil Whitaker, Madras Crocodile Bank; Bivash Pandav, Wildlife Biologist and turtle researcher; Basudev Tripathy, Member, MTSG; Wesley Sunderraj, Member, MTSG; Todd Steiner, Sea Turtle Restoration Project and Member, MTSG; Scott Eckert, WIDECAST and Member, MTSG; Karen Eckert, WIDECAST and Member, MTSG; Roldán A. Valverde, Southeastern Louisiana University, Member, MTSG; M. Zahirul Islam, MarineLife Alliance and Member, MTSG; John G. Frazier, Member, MTSG; Lily Venizelos, MEDASSET, Member, MTSG; Edward Aruna, Conservation Society of Sierra Leone, Member, MTSG; Dimitris Margaritoulis, Member, MTSG; dated 20 April 2009.

38 Letter to National Geographic Channel from Sanjeev Gopal, Greenpeace and other conservationists and biologists (Aarthi Sridhar, Ashish Fernandes, Belinda Wright, Biswajit Mohanty, Janaki Lenin, Kartik Shanker, Rom Whitaker, Sudarshan Rodriguez).

39 Letter to IUCN–India from Bivash Pandav, Kartik Shanker, Basudev Tripathy and Rom Whitaker, dated 19 November 2008.

40 Open letter from the MTSG–India members to the MTSG Co–Chairs on the Dhamra Turtle issue; letter dated 8 May 2009 and signed by BC Choudhury, Bivash Pandav, Kartik Shanker, Basudev Tripathy and Rom Whitaker.

41 Lenin, J et al. (2009) Editorial. The IUCN's new clothes: an update on the Dhamra – turtle saga. *Marine Turtle Newsletter* 126: 2–5.

42 Wright, B. (1996) *The Night of the Sea Turtles*. In: *Through The Tiger's Eyes; A Chronicle of India's Wildlife* (Eds. S Breeden and B Wright), pp. 120–122. Ten Speed Press, Berkeley, USA.

43 Anonymous (2004) Greenpeace ship SV Rainbow Warrior to tour the East Coast of India. Greenpeace.org, October 15, 2004 (http://www.greenpeace.org/india/en/news/greenpeace–ship–sv–rainbow–war/).

44 Anonymous (2006) Turning Turtle, Greenpeace.org, 17 February 2006 (http://www.greenpeace.org/international/en/news/features/turning–turtle/).

45 Anonymous (2006) Witnesses arrested, accused walks free. Greenpeace.org, 14 April 2006 (http://www.greenpeace.org/india/en/news/witnesses–arrested–accused–wa–2/).

46 Gopal, S (2005) A Journey to Oblivion? Orissa and its Olive Ridleys! Greenpeace.org, March 10, 2005 (http://www.greenpeace.org/india/en/news/a–journey–to–oblivion–orissa/).

47 Mrosovsky, N (2002) Editorial: Hype. *Marine Turtle Newsletter* 96: 1–4.

48 Mace, GM & R Lande (1990) Assessing extinction threats: Toward a reevaluation of IUCN threatened species categories. *Conservation Biology* 5:148–157.

49 See www.iucn.org.

50 See http://www.iucn.org/about/.

51 Chapin, M (2004) A challenge to conservationists. *Worldwatch magazine*. November/December: 17–31.

52 McDonald, CC (2008) *Green, Inc.: An Environmental Insider Reveals How a Good Cause Has Gone Bad*. The Lyons press, Connecticut, USA.
53 Dowie, M (2005) Conservation Refugees: When protecting nature means kicking people out. *Orion* November/December: 16–27.
54 Brockington, D (2002) *Fortress conservation: the preservation of the Mkomazi Game Reserve, Tanzania*. Indiana University Press, USA.
55 Brockington, D & J Igoe (2006) Eviction for conservation: A global overview. *Conservation and society* 4: 424.
56 Brockington, D (2011) A Brief Guide to (Conservation) NGOs. *Current Conservation* 5 (1): 30–31.
57 Shanker, K, B Pandav & BC Choudhury (2004) An assessment of the olive ridley turtle (*Lepidochelys olivacea*) nesting population in Orissa, India. *Biological Conservation*, 115(1): 149–160.
58 Brijnath, R & R Banerjee (1994) Turning the turtles away. *India Today*, February 28, 1994.
59 Silas, EG, M Rajagopalan and AB Fernando (1983) Sea turtles of India – Need for a crash programme on conservation and effective management of the resource. *Marine Fisheries Information Service T & E Series* 50: 1–12.
60 Shanker, K (2008) My way or the highway!!! Where corporations and conservationists meet. *Indian Ocean Turtle Newsletter* 8: 10–14.
61 Singh, C (1996) Bhitarkanika Ecosystem Protected by Court's Decision. *Marine Turtle Newsletter* 73: 1–2.
62 Rothenberg, D and Ulvaeus, M (1999) The New Earth Reader: The Best of Terra Nova. The MIT Press, USA.
63 Email from Kartik Shanker to the MTSG Listserver, dated 26 September 2007.
64 Email from Kartik Shanker, MTSG Regional Vice Chair; posted on MTSG listserver, 16 June 2008.
65 Email from Kartik Shanker, MTSG member; posted on MTSG listserver, 9 July 2008.
66 *Marine Turtle Newsletter* 121 and *Indian Ocean Turtle Newsletter* 8.
67 Hykle, DJ (2008) India's Dhamra Port controversy heats up again. *Indian Ocean Turtle Newsletter* 8: 2–3.
68 Mrosovsky, N (2008) Continuing controversy over ridleys in Orissa: Cui bono? *Indian Ocean Turtle Newsletter* 8: 6–9.
69 Fernandes, A (2008) IUCN–Tata partnership – undermining conservation. *Indian Ocean Turtle Newsletter* 8: 17–18.
70 Rodriguez, S and A Sridhar (2008) Dhamra Port: How environmental regulatory failure fuels corporate irreverence. *Indian Ocean Turtle Newsletter* 8:19–23.
71 Dublin, HT (2008) The Dhamra Port issue: Some views from the Chair of the IUCN SSC. *Indian Ocean Turtle Newsletter* 8: 25–27.
72 Dutta, A (2008) Dhamra Port: The other perspective. *Indian Ocean Turtle Newsletter* 8: 27–28.
73 Current Conservation Issue 3.4 (www.currentconservation.org).

Ridleys in the Big Idly

1 Jones, S and AB Fernando (1973) The present status of the turtle fishery in the Gulf of Mannar and Palk Bay. In: *Proceedings of the Symposium on Living Resources of the Seas Around India*, pp. 712–715. Central Marine Fisheries Research Institute, Cochin, India.
2 Valliappan, S & R Whitaker (1975) Olive ridleys on the Coromandel coast of India. *Herpetological Review* 6(2): 42–43.
3 Valliappan, S & R Whitaker (1974) *Olive ridleys on the Coromandel Coast*. Madras Snake Park Trust, Madras. 14 p.
4 Ibid.
5 Whitaker, R (1979) Captive rearing of marine turtles. *Journal of the Bombay Natural History Society* 76(1): 163–66.
6 Anonymous (1977) News from the Madras Snake Park and Madras Crocodile Bank. Hamadryad 2(1):2–3.
7 Whitaker, R (1979) Captive rearing of marine turtles. *Journal of Bombay Natural History Society* 76(1): 163–66.
8 Whitaker, R & JG Frazier (1993) Growth of a captive hawksbill turtle in India. *Hamadryad* 18: 47–48.
9 Paulraj, S, S Subbarayalu Naidu & J Pakkiaraj (1986) Rearing the olive ridley *Lepidochelys olivacea* in artificial sea water. *Aquatic Exhibits*: 90–94.
10 Anonymous (1982) WWF sea turtle hatchery. *Hamadryad* 7(2): 2.
11 Silas, EG (1984) Sea turtle research and conservation. *Bulletin of Central Marine Fisheries Research Institute* 35: 1–8.
12 Silas, EG, M Vijayakumaran & M Rajagopalan (1984) Yolk utilization in the egg of the olive ridley *Lepidochelys olivacea*. *Bulletin of Central Marine Fisheries Research Institute* 35: 22–33.
13 Vijayakumaran, M, M Rajagopalan & EG Silas (1984) Food intake and conversion in hatchlings of olive ridley *Lepidochelys olivacea* fed animal and plant food. *Bulletin of Central Marine Fisheries Research Institute* 35: 41–48.
14 Rajagopalan, M (1984) Studies on the growth of olive ridley *Lepidochelys olivacea* in captivity. *Bulletin of Central Marine Fisheries Research Institute* 35: 49–54.
15 Anonymous (1978) The vanishing turtle. *Central Marine Fisheries Research Institute Newsletter* 7: 5–6.
16 Silas, EG & M Rajagopalan (1984) Recovery programme for olive ridley *Lepidochelys olivacea* (Eschscholtz, 1829) along Madras Coast. *Bulletin of Central Marine Fisheries Research Institute* 35: 9–21.
17 Venkatramani, SH (1984) Hatching a success. *India Today* May 31: 155.
18 Abraham, C (1989) A *report on the conservation and management of sea turtles on the Madras coast 1988–89*. Students' Sea Turtle Conservation Network, Madras, India. 23 p.
19 Abraham, C (1990) Preliminary observations on the nesting of the olive ridley sea turtle (*Lepidochelys olivacea*) on the Madras coast, south India. *Hamadryad* 15: 10–12.

20 Arun, V (2013) Students Sea Turtle Conservation Network: 25 years of conservation. *Indian Ocean Turtle Newsletter* 18: 18–21.
21 Abraham, C, K Shanker & Y Thiruchelam (1990) *Conservation and management of sea turtles on the Madras coast, 1989–1990 report.* Students' Sea Turtle Conservation Network, Madras, India. 30 p.
22 Abraham, C (1989) *A report on the conservation and management of sea turtles on the Madras coast, 1988–89.* Students' Sea Turtle Conservation Network, Madras, India. 23 p.
23 Anonymous (1990) Students' Sea Turtle Conservation Network Newsletter 2.
24 Anonymous (1989) Students' Sea Turtle Conservation Network Newsletter 1.
25 Anonymous (1993) *A report on the sea turtle programme, 1992–1993.* Students' Sea Turtle Conservation Network, Madras, India. 16 p.
26 Sivasundar, A, C Bharadwaj, J Mathew, G Vivek & M Anand. *Conservation of Sea Turtles on the Madras Coast, 1993–94 report.* Students' Sea Turtle Conservation Network, Madras, India. 28 p.
27 Sivasundar, A, C Bharadwaj, G Vivek & M Anand. *Conservation of Sea Turtles on the Madras Coast, 1994–95 report.* Students' Sea Turtle Conservation Network, Madras, India. 30 p.
28 Shanker, K (2003) Thirty years of sea turtle conservation on the Madras coast: A review. *Kachhapa* 8: 16–19.
29 Arun, V (2011) Student's Sea Turtle Conservation Network: A victory for volunteerism! *Indian Ocean Turtle Newsletter* 14: 21–25.
30 See n. 20.
31 Mukherjee, N, B Muthuraman and K Shanker (2009) Bioshields and ecological restoration in tsunami affected areas in India. In: *Integrated Coastal Zone Management* (Eds. E Dahl, E Moskness and J Stottrup). Wiley Blackwell Publishing, UK.
32 Pandav, B (2000) *Post cyclone impact in Orissa with reference to marine turtle conservation – A GOI–UNDP sea turtle project report.* Wildlife Institute of India, Dehradun, India.
33 Mukherjee, N, K Shanker, N Koedam & F Dahdouh–Guebas (2014) Governance of coastal plantations in southern India: long term ecological vision or short term economic opportunity? *Acta Oecologica*.
34 Feagin, RA, N Mukherjee, K Shanker et al. (2010) Shelter from the storm? The use and misuse of 'bioshields' in managing for natural disasters on the coast. *Conservation Letters* 3:1–11.
35 Mukherjee, N, A Sridhar, M Menon, S Rodriguez and K Shanker (2008) *Policy Brief: Bioshields.* UNDP/UNTRS, Chennai and ATREE, Bangalore, India, 8 p.
36 Chaudhari, S, KVD Prasad, & K Shanker (2009) *Impact of Casuarina Plantations on Olive Ridley Turtle Nesting along the Northern Tamil Nadu Coast, India.* ATREE, Bangalore and MCBT, Mammalapuram, India.
37 Namboothri, N, D Subramanian, A Sridhar, S Rodriguez, M Menon, and K Shanker (2008) *Policy Brief: Sand Dunes.* UNDP/UNTRS, Chennai and ATREE, Bangalore, India, 12 p.

38 Balu, A (2008) The World Bank funded coastal shelterbelt project threatens sea turtle nesting habitats in Tamil Nadu, India. *Indian Ocean Turtle Newsletter* 7: 23–25.
39 Dattatri, S (2012) Freeing sea turtle nesting beaches from Casuarina plantations – A case study from Tamil Nadu (http://www.conservationindia.org/case–studies/freeing–sea–turtle–nesting–beaches–from–casuarina–plantations—–a–case–study–from–tamil–nadu).
40 Arun, V (2011) Student's Sea Turtle Conservation Network: A victory for volunteerism! *Indian Ocean Turtle Newsletter*14: 21–25.
41 Whitaker, Z (1989) *Snakeman: the story of a naturalist*. Penguin, India.
42 Valliapan, S & S Pushparaj (1973) Sea turtles in Indian waters. *Cheetal* 16(1): 26–30.
43 Mahapatra, A and S Valliapan (197X) IIT Newsletter (other details not available).
44 Dattatri, S & D Samarajiva (1982) *The status and conservation of sea turtles in Sri Lanka*. Project of the sea turtle rescue fund, Washington DC, USA.
45 Email to Kartik Shanker dated August 13, 2014.
46 See n. 29.
47 Caillouet, CW Jr (1987) *Report on efforts to prevent extinction of Kemp's ridley sea turtle through head starting*. NOAA Technical Memorandum NMFS–SEFC–188, i plus 20 p.
48 Yntema, CL & N Mrosovsky (1982) Critical periods and pivotal temperatures for sexual differentiation in loggerhead sea turtles. *Canadian Journal of Zoology* 60(5): 1012–1016.
49 Mrosovsky, N & CL Yntema (1980) Temperature dependence of sexual differentiation in sea turtles: implications for conservation practices. *Biological Conservation* 18: 271–280.
50 Wibbels (2003) Critical approaches to sex determination in sea turtles. In: *The Biology of Sea Turtles – Volume II* (Eds. PL Lutz, JA Musick and J Wyneken), pp. 103–134. CRC Press, USA.
51 Shine, R (1999) Why is sex determined by nest temperature in many reptiles? *Trends in Ecology & Evolution* 14: 186–189.
52 Hawkes, LA, AC Broderick, MH Godfrey & BJ Godley (2007) Investigating the potential impacts of climate change on a marine turtle population. *Global Change Biology* 13: 923–932.
53 Shaver, DJ (1996) Head–Started Kemp's Ridley Turtles Nest in Texas. *Marine Turtle Newsletter* 74: 5–7.
54 Mrosovsky, N (1983) *Conserving sea turtles*. The British Herpetological Society, London, UK.
55 Mathew, J, R Rajagopal & K Shanker (1991) *Conservation and management of sea turtles on the Madras coast, 1990–1991 report*. Students' Sea Turtle Conservation Network, Madras, India. 39 p.
56 Shanker (1994) Conservation of Sea Turtles on the Madras Coast. *Marine Turtle Newsletter* 64: 3–6.
57 Bézy, V.S., R.A. Valverde & C.J. Plante (2014) Olive Ridley Sea Turtle Hatching

Success as a Function of Microbial Abundance and the Microenvironment of In Situ Nest Sand at Ostional, Costa Rica. *Journal of Marine Biology*.

Small beaches, big battles

1 Bhaskar, S (1978) Notes from the Gulf of Kutch. *Hamadryad* 3(3): 9–10.
2 Bhaskar, S (1979) Preliminary report on sea turtles in the Gulf of Kutch. *Marine Turtle Newsletter* 11: 3–4.
3 Bhaskar, S (1982) Turtle tracking in Gujarat. *Hamadryad* 7(1): 13–14.
4 Firdous, F (1991) A turtle's journey from Pakistan (Karachi) to India (Gujarat). *Marine Turtle Newsletter* 53: 18–19.
5 Frazier, JG (1987) *Report on a trip to Mangrol, Gujarat*. Unpublished report.
6 Frazier, JG (1989) Observations of stranded green turtles (*Chelonia mydas*) in the Gulf of Kutch. *Journal of the Bombay Natural History Society*, *86*, 250–252.
7 Frazier, JG (1987) *Tata nature reserves in the Mithapur area*. Unpublished report.
8 Nareshwar, EK (1998) Evaluation of sea turtle nesting beaches for promoting participatory conservation at Sundervan Beyt Dwaraka, India. *Hamadryad* 22(2): 121–122.
9 Sunderraj, SFW et al. (2002) *Status of the breeding population of sea turtle along the Gujarat coast – A GOI–UNDP sea turtle project report*. Gujarat Institute of Desert Ecology, Bhuj, India. 44 p.
10 Anonymous (2011) Prakruti Nature Club. *Indian Ocean Turtle Newsletter* 13: 8–9.
11 Karbari, JP (1985) Leatherback turtle caught off Devbag near Malvan. *Marine Fisheries Information Service T & E Series* 64: 23.
12 Karbari, JP (1981) On the rare occurrence of a giant sized hawksbill turtle off Elephanta Caves (near Bombay). *Marine Fisheries Information Service T & E Series* 33: 17.
13 Shaikh, KA (1984) Distribution and nesting sites of sea turtles in Maharashtra. In: *Proceedings of the Workshop on Sea Turtle Conservation* (Ed. EG Silas), pp. 109–115. Central Marine Fisheries Research Institute Special Publication No. 18, Kochi, India.
14 Giri, V & N Chaturvedi (2006) Sea turtles of Maharashtra and Goa. In: *Marine Turtles of the Indian subcontinent* (Eds. K Shanker and BC Choudhury), pp. 147–155.Universities Press, Hyderabad, India.
15 Katdare, B (2012) An update on olive ridley nesting along the west coast of Maharashtra, India in 2011–2012. *Indian Ocean Turtle Newsletter* 15: 3–4.
16 Katdare, B (2009) An update on Sahyadri Nisarga Mitra activities during 2007–2008. *Indian Ocean Turtle Newsletter* 9: 20–21.
17 Shanker, K & R Kutty (2005) Sailing the flagship fantastic: myth and reality of sea turtle conservation in India. *Maritime Studies* 3(2) and 4(1): 213–240.
18 Dongre, SM and MK Shambhu (2008) Achieving sea turtle conservation with

the help and involvement of local communities in Morjim, Goa, India. *Indian Ocean Turtle Newsletter* 7: 21–22.
19 Bhaskar, S (1984) The distribution and status of sea turtles in India. In: *Proceedings of the Workshop Sea Turtle Conservation* (Ed. EG Silas), pp. 21–35. Central Marine Fisheries Research Institute Special Publication No. 18, Kochi, India.
20 Chandarji, SL (1984) Status of sea turtle conservation in Karnataka state. In: *Proceedings of the Workshop on Sea Turtle Conservation* (Ed. EG Silas), pp. 109. Central Marine Fisheries Research Institute Special Publication No. 18, Kochi, India.
21 Frazier, JG (1987) *Survey of the marine turtle hatcheries of Karnataka*. Unpublished report.
22 Frazier, JG (1989) Survey of the marine turtle hatcheries of Karnataka. *Hamadryad* 14(2): 8–18.
23 Anonymous (2011) Canara Green Academy. *Indian Ocean Turtle Newsletter* 13: 9–11.
24 Pandit, R (2007) Release of juvenile green turtles in North Karnataka. *Indian Ocean Turtle Newsletter* 6: 28.
25 Anonymous (2011) Field Services and Inter–Cultural Learning. *Indian Ocean Turtle Newsletter* 13: 11–12.
26 See n. 19.
27 Shanmugasundaram, P (1968) Turtle industry. *Indian Seafood* 6: 18–19.
28 Dileepkumar, N & C Jayakumar (2002) *Field study and networking for turtle conservation in Kerala – A GOI–UNDP sea turtle project report*. Thanal Conservation Action & Information Network, Trivandrum, India.
29 Pillai, SK (2003) Instance of meat of Leatherback Turtle *Dermochelys coriacea* used as food. *Fishing Chimes* 23 (3): 46–47.
30 Anonymous (1981) Turtle Sanctuary. *Hamadryad* 6(2): 2.
31 Anonymous (1997) Priority Action: Survey of marine turtles along the Kerala coast. *Marine Turtle Newsletter* 76: 20–21.
32 Based on Shanker, K & R Kutty (2005) Sailing the flagship fantastic: myth and reality of sea turtle conservation in India. *Maritime Studies* 3(2) and 4(1): 213–240.
33 Ibid.
34 Kumar, S (2007) Sea turtle conservation in Kasargod, Kerala. *Indian Ocean Turtle Newsletter* 6: 28–29.
35 James, NJ (2011) Green Habitat, Guruvayur, Kerala. *Indian Ocean Turtle Newsletter* 14: 29–30.
36 Whitaker, R (2006) Turtle Trekker: Satish Bhaskar. In: *Marine Turtles of the Indian subcontinent* (Eds. K Shanker and BC Choudhury), pp. 271–290. Universities Press, Hyderabad, India.
37 Bhaskar, S (1984) The distribution and status of sea turtles in India. In: *Proceedings of the Workshop Sea Turtle Conservation* (Ed. EG Silas), pp. 17–21. Central Marine Fisheries Research Institute Special Publication No. 18, Kochi, India.
38 Tripathy, B, K Shanker & BC Choudhury (2006) The status of sea turtles and

their habitats in the Lakshadweep archipelago, India. *Journal of the Bombay Natural History Society* 103(1): 33–43.

39 Tripathy, B, BC Choudhury & K Shanker (2002) A *survey of marine turtles and their nesting habitats in the Lakshadweep islands, India – A GOI–UNDP sea turtle project report.* Wildlife Institute of India, Dehradun, India. 62 p.

40 Arthur, R, N Kelkar, T Alcoverro and MD Madhusudan (2013) Complex ecological pathways underlie perceptions of conflict between green turtles and fishers in the Lakshadweep Islands. *Biological Conservation* 167: 25–34.

41 Lal, A et al. (2010) Implications of conserving an ecosystem modifier: Increasing green turtle (*Chelonia mydas*) densities substantially alters seagrass meadows. *Biological Conservation* 143: 2730–2738.

42 Kelkar, N, R Arthur, N Marbà, & T Alcoverro (2013) Greener pastures? High-density feeding aggregations of green turtles precipitate species shifts in seagrass meadows. *Journal of Ecology* 101: 1158–1168.

43 Kumar, S and BC Choudhury (2013) Sea turtle nesting trends in the Lakshadweep Islands. In: *Proceedings of the Thirtieth Annual Symposium on Sea Turtle Biology and Conservation* (Compilers: J Blumenthal, A Panagopolou and A Rees), pp. 141. NOAA Technical Memorandum NMFS–SEFSC–640, 177p.

44 Anonymous (2011) Lakshadweep Marine Research and Conservation Centre. *Indian Ocean Turtle Newsletter* 13: 14.

45 Bhupathy, S and S Saravanan (2006) Marine turtles of Tamil Nadu. *Marine Turtles of the Indian subcontinent* (Eds. K Shanker and BC Choudhury), pp. 58–67. Universities Press, Hyderabad, India.

46 Bhupathy, S & S Saravanan (2003) Exploitation of sea turtles along the south–east coast of Tamil Nadu, India. *Journal of Bombay Natural History Society* 100(2&3): 628–631.

47 Mohanraj, T (2011) Observations on the exploitation of sea turtles along the Tuticorin coast, Tamil Nadu, India. *Indian Ocean Turtle Newsletter* 14: 9–11.

48 Bhupathy, S and S Saravanan (2006) Status of marine turtles in the Gulf of Mannar, India. *Chelonian Conservation and Biology* 5(1): 139–141.

49 Rahaman, A, PK Ponnuswamy & K Rajendran (1985) Recovery plan for olive ridley *Lepidochelys olivacea* at Point Calimere, Tamil Nadu. In: *Proceedings of the Symposium on Endangered Marine Animals and Marine Parks, Cochin* (Ed. EG Silas), pp. 250–254. Marine Biological Association of India, Cochin, India.

50 Ponnuswamy, PK & AA Rahaman (1985) Captive rearing of hatchlings of olive ridley *Lepidochelys olivacea* at Point Calimere, Tamil Nadu. In: *Proceedings of the Symposium on Endangered Marine Animals and Marine Parks, Cochin* (Ed. EG Silas), pp. 248–249. Marine Biological Association of India, Cochin, India.

51 Baruah, AD (2001) Olive ridley hatchery program of Point Calimere wildlife sanctuary, Tamil Nadu. *Kachhapa* 5: 18.

52 Velusamy, T and R Sundararaju (2009) Olive ridley turtle conservation activities along the Nagapattinam coast, Tamil Nadu, India. *Indian Ocean Turtle Newsletter* 10: 21–24.

53 Dharani, S (2003) Turtle conservation by local communities in Madras. *Kachhapa* 8: 22.
54 Currimboys, S (2005) Turtle conservation & community development: Activities of the TREE Foundation. *Indian Ocean Turtle Newsletter* 2: 4–5.
55 Dharini, S (2007) Interaction between the Sea Turtle Protection Force and trawler owners and workers at Kasimedu fishing harbour, Chennai. *Indian Ocean Turtle Newsletter* 5: 22.
56 Dharini, S (2012) Capacity building for trawl/commercial fishers to reduce bycatch mortality of sea turtles along the Chennai coast, India. *Indian Ocean Turtle Newsletter* 15: 5–7.
57 Anonymous (2011) TREE Foundation. *Indian Ocean Turtle Newsletter* 13: 25–28.
58 Dharini, S and R Muralidharan (2012) Tree Foundation – Sea Turtle Rescue and Rehabilitation Centre. *Indian Ocean Turtle Newsletter* 15: 8–9.
59 See n. 19.
60 Ibid.
61 Anonymous (2011) Visakha Society for Protection and Care of Animals (VSPCA). *Indian Ocean Turtle Newsletter* 13: 6–7.
62 Writ petition No. 4373 of 2000, High court of Judicature, Andhra Pradesh, Hyderabad dated 10.11.2000.
63 Multiple emails to Kartik Shanker in 2000 and 2001.
64 Letter from SD Mukherji, IFS (Principal Chief Conservator of Forests, Forest Department, Government of Andhra Pradesh) to Joint Director, Ministry of Environment and Forests, dated 24.6.2000.
65 Shanker, K (2001) Guest editorial: The swampland of sea turtle conservation: In search of a philosophy. *Marine Turtle Newsletter* 95: 1–4.
66 Shanker, K (2001) Opinion: Sea turtles and submarines – sinking the wrong ship? *Kachhapa* 4: 27–28.
67 See n. 61.
68 Ramana Murthy, KV (2001) Conservation of sea turtles in northern Andhra Pradesh. *Kachhapa* 4: 18–19.
69 Anonymous (2000) Operation Angel. *Kachhapa* 2: 16–17.
70 Dharini S and R Muralidharan (2012) Tree Foundation – Sea Turtle Rescue and Rehabilitation Centre. *Indian Ocean Turtle Newsletter* 15: 8–9.
71 Anonymous (2009) NGO Profile: The Rushikulya Sea Turtle Protection Committee. *Indian Ocean Turtle Newsletter* 9: 28–29.
72 Anonymous (2011) Rushikulya Sea Turtle Protection Committee (RSTPC), Orissa. *Indian Ocean Turtle Newsletter* 13: 22–23.
73 Tripathy, B (2005) Letter to the Editor: Lighting and sea turtle hatchlings in Rushikulya. *Indian Ocean Turtle Newsletter* 1: 26.
74 Anonymous (2005) Profile of NGOS working on sea turtle conservation and fisheries in Orissa. *Indian Ocean Turtle Newsletter* 1: 18–25.
75 Anonymous (2011) Podampeta Ecotourism and Olive Ridley Protection Club (PEORPC), Orissa. *Indian Ocean Turtle Newsletter* 13: 20–21.
76 See n. 74.

77 Anonymous (2011) Green Life Rural Association. *Indian Ocean Turtle Newsletter* 14:26–28.
78 Anonymous (2011) Action for Protection of Wild Animals (APOWA), Orissa. *Indian Ocean Turtle Newsletter* 13:16–18.
79 Anonymous (2011) Alacrity, Orissa. *Indian Ocean Turtle Newsletter* 13:18–20.
80 Behera, C (2006) Beyond TEDs: The TED controversy from the perspective of Orissa's trawling industry. In: *Marine turtles of the Indian Subcontinent* (Eds. K Shanker & BC Choudhury), pp. 238–243. Universities Press, Hyderabad, India.
81 Silas, EG, M Rajagopalan & SS Dan (1983) Marine turtle conservation and management: A survey of the situation in West Bengal 1981/82 & 1982/83. *Marine Fisheries Information Service T & E Series* 50: 24–33.
82 Kar, CS and S Bhaskar (1982) Status of sea turtles in the eastern Indian Ocean. *Biology and conservation of sea turtles* (Eds. K Bjorndal), pp. 365–372. Smithsonian Institution Press, Washington DC, USA.
83 Das, I (1986) Marine turtle conservation: The tribal connection. *Marine Turtle Newsletter* 36: 2–3.
84 GOI UNDP Sea turtle project.
85 See http://www.marinelifealliance.org.
86 This section is derived from Shanker, K (2007) Deconstructing sea turtle conservation in India. In: *Making Conservation Work* (Eds. G Shahabuddin and M Rangarajan), pp. 89–110. Permanent Black, New Delhi, India.
87 Troëng, S, & E Rankin (2005) Long–term conservation efforts contribute to positive green turtle (*Chelonia mydas*) nesting trend at Tortuguero, Costa Rica. *Biological Conservation* 121: 111–116.
88 Marcovaldi, MÂ & GG Dei Marcovaldi (1999) Marine turtles of Brazil: the history and structure of Projeto TAMAR–IBAMA. *Biological Conservation* 91: 35–41.
89 Kapurusinghe, T (2006) Status and conservation of marine turtles in Sri Lanka. In: *Marine Turtles of the Indian subcontinent* (Eds. K Shanker and BC Choudhury), pp. 173–187. Universities Press, Hyderabad, India.
90 Richardson, PB (2013) Satellite telemetry reveals behavioural plasticity in a green turtle population nesting in Sri Lanka. *Marine Biology* 160: 1415–1426.
91 See n. 19.
92 Ibid.
93 Fernando, A B (1983) Nesting site and hatching of the hawksbill turtle along Thirunelveli coast of Tamil Nadu. *Marine Fisheries Information Service T & E Series* 50: 33–34.
94 Agastheesapillai, A & R Thiagarajan (1979) Biology of the green turtle *Chelonia mydas* (Linnaeus) in the Gulf of Mannar and Palk Bay. *Journal of Marine Biological Association of India* 21(1&2): 45–60.
95 Kannan, P (2008) Studies on the green turtle (*Chelonia mydas*) in the Gulf of Mannar Biosphere Reserve, Tamil Nadu, India. *Indian Ocean Turtle Newsletter* 7: 12–15.
96 Raja Sekhar, PS & MV Subba Rao (1993) Conservation and management of

the endangered olive ridley sea turtle, *Lepidochelys olivacea* (Eschscholtz), along the northern Andhra Pradesh coastline, India. *B.C.G. Testudo* 3(5): 35–53.

97 Priyadarshini, KVR (1998) *Status, ecology and management of olive ridley sea turtles and their nesting habitats along north coastal Andhra Pradesh*. A WWF–India, Conservation Corps Volunteer Annual Report (Jan 1997 to June 1998). 51 pp.

98 Tripathy, B & BC Choudhury (2001) *Sea turtles and their nesting beaches along the Andhra Pradesh coast, India: A status survey – A GOI–UNDP Sea turtle project report*. Wildlife Institute of India, Dehradun.

99 Tripathy, B, K Shanker & BC Choudhury (2003) Important nesting habitats of olive ridley turtles (*Lepidochelys olivacea*) along the Andhra Pradesh coast of eastern India. *Oryx* 37(4): 454–463.

100 See n. 96.

101 See n. 99.

102 See profiles in Kachhapa 4, pp 23–28 (http://www.iotn.org/pdf/kachhapa/kachhapa4.pdf).

Islands of hope

1 Abbi, A (2009) Is Great Andamanese genealogically and typologically distinct from Onge and Jarawa? *Language Sciences* 31: 791–812.

2 Thangaraj, K (2005) Reconstructing the origin of Andaman Islanders. *Science* 308: 996–996.

3 Sastry, KAN (1935) The Cholas. University of Madras, India.

4 Ripley, SD & BM Beehler (1989) Ornithogeographic affinities of the Andaman and Nicobar Islands. *Journal of Biogeography* 16: 323–332.

5 Blyth, E (1846) Notices and descriptions of various new or little known species of birds. *Journal of the Asiatic Society of Bengal* 15(169): 1–54.

6 Mouat, FJ (1863) *Adventures and researches among the Andaman Islanders*. Hurst and Blackett, London, UK.

7 Blyth, E (1863) The Zoology of the Andaman Islands. Appendix in: *Adventures and researches among the Andaman Islanders* (FJ Mouat) Hurst and Blackett, London, UK.

8 Annandale, N (1915) Notes on some Indian Chelonia. *Records of the Indian Museum* 11: 189–195.

9 Smith, MA (1931) *The fauna of British India, including Ceylon and Burma. Reptilia and Amphibia. Vol. 1. Loricata, Testudines*. Taylor & Francis, London, UK. 185 pp.

10 Mackey, S (1847) Extract 'Notice of the Nicobars'; sent to the Directors of the East India Company, Calcutta, 6 February 1847, authored by Rev. Dr. P Barbe.

11 See n. 6.

12 In a letter from Henry Nottidge Moseley, Linacre Professor of Human Anatomy at the University of Oxford to E.B. Tylor, Keeper of the Oxford University Museum

[Tylor Papers/E. H. Man Manuscript Collection no. 3].
13 Man, EH (1883) The Aboriginal Inhabitants of the Andaman Islands. Reprinted in 2001 by Mittal Publications, New Delhi, India.
14 Pandya, V (1993) *Above the Forest: A Study of Andamanese Ethnoanemology, Cosmology, and the Power of Ritual*. Oxford University Press, New Delhi, India.
15 Alcock, A (1902) A *Naturalist in Indian Seas: Or, Four Years with the Royal Indian Marine Survey Ship 'Investigator'*. Dutton, London, UK.
16 Portman, MV (1899) A *history of our relations with the Andamanese* (Vol. 1). Office of the Superintendent of Government Print, India.
17 Kloss, CB (1902) *Andaman and Nicobars*. Reprint (1971) Vivek, New Delhi, India.
18 Cutting, CS (1932) Natives of the Andaman Islands. *Journal of the American Museum of Natural History* 32: 521–530.
19 Bonington, MCC (1931) *Census of India: The Andaman and Nicobar Islands.*
20 Biswas, S and DP Sanyal (1977) Notes on the Reptilia collection from the Great Nicobar Island during the Great Nicobar Expedition in 1966. *Records of the Zoological Survey of India* 72: 107–124.
21 Davis, TA and R Altevogt (1976) Giant turtles and robber crabs of the South Sentinel. *Yojana* 20: 75–79.
22 Bhaskar, S (1984) The distribution and status of sea turtles in India. In: *Proceedings of the Workshop on Sea Turtle Conservation* (Ed. E.G. Silas), pp. 21– 35. Central Marine Fisheries Research Institute Special Publication 18, Cochin, India.
23 Whitaker, R and J Lenin (2010) Satish 'Batagur' Bhaskar. *Indian Ocean Turtle Newsletter* 12: 24–28.
24 Bhaskar, S (1985) *Management and Research of Marine Turtle Nesting sites on the North Vogelkop coast of Irian Jaya Progress Report. 5 Nov. 1984 – 30 April 1985*. WWF/IUCN Project 1528. 14 pp.
25 Whitaker, Z (1979) Editors note. *Hamadryad* 4(3): 1.
26 Bhaskar, S (1979) Sea turtle survey in the Andaman and Nicobars. *Hamadryad* 4(3): 2–26.
27 Bhaskar, S (1979) Andamans. *Hamadryad* 4(1): 3.
28 Bhaskar, S (1979) Sea turtles in the South Andaman Islands. *Hamadryad* 4(1): 3–5.
29 Ibid.
30 Bhaskar, S (1979) Letters from the Andamans. *Hamadryad* 4(2): 3–6.
31 Bhaskar, S and R Whitaker (1983) Sea turtle resources in the Andamans – Mariculture potential in Andaman and Nicobar Islands, An indicative survey. *Bulletin of Central Marine Fisheries Research Institute* 34: 94–97.
32 See n. 30.
33 See n. 31.
34 Ibid.
35 See n. 30.
36 See n. 26.

37 Bhaskar, S (1981) Travels in the Andaman and Nicobar Islands. *Hamadryad* 6(1): 2–7.
38 See n. 26.
39 See n. 30.
40 See n. 26.
41 Ibid.
42 Bhaskar, S (1981) *Sea turtle surveys of Great Nicobar and Little Andaman Islands*. Report to WWF–India. 5 pp.
43 Bhaskar, S (1984) *Sea turtles in North Andaman and other Andaman Islands*. Report to WWF– India. 46 pp.
44 See n. 22.
45 Ibid.
46 Bhaskar, S and M Tiwari (1992) *Andaman and Nicobar Sea Turtle Project. Phase–I. Great Nicobar Island*. Unpublished report for the Centre for Herpetology Madras Crocodile Bank Trust, Tamil Nadu, India.
47 Bhaskar, S (1996) Re–nesting intervals of the hawksbill turtle (*Eretmochelys imbricata*) on south Reef Island, Andaman Islands, India. *Hamadryad* 21: 19–22.
48 Shanker, K (2010) Special profile: Satish Bhaskar. *Indian Ocean Turtle Newsletter* 12: 23.
49 Whitaker, R and J Lenin (2010) Satish 'Batagur' Bhaskar. *Indian Ocean Turtle Newsletter* 12: 24–28.
50 Shanker, K and J Lenin (2010) Satish Bhaskar's publications and surveys. *Indian Ocean Turtle Newsletter* 12: 29–32.
51 Andrews, HV, S Krishnan and P Biswas (2006) Distribution and status of marine turtles in the Andaman and Nicobar Islands. In: *Marine turtles of the Indian subcontinent* (Eds. K Shanker and BC Choudhury), pp. 33–57. Universities Press, Hyderabad, India.
52 Andrews, HV and K Shanker (2000) A significant population of leatherback turtles in the Indian Ocean. *Kachhapa* 6: 17.
53 Andrews, HV et al. (2006) Marine turtle status and distribution in the Andaman and Nicobar Islands after the 2004 M9 quake and tsunami. *Indian Ocean Turtle Newsletter* 4: 3–11.
54 Chandi, M (2007) Traditional sensibility in the Andamans. *SWOT Report* 3: 21.
55 Andrews, HV, A Tripathy, S Aghue, S Glen, S John & K Naveen (2006) The status of sea turtle populations in the Andaman and Nicobar Islands (Eds. K Shanker & HV Andrews) *Towards an Integrated and Collaborative Sea Turtle Conservation Programme in India: A UNEP/CMS–IOSEA Project Report*. Centre for Herpetology/ Madras Crocodile Bank Trust, Tamil Nadu, India.
56 Sivasundar, A and KV Devi Prasad (1996) Placement and predation of nests of leatherback sea turtles in the Andaman Islands, India. *Hamadryad* 21: 36–42.
57 Whitaker, Z (1979) Editor's note. *Hamadryad* 4(3): 1.
58 Tiwari, M (2012) Sea turtles in the southern Nicobar Islands: results of surveys from February–May 1991. *Indian Ocean Turtle Newsletter* 16: 14–18.
59 Chandi, M (2009) Surviving the tsunami at the Galathea Bridge, Great Nicobar

Island. *Current Conservation* 3(1): 16–19 (See also http://madrascrocbank.blogspot.in/2007/12/surviving–tsunami–at–galathea–bridge.html).

60 MacArthur, RH and EO Wilson (1967) *The Theory of Island Biogeography*. Princeton University Press, Princeton, New Jersey, USA.

61 Diamond, JM (1975) The Island Dilemma: Lessons of Modern Biogeographic Studies for the Design of Natural Reserves. *Biological Conservation* 7: 129–146.

62 Simberloff, DS and LG Abele (1982) Refuge design and island biogeograpic theory – effects of fragmentation. *American Naturalist* 120: 41–56.

63 Wilcox, BA and DD Murphy (1985) Conservation strategy – effects of fragmentation on extinction. *American Naturalist* 125: 879–887.

64 Levins, R (1969) Some demographic and genetic consequences of environmental heterogeneity for biological control. *Bulletin of the Entomological Society of America* 15: 237–240.

65 Hanski, IA and ME Gilpin (1997) *Metapopulation biology: Ecology, Genetics and Evolution*. Elsevier Science, USA.

66 See n. 26.

67 Oommen MA (2009) Mathe Budda. *Current Conservation* 3(1): 20.

68 Bhaskar, S (1993) *The Status and Ecology of Sea Turtles in the Andaman and Nicobar Islands. ST 1/93*. Centre for Herpetology/Madras Crocodile Bank Trust, Tamil Nadu, India, 37 pp.

69 Bhaskar, S (1994) *Andaman & Nicobar Sea Turtle Project, Phase V*. Centre for Herpetology/Madras Crocodile Bank Trust, Tamil Nadu, India (unpublished report).

70 Bhaskar, S (1995) *Andaman & Nicobar Sea Turtle Project, Phase VIII*. Centre for Herpetology/Madras Crocodile Bank Trust, Tamil Nadu, India (unpublished report).

INDEX

COMMONLY USED ACRONYMS

APOWA	Action for the Protection of Wild Animals
ARRS	Agumbe Rainforest Research Station
ANET	Andaman and Nicobar Islands Environmental Team
ATREE	Ashoka Trust for Research in Ecology and the Environment
BNHS	Bombay Natural History Society
CEC	Central Empowered Committee
CIFT	Central Institute of Fisheries Technology
CMFRI	Central Marine Fisheries Research Institute
CES	Centre for Ecological Sciences
CMC	Centre for Marine Conservation
CCDP	Coastal Community Development Programme
CI	Conservation International
CITES	Convention on International Trade in Endangered Species of Wild Flora and Fauna
CMS	Convention on the Conservation of Migratory Species
CMS-UNEP	Convention on the Conservation of Migratory Species – United Nations Environment Programme
DRDO	Defence Research and Development Organization
DPCL	Dhamra Port Company Limited
DGH	Directorate General of Hydrocarbons
FSL	Field Services and Learning
FAO	Food and Agricultural Organization
GIS	Geographic Information Systems
GLRA	Green Life Rural Association
ICAR	Indian Council of Agricultural Research

IISc	Indian Institute of Science
INSA	Indian National Science Academy
ISPL	International Seaports Private Limited
JBNHS	Journal of the Bombay Natural History Society
KAP	Kadal Aamai Paadukavalargal (Sea Turtle Protectors)
LMRCC	Lakshadweep Marine Resource Conservation Centre
MSSRF	M.S. Swaminathan Research Foundation
MPEDA	Marine Products Export Development Authority
MTSG	Marine Turtle Specialist Group
NSS	National Service Scheme
NEWS	Nature, Environment & Wildlife Society
NORAD	Norwegian Agency for Development Cooperation
OTFWU	Odisha Traditional Fishworkers' Union
OMFRA	Odisha Marine Fisheries Regulation Act
OMRCC	Odisha Marine Resources Conversation Consortium
PNC	Prakruti Nature Club
RSTPC	Rushikulya Sea Turtle Protection Committee
SNM	Sahyadri Nisarga Mitra
STAP	Sea Turtle Action Programme
SNIRD	Society for National Integration through Rural Development
SEAFDEC	Southeast Asian Fisheries Development Center
SSTCN	Students' Sea Turtle Conservation Network
IUCN	The International Union for Conservation of Nature and Natural Resources
TNC	The Nature Conservancy
TCDS	Tribal Community Development Society
TREE	Trust for Environmental Education
TAG	Turtle Action Group
TCP	Turtle Conservation Project
TED	Turtle Excluder Device
UAA	United Artists Association
UNDP	United Nations Development Programme
VSPCA	Visakha Society for the Prevention of Cruelty to Animals
WCS	Wildlife Conservation Society
WII	Wildlife Institute of India
WPSI	Wildlife Protection Society of India
WWF	World Wide Fund for Nature

ACKNOWLEDGEMENTS

THIS BOOK WAS, FITTINGLY, instigated and supported by two of India's leading historians. Ramachandra Guha has been encouraging my writing from the time I was a Ph.D student at the Indian Institute of Science in the 1990s. In 2010, I received a fellowship from the New India Foundation, of which he is a Trustee, without which this book would not have been possible. The idea for this particular narrative took root nearly a decade ago when I wrote a couple of articles on the politics of sea turtle conservation in Odisha and Chennai. One of these was published in a collection edited by Mahesh Rangarajan. Mahesh's comment that the chapter could be fleshed out into a book set me thinking about it.

I am particularly grateful for the freedom I received as faculty at the Indian Institute of Science. R. Sukumar, my Ph.D supervisor and now colleague at the Centre for Ecological Sciences, is an author of many books himself, and understood the travails of writing. My colleagues and students at CES and Dakshin Foundation put up with my frequent disappearances and provided support whenever I needed it. And Deepa's efficiency in managing the office kept things running smoothly.

A great amount of information and insight came from interviews with numerous individuals who freely gave their time and shared their memories. For their long walks down memory lane, thanks in particular to Ann and Preston Ahimaz, Harry Andrews, Satish Bhaskar, Manish Chandi, B.C. Choudhury, Shekar Dattatri, Jack Frazier, C.S. Kar (late), Biswajit Mohanty, Bivash Pandav, E.G. Silas, Samar Singh, Rom Whitaker, Sejal

Worah and Anne and Belinda Wright. Thanks also to Amala Akkineni, Indraneil Das, Nirmal Kulkarni, Colin Limpus, Amukta Mahapatra, Anil Panwar, Mahesh Rangarajan, Manjula Tiwari, Pranav Trivedi, Lotika Varadarajan, E. Vivekanandan and Zai Whitaker. The leaders of several NGOs – Arun V., Sovakar Behera, Bichi Biswal, Jignesh Gohil, Dinesh Goswamy, Bhau Katdare, Pradeep Nath, Mangaraj Panda and Rabindranath Sahu – also provided information for the book. In addition, for endless heated conversations about sea turtles, thanks to Rohan Arthur, Ashish Fernandes, Sanjiv Gopal, Janaki Lenin, Sudarshan Rodriguez, Aarthi Sridhar, Wesley Sunderraj and Basudev Tripathy. Gautami Balasubramaniam, Tito Chandy, John Mathew, Savita Moorthy, Tharani Selvam (late), Shabbar Sheerazi, Tara Thiagarajan, Venkat and Yohan Thiruchelvam were the first in that long line of friends during our turtle walks in Chennai in the 1980s.

A number of colleagues and friends read the book and provided comments. Ramachandra Guha's review of the entire manuscript was invaluable. Various others read one or more chapters each: Manish Chandi, Matthew Godfrey, Jack Frazier, Madhuri Ramesh, Aarthi Sridhar and Rom Whitaker. I also received inputs from Robert Bustard, B.C. Choudhury, Indraneil Das, Shekar Dattatri, Ashish Fernandes, Sudhakar Kar, Nirmal Kulkarni, Janaki Lenin, Biswajit Mohanty, Manjula Tiwari, Basudev Tripathy and Belinda Wright. Thanks also to Somak Ghoshal at HarperCollins India for his careful reading and editing of the final manuscript.

Seema Shenoy, who worked with me for several years on sea turtle projects, transcribed most of the interviews, gathered information from NGOs and was generally an invaluable source of support. Muralidharan, who now coordinates my sea turtle projects in India, transcribed several interviews and helped with referencing the entire manuscript. And Mallika Sardeshpande painstakingly went through the proofs – twice.

My work on sea turtles was inspired by Satish Bhaskar, my first mentor, and Rom Whitaker. B.C. Choudhury, my postdoctoral advisor, supported my return to sea turtle research and seeded the idea of building civil society networks for sea turtles in India. Brian Bowen sparked my first postdoctoral project on ridley genetics, Peter Dutton and Earl Possardt have supported our sea turtle work for several years, and Jack Frazier has been the constant voice of wisdom (and occasionally mischief) in the background. The late Nicholas Mrosovsky kept us all on our toes with his thought provoking ideas and opinions.

Matthew Godfrey (my turtle encyclopaedia) and Lisa Campbell, Brendan Godley and Annette Broderick, Mark Hamann and Chloe Schauble, Peter Richardson and Sue Ranger, Jeff Seminoff, J. Nichols, Michael Coyne, Andrea Phillott, Ana Barragan, Emma Harrison, Zoe Meletis, Nicolas Pilcher and Rod Mast are friends and colleagues around the world who motivate me to continue my work on sea turtles.

Over the last few years, my researchers and project coordinators – Ema Fatima, Sajan John, Nupur Kale, Divya Karnad, M. Muralidharan, Naveen Namboothri, Chetan Rao, Adhith Swaminathan, Amrita Tripathy and others – have kept various projects going which has allowed me to stay engaged with these narratives. Amukta and Aditya Mahapatra provided a home in Bhubaneshwar during the early years of my research in Odisha.

I would also like to fondly recall Ambika Tripathy, who was conducting surveys for our sea turtle project in Great Nicobar, and lost his life when the tsunami struck in December 2004; my student, Archana Bali, lost her long battle with cancer last year; Archana worked on the nuances of conservation through engagement with communities – her Ph.D was on Inuit communities in Alaska – and would have enjoyed this book, I think; and Ravi Sankaran, a dynamic researcher and conservationist who spent much time in the islands and passed away suddenly in 2009.

And a brief mention of my friends and colleagues who indulge my endless thirst for conversations about life, the universe and everything – Aarthi, Meera and Naveen (at Dakshin), Kavita, Maria, Praveen, Rohini, Siddhartha, Sumanta and Vishu (at IISc), Abi, Ankila, Jagdish and Nitin (at ATREE), Ajith and Mahesh (at NCBS), Giri and Rekha, Arundhati and Dev, Bharath, Jagdeesh and Tasneem.

Thanks to my mother, Uma, who is an irrepressible historian by nature, and my father, Shanker, for timely infusions of spirit, including of the Scottish variety; to Jaya, who makes my endless cups of tea and coffee; to Meera, my editor for everything, and Vishak, the little prince and my favourite critic.

Published articles

I have obviously been mulling over turtles and the ideas in this book for several years now, and they have appeared in my writing in the past. Many of these ideas are revisited here, and a small amount of text has also been extracted from previous articles. I thank the publishers and my co-authors for letting me use this material. The sources are credited below.

The ideas and narratives in several chapters have appeared before in:

- Shanker, K. & Kutty, R., 'Sailing the flagship fantastic: myth and reality of sea turtle conservation in India,' *Maritime Studies* 3(2) and 4(1) (2005): pp. 213–240.
- Shanker, K., 'Deconstructing sea turtle conservation in India,' *Making Conservation Work* (eds. G. Shahabuddin and M. Rangarajan), New Delhi: Permanent Black (2007): pp. 89–110.

The description and discussion of the drama over Dhamra in 'Hype and Hypocrisy in Las Tortugas' is based on:

- Shanker, K., 'My way or the highway: where conservationists and corporations meet,' *Indian Ocean Turtle Newsletter* 8 (2008): pp. 10–14. (*Marine Turtle Newsletter* 121 (2008): pp. 16–18)

In 'Islands of Hope', the introduction and conclusion are largely drawn from:

- Shanker, K. and Oommen, M.A., 'The edge of the world,' *Sanctuary Asia*, December 2003.
- Shanker, K., 'Voyages of the Leatherback,' *Sanctuary Asia*, April 2003.

In addition, some lines have been extracted for various chapters from:

- Shanker, K. & Mohanty, B., 'Operation Kachhapa: In search of a solution for the Olive Ridleys of Orissa' (*Guest Editorial*) *Marine Turtle Newsletter* 86 (1999): pp. 1–3.
- Shanker, K., 'Solving the ridley riddle,' *Sanctuary Asia*, August 2001.
- Shanker, K., 'Sea turtles and Submarines – Sinking the wrong ship?' *Kachhapa* 4 (2001): pp. 29-30.

- Shanker, K., 'Tracking turtles through time and space,' *Resonance*, June 2002.
- Shanker, K., 'The swampland of sea turtle conservation: in search of a philosophy,' (*Guest Editorial*). *Marine Turtle Newsletter* 95 (2002): pp. 3–6.
- Shanker, K., 'What ails the ridley?' *The Hindu*, June 8, 2003.
- Shanker, K., Hiremath, A., and Bawa, K.S., 'Linking Biodiversity Conservation and Livelihoods in India,' *PLOS Biology* 3 (2005): pp. 1878–1880.
- Shanker, K., 'An island away,' *Down to Earth*, April 30, 2006.
- Shanker, K., 'Small empty patches are important,' *Down to Earth*, June 15, 2006.
- Shanker, K., 'Special profile: Satish Bhaskar,' *Indian Ocean Turtle Newsletter* 12 (2010): p. 23.
- Shanker, K., 'Warm turtle in cold waters: the leatherback's journey,' in *Nature Chronicles of India: Essays on Wildlife*, ed. by Ananda Banerjee (New Delhi: Rupa Publications, 2014).

ABOUT THE AUTHOR

KARTIK SHANKER WAS inspired to a career in ecology by an ancient reptile, a sea turtle that crawled ashore late one night in Madras (now Chennai). As faculty at the Centre for Ecological Sciences, Indian Institute of Science, Bengaluru, he now indulges his fascination for ecology and evolution, working with students on frogs, reptiles, birds, plants, reef fish and other marine fauna.

But that initial encounter with the turtle, an olive ridley, hooked him for life. In the twenty-five or so years since then, he has helped establish a students' group for sea turtle conservation in Chennai, conducted research on olive ridleys in Odisha and leatherback turtles in the Andaman and Nicobar Islands, started newsletters and websites, conducted international symposiums on sea turtles, and established regional and national networks for coastal and marine conservation. He has also served as the president of the International Sea Turtle Society and regional vice-chair of the IUCN/SSC Marine Turtle Specialist Group.

Shanker is a founding trustee of Dakshin Foundation, which works largely with coastal communities on natural resource conservation and management, and a founding editor of the magazine, *Current Conservation*. In his spare time, he seeks to distract young minds from more serious pursuits with books such as *Turtle Story* and *The Adventures of Philautus Frog*. Kartik loves hanging out in the islands, diving at reefs and looking for turtles, a passion that he shares with his family. At home, in Bengaluru, sadly, he is a slave to Meera, Vishak and two cats, which leaves little time for important activities like playing the guitar, basketball and going to the gym.